Springer Proceedings in Business and Economics

Springer Proceedings in Business and Economics brings the most current research presented at conferences and workshops to a global readership. The series features volumes (in electronic and print formats) of selected contributions from conferences in all areas of economics, business, management, and finance. In addition to an overall evaluation by the publisher of the topical interest, scientific quality, and timeliness of each volume, each contribution is refereed to standards comparable to those of leading journals, resulting in authoritative contributions to the respective fields. Springer's production and distribution infrastructure ensures rapid publication and wide circulation of the latest developments in the most compelling and promising areas of research today.

The editorial development of volumes may be managed using Springer Nature's innovative EquinOCS, a proven online conference proceedings submission, management and review system. This system is designed to ensure an efficient timeline for your publication, making Springer Proceedings in Business and Economics the premier series to publish your workshop or conference volume.

This book series is indexed in SCOPUS.

Mihail Busu
Editor

Digital Economy and New Value Creation

15th International Conference on Business Excellence, ICBE 2021, Bucharest, Romania, March 18–19, 2021

Springer

Editor
Mihail Busu
Bucharest University of Economic Studies
Bucharest, Romania

ISSN 2198-7246 ISSN 2198-7254 (electronic)
Springer Proceedings in Business and Economics
ISBN 978-3-031-07264-2 ISBN 978-3-031-07265-9 (eBook)
https://doi.org/10.1007/978-3-031-07265-9

This Springer imprint is published by the registered company Springer Nature Switzerland AG
The registered company address is: Gewerbestrasse 11, 6330 Cham, Switzerland

Preface

This book consists of selected papers from the 15th edition of the International Conference on Business Excellence (ICBE), March 18–19, 2021, Bucharest, Romania. The International Conference on Business Excellence is an annual international scientific event organized by the Faculty of Business Administration in foreign languages (UNESCO Department for Business Administration) and the Society for Business Excellence and takes place at Bucharest University (http://www.bizexcellence.ro/icbe/call of for Economic papers/Studies).

The goal of this international event was to reunite specialists from the triple helix fields: university and research, consultancy and business, and policy making pursuant to optimizing the business added value by providing a stimulating environment for knowledge and know how business excellence transfers and alliance formation, and by ensuring a better understanding of new challenges, opportunities and threats generated by sustainability and innovation in general and particularly in the fields like energy, social entrepreneurship, finance, marketing, digitization and the list may continue.

The previous editions of the International Conference on Business Excellence attracted the interest of more than 1500 international authors. The geographical distribution of the guest speakers, authors and members of the scientific committee covers the following countries: Bulgaria, France, Germany, Hungary, Iran, Italy, Mexico, Poland, Portugal, Romania, Serbia, Switzerland, Turkey, Czech Republic, Singapore, Slovakia, Finland, United Kingdom and USA. The Proceedings of the International Conference on Business Excellence were indexed in the ISI Web of Science (WOS) or published as special issues of the international journal Management & Marketing, indexed in various international databases (ESCI, Scopus, etc).

Primarily, after the review process, selected papers from the International Conference on Business Excellence (ICBE 2021) will be considered for publication. The papers submitted and selected for this book should neither have been previously published nor be under consideration for publication elsewhere, and are subject to a very rigorous peer review process. The topics of the conference include, but are not limited to the following aspects:

- Accounting for the future: the impact of digital economy on the financial profession: what's new?
- Applied statistics in economics and management
- Business digitization
- Business internationalization and globalization
- Business Law
- Clean energy, climate change and circular economy—building bridges towards a climate friendly future
- Complexity
- Driving marketing performance through Innovation, Creativity and Entreprenurial Focus
- DS2 Data Science & Digital Society & Fintech
- Human resources: A game changer of the Business Strategy
- International Business Finances in a Globalized World
- Knowledge Economy
- New Challenges For Macroeconomic Modeling
- Rethinking leadership in challenging times
- Social innovation and entrepreneurship ecosystem
- Strategic options for an uncertain future
- Technology entrepreneurship.

The topic of this conference is strongly correlated with the new challenges of digital economy and research systems, and the papers presented during this conference, the results and ideas exchanged are useful and relevant in the present context of a global crisis and they might offer solutions and inputs for the real challenges we face today.

Bucharest, Romania

Associate Professor Dr. Mihail Busu
Editor-in-Chief

Contents

Study Regarding the Impact of ICT Development on Romania's Economic Growth

Vanesa Madalina Vargas, Sorin Cristian Nita, and Razvan Hoinaru

Abstract Digitalization is a reality that happens all around the world and people enjoy the benefits it offers. After several analysis of the literature review, we have discovered there might be a connection between the development of the Information and Communication Technologies and the Gross Domestic Product. If such a hypothesis holds true, a source of support for countries' economic development might have been found. In order to test it, the Augmented Dickey Fuller test was used to check whether the time series have unit roots, followed by using the Granger causality test for analyzing the causality between the time series. The variables tested for this article are: level of internet access, internet purchases by individuals and gross domestic product, the latter being the dependent variable. Results showed us that while level of internet access has a positive, direct causal relationship, the variable purchases by individuals cannot be connected to the evolution of the GDP.

Keywords Digitalization · Internet access · Internet purchases · GDP

1 Introduction

Digital innovations, such as the Internet, smartphones, and other apps and technologies that capture, store, interpret, and exchange data, are changing the global economy. Entrepreneurs today have new opportunities to start companies and market their goods and services all over the world thanks to digital technology (Elia et al. 2016). The Gross Domestic Product (hereinafter GDP) and Information and Communication Technologies have been considered as connected in the scientific literature (Vu 2011; Ho et al. 2011; Warr and Ayres 2012). As a result, the influence of Information and Communication Technologies (from now on ICT) will become a source of support for countries' economic growth (Jorgenson and Vu 2016).

V. M. Vargas (✉) · S. C. Nita
Bucharest University of Economic Studies, Bucharest, Romania
e-mail: vanesa.vargas@fabiz.ase.ro

R. Hoinaru
Queen Mary University of London, London, UK

M. Busu (ed.), *Digital Economy and New Value Creation*, Springer Proceedings in Business and Economics, https://doi.org/10.1007/978-3-031-07265-9_1

At the macroeconomic level (Chaves et al. 2014; Kumar et al. 2016), as well as at the company level (Rowe, 2013; Eze et al. 2018), there is a huge amount of data proving this statement. In this regard, studies suggest a correlation between ICT and technological change that leads to economic growth, either as a result of changes in productivity (Venturini 2015) or as a result of externalities related to information and innovation dissemination (Albiman and Sulong 2017).

While the majority of research on the relationship between ICT and economic growth indicate a positive relationship (Vu et al. 2020; Mayer et al. 2020), others report null relationships in terms of importance, so the findings are not homogenous. No major impact was found between business productivity and ICT variables by Thompson and Garbacz (2011) or Haller and Lyons (2015). Even negative relationship was suggested by some researchers. The effect of ICT on economic performance is not consistent across countries; in fact, it varies depending on economic status (Yazdan and Hossein 2013). This situation indicates that it will be more realistic to research the relationships between ICT and economic growth separately for each country based on its level of development. However, different types of human progress were led thanks to the advances in technology (Isaic et al. 2019; Milićević et al. 2020).

2 Literature Review

Scuotto et al. (2017) found a positive and vital correlation between the use of unique ICTs and the innovation outcomes of SMEs, in line with the majority of previous research in this area (Fernández-Mesa et al. 2014; Joshi et al. 2010; Soto-Acosta et al. 2014).

There are two ways to look at the effect of digitalization on entrepreneurship intentions (Bryniolfsson and McAfee 2014; World Bank 2016). To begin with, the context of future entrepreneurial activity has a substantial economic impact. It is believed, according to Mansfield (1962), that there is a long line of well-informed aspiring entrepreneurs willing to enter the market. The predicted profit level is thought to be the catalyst for entry. Exogenous entry barriers, such as the required expenditure level (Geroski and Schwalbach 1991) and bureaucratic entry laws (Djankov et al. 2002; Youssef et al. 2018), as well as endogenous entry barriers, such as advertisement and R&D costs (Sutton 1991), may impede market entry.

The Endogenous Theory (Romer 1986; Lucas 1993) and the Exogenous Theory (Solow 1956; Swan 1956) are two theoretical frameworks that have been used by scientists to link Information and Communication Technology and economic development. Both of them illustrate the importance of technological change in financial development, but their essence differs. In the case of the Endogenous Hypothesis, it describes technological transition while ignoring the exogenous variable's status (outside the model). The Exogenous Theory, on the other hand, is not clarified within the model, resulting in the "Solow residual," or the aspect of economic growth that is not explained by the model, namely technological change. Exogenous models

(Jorgenson and Vu 2016) and endogenous models (Chaves et al. 2014) are used to study ICT from an entirely empirical viewpoint (Bratianu 2020). In this regard, it is important to note that the works cited with an exogenous perspective only use this theory for analytical purposes, and do not explicitly consider an exogenous technological change relationship in the context of the Solow residual.

Different ways for ICT to produce economic growth are discussed in scientific literature. Despite the fact that the proposals are not fully homogeneous, there are traces of alignment. Direct and indirect effects are the two most common forms of effects proposed (Dima et al. 2019). Indirect effects are the materialization of externalities arising from the use and production of ICT, while direct effects are generally productivity benefits that result directly from the use of ICT. Skorupinska and Torrent-Sellens (2017) suggest that ICT improves efficiency and economic growth both directly and indirectly by promoting complementary technologies that affect the overall productivity factor.

Albiman and Sulong (2017) define several theoretical aspects in which information and communication technology affects the economy. On one hand, the use of e-commerce and e-business has increased the efficiency and versatility of banking operations and staff transactions, which, combined with enhanced personal communications, has increased productivity and economic development. Dima and Maassen (2018) outline how traditional models of software development influenced in the direction of more complex organizations and a better cooperation between several layers of management. In addition, as a result of ICT progress, inventions and R&D have a positive effect on economic growth.

Kumar et al. (2016) demonstrate how investments in ICT improve total factor efficiency at a macroeconomic level, as long as there is a skilled workforce with the necessary experience. It is without a doubt that there are economical differences between one country's regions and diversified public policies must be taken for a direct support of these geographic areas (Miron et al. 2009) Without ignoring that ICT investments have a multiplier effect due to the complementary developments that arise as a result of them.

Vu (2011) recognizes three channels by which ICT will improve economic growth. The first channel has an influence on information and innovation sharing, and is largely spread from developed to developing countries due to ICT penetration. Since ICT enhances decision-making, the second channel deals with resource allocation effectiveness and performance. The final channel has an impact on the reduction of costs of production, the increase of demand, and the investment that ICT penetration produces.

3 Methodology

The objective of this study is to determine if the selected variables has a Granger causality over the GDP. To analyze the dynamic interactions between the selected variables, a two-step statistical analysis will be performed. First, it will be checked whether the time series have unit roots by using the Augmented Dickey Fuller (henceforth, ADF) test, Dickey Fuller (1981) and secondly, we analyze the causality between the time series by performing the Granger causality test.

It is assumed that there is a close correlation between the Level of internet access among households, Internet purchases by individuals and the Gross domestic product. This means that the long-term trend is going in the same direction, as seen in the tables below, the general trend being upward. In this sense, the Granger causality test helps us to calibrate the potential relationship between the Level of internet access among households, Internet purchases by individuals and the Gross domestic product. Thus, the question is whether both variables move in the same direction, whether the differences between them are stable and whether there is an explanatory cause-effect relationship. The stationarity of the series is especially important because when this condition is not fulfilled, two completely independent variables can appear to be closely correlated in one regression, both showing the same trend and both increasing / decreasing over time, which may lead to unrealistic conclusions in the analysis. To this end, the time series will be analyzed to test whether they are stationary or non-stationary.

4 Results and Discussions

Before the actual start of the empirical analysis, the series of observations must be transformed into a stationary series, in other words, into series in which the mean, variance and autocorrelations do not depend on time. Following the ADF test, it is found that the GDP is not stationary at the level. However, the data series is stationary at the first level, the probability being of 0.2%, lower than 5% allowing us to reject the null hypothesis according to which the variable has a unit root as shown in Table 1.

Following the ADF test, it is found that the Internet purchases by individuals is stationary at the level. The probability being of 4.5%, lower than 5% allowing us to reject the null hypothesis according to which the variable has a unit root as shown in Table 2. Once the data series is stationary, we are allowed to draw conclusions about the effects that a short-term economic change has on the behavior of long-term data series.

Following the ADF test, it is found that the Internet purchases by individuals is stationary at the level. the probability being of 4.5%, lower than 5% allowing us to reject the null hypothesis according to which the variable has a unit root as shown in Table 3. Once the data series is stationary, we are allowed to draw conclusions about the effects that a short-term economic change has on the behavior of long-term data series.

For causality testing, the Granger causality test will be used. The purpose of this statistical analysis is to describe the dynamic interactions between time series as well as to reveal their independent movements.

Table 1 Dickey-fuller stability test (ADF) for GDP

Null Hypothesis: DGDP has a unit root				
Exogenous: Constant				
Lag Length: 1 (Automatic - based on SIC, maxlag = 1)				
			t-Statistic	Prob.*
Augmented Dickey-Fuller test statistic			−6.710874	0.0027
Test critical values:	1% level		−5.119808	
	5% level		−3.519595	
	10% level		−2.898418	
*MacKinnon (1996) one-sided p-values				
Warning: Probabilities and critical values calculated for 20 observations and may not be accurate for a sample size of 6				
Augmented Dickey-Fuller Test Equation				
Dependent Variable: D(DGDP)				
Method: Least Squares				
Date: 02/28/21 Time: 22:09				
Sample (adjusted): 2014 2019				
Included observations: 6 after adjustments				
Variable	Coefficient	Std. Error	t-Statistic	Prob
DGDP(-1)	−3.164630	0.471568	−6.710874	0.0068
D(DGDP(-1))	1.535821	0.330453	4.647620	0.0188
C	1.694243	0.490455	3.454431	0.0408
R-squared	0.951771	Mean dependent var		-0.366667
Adjusted R-squared	0.919619	S.D. dependent var		2.979038
S.E. of regression	0.844605	Akaike info criterion		2.806958
Sum squared resid	2.140073	Schwarz criterion		2.702837
Log likelihood	−5.420873	Hannan-Quinn criter		2.390156
F-statistic	29.60174	Durbin-Watson stat		3.177036
Prob(F-statistic)	0.010592			

Source Authors' own research

Table 4 shows the result of the Granger test. Following this test, the probability associated with the ipurchases2 is 68% being above the 5% threshold while the dgdp variable has a probability of 75.9% being above the 5% threshold. The test demonstrates that the two variables do not have a Granger causality.

With regards to the variable DIaccess and dgdp, the probability associated with DIaccess is 73% being above the 5% threshold while the dgdp variable has a probability of 4.5% being below the 5% threshold. Thus, in the case of the variable dgdp it can be said that the null hypothesis according to which the variable does not have a Granger causality on the level of internet access is rejected and the alternative hypothesis according to which the GDP has a Granger causality on the level

Table 2 Dickey-Fuller Stability Test (ADF) for the Level of internet access among households

Null Hypothesis: DIACCES has a unit root				
Exogenous: Constant				
Lag Length: 1 (Automatic - based on SIC, maxlag = 1)				
			t-Statistic	Prob.*
Augmented Dickey-Fuller test statistic			−4.015060	0.0297
Test critical values:	1% level		−5.119808	
	5% level		−3.519595	
	10% level		−2.898418	
*MacKinnon (1996) one-sided p-values				
Warning: Probabilities and critical values calculated for 20 observations and may not be accurate for a sample size of 6				
Augmented Dickey-Fuller Test Equation				
Dependent Variable: D(DIACCES)				
Method: Least Squares				
Date: 02/28/21 Time: 22:14				
Sample (adjusted): 2014 2019				
Included observations: 6 after adjustments				
Variable	Coefficient	Std. Error	t-Statistic	Prob
DIACCES(-1)	−2.551205	0.635409	-4.015060	0.0277
D(DIACCES(-1))	0.649096	0.329475	1.970091	0.1434
C	11.53012	2.985440	3.862118	0.0307
R-squared	0.876962	Mean dependent var		-0.166667
Adjusted R-squared	0.794937	S.D. dependent var		2.483277
S.E. of regression	1.124526	Akaike info criterion		3.379453
Sum squared resid	3.793675	Schwarz criterion		3.275332
Log likelihood	−7.138358	Hannan-Quinn criter		2.962651
F-statistic	10.69135	Durbin-Watson stat		0.906864
Prob(F-statistic)	0.043158			

Source Authors' own research

of internet access is accepted. Regarding the internet access and internet purchases, according to the Granger causality test, the probability associated with the two variables is above 5%, thus accepting the null hypothesis according to which there is no causality between the two variables. The relevance of this study is given by the fact that by studying the variables that may impact the GDP a national strategy may be developed in order to obtain a sustainable economic growth. Also, considering the fact that Romania has a comparative advantage on the IT sector, we may draw the conclusion that if Romania is able to maintain a sustainable economic growth the comparative advantage that Romania has will increase. Also, considering the fact that an increase in the level of internet purchases does not necessarily Granger cause

Table 3 Dickey-Fuller Stability Test (ADF) for the Level of internet access among households

Null Hypothesis: IPURCHASES2 has a unit root				
Exogenous: Constant				
Lag Length: 0 (Automatic - based on SIC, maxlag = 1)				
			t-Statistic	Prob.*
Augmented Dickey-Fuller test statistic			−3.395284	0.0453
Test critical values:	1% level		−4.582648	
	5% level		−3.320969	
	10% level		−2.801384	
*MacKinnon (1996) one-sided p-values				
Warning: Probabilities and critical values calculated for 20 observations and may not be accurate for a sample size of 8				
Augmented Dickey-Fuller Test Equation				
Dependent Variable: D(IPURCHASES2)				
Method: Least Squares				
Date: 02/28/21 Time: 22:24				
Sample (adjusted): 2012 2019				
Included observations: 8 after adjustments				
Variable	Coefficient	Std. Error	t-Statistic	Prob
IPURCHASES2(−1)	−0.385542	0.113552	−3.395284	0.0146
C	22.09639	5.959737	3.707611	0.0100
R-squared	0.657690	Mean dependent var		2.000000
Adjusted R-squared	0.600638	S.D. dependent var		3.116775
S.E. of regression	1.969649	Akaike info criterion		4.405906
Sum squared resid	23.27711	Schwarz criterion		4.425766
Log likelihood	−15.62362	Hannan-Quinn criter		4.271956
F-statistic	11.52795	Durbin-Watson stat		2.058688
Prob(F-statistic)	0.014580			

Source Authors' own research

an increase in the GDP according to our results, we may draw the conclusion that most of the purchases that the Romanian people do online come from imports, thus an encouragement of the autochthone industry is desirable.

5 Conclusion

In this paper, the relationship between level of internet access among households, internet purchases by individuals and the gross domestic product in Romania was analyzed. In this regard, statistical evidence was identified using the Granger test,

Table 4 Granger causality tests

Pairwise granger causality tests			
Date: 02/28/21 Time: 21:55			
Sample: 2011 2019			
Lags: 2			
Null Hypothesis:	Obs	F-Statistic	Prob
IPURCHASES2 does not Granger Cause DGDP	6	0.57286	0.6827
DGDP does not Granger Cause IPURCHASES2		0.36647	0.7596
DIACCES does not Granger Cause DGDP	6	0.42115	0.7367
DGDP does not Granger Cause DIACCES		243.540	0.0453
DIACCES does not Granger Cause IPURCHASES2	6	2.24506	0.4268
IPURCHASES2 does not Granger Cause DIACCES		0.94406	0.5884

Source Authors' own research

according to which an increase in the GDP causes changes in the level of internet access by households. With an increasing GDP the wealth of the population increases and by having more financial resources available, some of those financial resources are used in order to increase the standard of living.

Our study showed us that there is a close direct correlation between the level of internet access among households and the gross domestic product. This means that the long-term trend is going in the same direction. It is somehow expected, given that a higher level of internet access means a better infrastructure, connecting household with the e-commerce and digitalizing institutions, all these bringing more wealth to one's country. However, no direct link was found between the variable Internet purchases by individuals and GDP. One reason would be that people are not buying online from local stores and local products or even the fact that not enough national businesses sell their services and goods online. After more local enterprises will have a greater presence in the electronic environment, the GDP of the country might be correlated to the number of internet purchases. Of course, the study has its own limitation of research, being conducted for Romania only. More countries should be taken into consideration in the next studies in order to sustain our hypothesis and check it for several economies.

References

Albiman, M.M., Sulong, Z.: The linear and non-linear impacts of ICT on economic growth, of disaggregate income groups within SSA region. Telecommun. Pol. **41**, 555–572 (2017)

Bratianu, C.: Toward understanding the complexity of the COVID-19 crisis: a grounded theory approach. Manag. Market. Chall. Knowl. Soc. **15**(s1), 410–423 (2020). https://doi.org/10.2478/mmcks-2020-0024

Bryniolfsson, E., McAfee, A.: The Second Machine Age: Work, Progress, and Prosperity in a Time of Brilliant Technologies. W. W. Norton & Company (2014)

Chaves, R., Bernal, E., Mozas, A., Puentes, R.: Improving e-economy by regional governments. Manag. Decis. **52**, 559–572 (2014)

Dima, A.M., Tuclea, C.E., Vrânceanu, D.M., Tigu, G.: Sustainable social and individual implications of telework: a new insight into the Romanian labor market. Sustainability **11**(13), 3506 (2019). https://doi.org/10.3390/su11133506

Dima, A.M., Maassen, M.A.: From waterfall to agile software: development models in the IT sector, 2006 to 2018. impacts on company management. J. Int. Stud. **11** (2), 315–326 (2018)

Djankov, S., La Porta, R., Lopez-de-Silanes, F., Shleifer, A.: The regulation of entry. Q. J. Econ. **117**(1), 1–37 (2002). https://doi.org/10.1162/003355302753399436

Elia, G., Margherita, A., Petti, C.: An operational model to develop technology entrepreneurship EGO-system. Int. J. Innov. Technol. Manag. **13**(5), 1640008 (2016). https://doi.org/10.1142/S0219877016400083

Eze, S.C., Chinedu-Eze, V.C., Bello, A.O.: Actors and emerging information, communications and technology (EICT) adoption: a study of UK small and medium services enterprises. Cogent Bus. Manage. **5**, 1480188 (2018). https://doi.org/10.1080/23311975.2018.1480188

Fernández-Mesa, A., Ferreras-Méndez, J. L., Alegre, J., Chiva, R.: IT competency and the commercial success of innovation. Indust. Manage. Data Syst. (2014)

Geroski, P.A., Schwalbach, J.: Entry and Market Contestability: An International Comparison. Blackwell Publishing (1991)

Haller, S.A., Lyons, S.: Broadband adoption and firm productivity: evidence from Irish manufacturing firms. Telecommun. Pol. **39**, 1–13 (2015)

Ho, S.C., Kauffman, R.J., Liang, T.P.: Internet-based selling technology and ecommerce growth: a hybrid growth theory approach with cross-model inference. Inf. Technol. Manag. **12**, 409–429 (2011)

Isaic, R., Smirna, T., Paun, C.: A critical view on the mainstream theory of economic cycles. Manage. Market. Chall. Knowl. Soc. **14**(1), 48–58 (2019). https://doi.org/10.2478/mmcks-2019-0004

Jorgenson, D.W., Vu, K.M.: The ICT revolution, world economic growth, and policy issues. Telecommun. Pol. **40**, 383–397 (2016)

Joshi, K.D., Chi, L., Datta, A., Han, S.: Changing the competitive landscape: continuous innovation through IT-enabled knowledge capabilities. Inform. Syst. Res. **21**(3), 472–495 (2010)

Kumar, R.R., Stauvermann, P.J., Samitas, A.: The effects of ICT on output per worker: a study of the Chinese economy. Telecommun. Pol. **40**, 102–115 (2016)

Lucas, R.E.: Making a miracle. Econometrica **61**, 251–272 (1993)

Mansfield, E.: Entry, Gibrat's law, innovation, and the growth of firms. Am. Econ. Rev. **52**(5), 1023–1051 (1962). Margo, T., 2017.

Mayer, W., Madden, G., Wu, C.: Broadband andeconomic growth: a reassessment. Inf. Technol. Develop. **26**, 128–145 (2020)

Milićević, S., Petrović, J., Đorđević, N.: ICT as a factor of destination competitiveness: the case of the republics of former Yugoslavia. Manage. Market. Chall. Knowl. Soc. **15**(3), 381–392 (2020). https://doi.org/10.2478/mmcks-2020-0022

Miron, D., Dima, A.M., Vasilache, S.: Indexes of regional economic growth in post-accession Romania. Rom. J. Econ. Forecast. **11**(3), 138–152 (2009)

Romer, P.M.: Increasing returns and long-run growth. J. Polit. Econ. **94**, 1002–1037 (1986)

Rowe, W.B.: Principles of Modern Grinding Technology. Elsevier Science, London (2013)

Scuotto, V., Santoro, G., Bresciani, S., Giudice, M.D.: Shifting intra-and interorganizational innovation processes towards digital business: an empirical analysis of SMEs. Creativity Innov. Manage. **26**(3), 247–255 (2017)

Skorupinska, A., Torrent-Sellens, J.: ICT, innovation and productivity: evidence based on eastern European manufacturing companies. J. Knowl. Econ. **8**, 768–788 (2017)

Solow, R.A.: Contribution to the theory of economic growth. Q. J. Econ. **70**, 65–94 (1956)

Soto-Acosta, P., Colomo-Palacios, R., Popa, S.: Web knowledge sharing and its effect on innovation: an empirical investigation in SMEs. Knowl. Manage. Res. Practice **12**(1), 103–113 (2014)

Sutton, J.: Sunk costs and market structure, price competition, advertising, and the evolution of concentration. MIT press, Cambrige (1991)

Swan, T.W.: Economic growth and capital accumulation. Econ. Rec. **32**, 334–361 (1956)

Thompson, H.G., Garbacz, C.: Economic impacts of mobile versus fixed broadband. Telecommun. Pol. **35**, 999–1009 (2011)

Venturini, F.: The modern drivers of productivity. Res. Pol. **44**, 357–369 (2015)

Vu, K.M.: ICT as a source of economic growth in the information age: empirical evidence from the 1996–2005 period. Telecommun. Pol. **35**, 357–372 (2011)

Vu, K., Hanafizadeh, P., Bohlin, E.: ICT as a driver of economic growth: a survey of the literature and directions for future research. Telecommun. Pol. **44**, 101922 (2020). https://doi.org/10.1016/j.telpol.2020.101922

Warr, B., Ayres, R.U.: Useful work and information as drivers of economic growth. Ecol. Econ. **73**, 93–102 (2012)

World Bank: World development report–digital dividends. World Bank (2016). https://doi.org/10.1017/CBO9781107415324.004.x

Yazdan, G.F., Hossein, S.S.M.: FDI and ICT effects on productivity growth. Proc. Soc. Behav. Sci. **93**, 1710–1715 (2013)

Youssef, A., Boubaker, S., Omri, A.: Entrepreneurship and sustainability: the need for innovative and institutional solutions. Technol. Forecast. Soc. Chang. **129**, 232–241 (2018)

EU Countries' Performance in Digitalization

Anca Maria Paraschiv, Narcisa Alexandra Panie, Tamara Maria Nae, and Laura Ciobanu

Abstract Digital transition represents one of the main objectives of the European Commission, and the current pandemic context is considered an 'opportunity' to accelerate the implementation of new digital technologies, which, in line with the climate neutrality goal, contribute to transforming today's Europe into a more resilient and functional one. However, evidence from across the globe shows that the digital transition process has different levels of implementation in EU countries. Based on the DESI data we analyzed the evolution of the digital transition in EU Member States and answered the question which Member States are performing best in terms of digitization and which are the least performing ones, in order to provide a macro-perspective regarding the structure of the digitalization process. In the second part, the analysis focuses on the current trends in digital transition in Finland, Sweden, Romania and Bulgaria, including pandemic period in order to see its impact. The results indicated that there is a gap between Member States regarding the level of digitalization. Each domain of DESI is important, there are countries that have good results in two areas but have poor results in the other three and overall they obtain a low score. The pandemic has slowed down the implementation of digitalization policies. However, there is a good improvement regarding e-commerce area and not only. This paper mainly contributes to the intensification of debates on digitalization in order to identify as many policies as possible that increase the level of digitalization and reduce the existing gap.

Keywords Digital Gap · Digital Economy · Society Index · Digital Transition

---

A. M. Paraschiv (✉) · N. A. Panie · T. M. Nae · L. Ciobanu
Bucharest University of Economic Studies, Bucharest, Romania
e-mail: anca.paraschiv@economie.ase.ro

T. M. Nae
e-mail: tamara.nae@economie.ase.ro

L. Ciobanu
e-mail: laura.ciobanu@eam.ase.ro

M. Busu (ed.), *Digital Economy and New Value Creation*, Springer Proceedings in Business and Economics, https://doi.org/10.1007/978-3-031-07265-9_2

1 Introduction

Ensuring a green and digital transition is currently one of the top priorities in Europe. Actions to achieve this include examining human implications, strengthening the digital skills of Europeans, and digitizing the public institutions.

The need to intensify the digitalization process was highlighted during the 2019 European Semester, where Member States received clear recommendations on stepping up the efforts to strengthen digital technologies and other components involved in the digitization process. The recommendations focused mainly on the development of citizens' digital skills as well as on the efforts to ensure an infrastructure that allows the use of new technologies, taking into account regional disparities.

Digitization can be perceived as a new dimension of globalization. Like any other transition, this process has both associated costs as well as numerous benefits. A higher degree of digitization can lead to better policy development, more sustainable business models, new jobs and income opportunities. To benefit from all these opportunities requires significant changes in infrastructure, education, regulation and governance. Clearly, this will require considerable financial resources but especially a concentrated effort from all socio-economic actors. In addition to the beneficial effects that digitalization can generate regarding the development and progress, it also raises a series of concerns about their impact on society, largely related to rising income inequalities and unemployment (by replacing human labor).

Regarding the current state of digitalization in the European Union, progress can be compared through the Digital Economy and Society Index (DESI) which is a composite index that summarizes relevant indicators on Europe's digital performance and tracks the evolution of EU Member States in digital competitiveness. DESI has been used since 2014 as a benchmarking tool in the digitalization process.

The aim of this paper is to highlight the evolution of the digitization process in EU Member States and to show there is a gap between them, using the most recent data, including the pandemic period. But the important result is to find which are the factors and premises that contribute in a negative way to digitalization process. We used DESI and its individual components, as well as other relevant indicators to capture the effects of the pandemic on the level of digitization. We also analyze the evolution of the two best performing countries, compared to the least two performing countries. This analysis will allow us to better understand, from the digital transition perspective, where are the non-functional areas and how this process can be accelerated to reduce the gaps in Europe.

Finally, we propose a series of recommendations to improve the level of digitization. We show how the best performing economies in the European Union behave compared to the least performing in terms of digitization. On the one hand, we must take into account the gaps present at the aggregate level, but at the same time the significant regional differences in our country force us to take into account the sub-national level, an analysis that will be carried out in further research.

2 Literature Review

The current European context is characterized by the beginning of a period of digital transition corresponding to the fourth industrial revolution. Also, the current pandemic crisis highlights the need to increase the level of digitalization both in production processes, but also in areas such as education and health.

In order to measure the level of digitalization of the economy and society, we have several indices, scores, indicators, and measurement units that indicate the evolution of this process. Regarding the European Union, DESI is considered the most appropriate and robust method for mapping Europe's progress on digitalization (Banhidi et al., 2020).

There are numerous authors who include the DESI index in them analyzes regarding digitization. There was analyzed the influence of the consumption index growth by the purchasing power parity and unemployment among the active population on the structural units of DESI (Stavytskyy et al., 2019). Using the panel regression, the authors have proven that a 1% increase in the consumption index results in about 0.2 increase in the DESI, and an increase in unemployment by 1% leads to about 0.2 DESI decline. It is also shown that the 98% value of DESI is actually determined by its previous trends, and therefore it is impossible to increase this index rapidly. Stoica and Bogoslov (2017) compared the data of the five areas of DESI for Romania and European Union. They conclude that a low performance in one of these areas is affecting the whole level of digitalization of Romania. At UE level they notice that the EU members should struggle for reducing the existing difference regarding their performance.

An important discussion when talking about a digital economy and society is related to the economic and social effects that this transformation brings. The changes that are taking place in the economy and society must have at their center the individual and its needs. The poverty rate, unemployment and economic and social inequalities are still high in the European Union, so European citizens are still waiting for measures to improve their daily lives. According to Deaton (2013), The Industrial Revolution in Britain (eighteenth-nineteenth centuries) was responsible for the economic progress and relieving from poverty hundreds of millions of people. Today people are better fed, more educated, and life expectancy grew. The industrial revolution has also created new jobs and better income opportunities.

There are studies that conclude that the digital revolution will have a positive effect on the economy and society, but there are also studies that emphasize that it will lead to job losses and significant changes in the structure of the economy and jobs. According to the growth strategy, the technological revolution is an important source to increase productivity and stimulate economic growth (Solow 1956).

Regarding the impact of technological revolution Arntz et al. (2016) analyzed the risk of automation for jobs in OECD countries. Their analysis suggests that 9% of OECD jobs are potentially automatable. However, there is a major difference between countries, while the share of automatable jobs is 6% in Korea, the corresponding share is 12% in Austria. An explanation for this could be the differences in workplace organization, differences in previous investments into automation technologies as well as differences in the education of workers across countries. They pointed out that the automation and digitalization process are unlikely to affect a major number of jobs. A few arguments are that the digitalization is a slow process, workers can adjust to changing technological endowments by switching tasks, thus preventing technological unemployment and additional jobs will be created. However, those who will be affected are part of the category of low-skilled workers. Even if the results are important, this study has its limitations. The analysis takes into consider technological capabilities rather than the actual utilization of such technologies, they consider only existing jobs, although new technologies are likely to create also new etc.

Another study with reference to the impact of digital technologies on the economy and society (Vasilescu et al., 2020) examine the hypothesis that digital divide leads to creation of vulnerable citizens or countries groups. Also, included the general perception of the respondents on the impact of recent digital technologies on the economy, on society, and the quality of life, using The Eurobarometer 87.1 Survey. They applied TwoStep Cluster Analysis (TSCA) with the aim to create homogenous groups of people in terms of three aspects related to digitalization: the attitude towards digitalization, the perception of EU citizens on their own digital skills and the actual use of technology. They found out interesting results, women tend to be more afraid than men about the implication of the workplace digitalization and the most vulnerable category in digitalization era is formed by the people over 55 years, with a low level of education and a low level of income and little internet use, mostly from the Hungary, Romania, Greece and Bulgaria. One of the major limitations of this study is the subjective perception of people that was taken into account.

Among the positive effects of the digital transition on the labor market, we mention the increase in the number of highly skilled workers (Acemoglu and Restrepo, 2017); environmental sustainability (more precisely in the manufacturing process - digitalization allows the development of ecological manufacturing processes) (De Sousa Jabbour, 2018); improving health technologies, etc. The new technology can help both society and the environment, but there are concerns that it may threaten the confidentiality of personal data, erode security, or even deepen income inequality.

Digitalization leads to rapid changes in the labor market, influencing the nature, quality, and productivity of labor. Thus, policy makers face the challenge of using digitalization to support economic growth and employment - while ensuring decent working conditions, social protection and equal opportunities for all (COM, 2019).

3 Methodology

In our analysis, we first established the hypothesis of the paper, namely that there is a gap in the process of digitization in the EU Member States. We wanted to determine which are the factors that place Romania on the bottom of the digital classmen. In order to understand the evolution of the digitalization process in the EU, and to test the hypothesis of the paper, we used the DESI indicator (The digital economy and society index), a composite measure that summarizes indicators related to the digital performance and digital competitiveness at the level of European Union. It comprises five sets of data related to: the degree of connectivity; human capital; use of internet services; integration of digital technologies; digital public services. The lack of DESI data for 2020 led us to identify other indicators that we considered substitutes for DESI components in order to capture the evolution of digitization during the pandemic.

We used both quantitative and qualitative methods. We analyzed the data for each of the 5 dimensions of the index for the 27 EU countries. The data series are annual. We interpreted the variables in a comparative manner that highlights the performing states in terms of digitization and the less performing ones. We took into consideration the structural dimension in terms of digital evolution but also the conjuncture dimension of this process.

We started by analyzing the DESI indicator at Member State level in 2020, and continued to explore each component of this indicator, taking into account the data series for 2019 and 2020. Thus, we monitored the evolution of each component in EU countries and presented in a comparative way that there is a gap between Member States and which are the best performing countries compared to the worst performing. We found that there are states that record very good results in some components and poor results in others.

In the last part of the analysis we chose two of the most performing countries Finland and Sweden and two of those with the worst results regarding the digitalization of Romania and Bulgaria, and we surprised the evolution in recent years, including the pandemic period. It was important to determine the causes of poor performance and also the capacity in terms of the digital process.

4 Results and Discussions

The need to digitize the European economy has been debated for several years, but the crisis caused by Covid 19 has led policymakers to decide on digitalization as a key pillar of the EU's recovery. More than ever, the European Union is determined to take the necessary steps for a digitized and green European future that supports sustainable development. Thus, EU wants 20% of the Recovery Plan fund to be used by Member States to implement public policies that support the digital transition.

Digital transition can respond both to the challenges the European community is facing and to bring added value to both companies and society. In addition to increasing competitiveness and innovation, a digital economy helps to create new jobs and to improve education and social inclusion. Thus, a digital economy refers not only to the way we communicate but also to the way we work and live. All these changes can increase economic well-being and generate social progress.

The areas of action established for the transition to a digital Europe are numerous and include important segments of everyday life, such as ensuring the protection of online operations, digitizing the public sector, ensuring greater connectivity for all European households, digitizing justice, the medical sector and last but not least the digitalization of education. By ensuring adequate public policies in these areas and by respecting European values and the fundamental rights of citizens, Europe can achieve its ambitious goals of creating a digital economy and society.

In terms of measures, EU Member States have as examples Finland and Sweden which have a high index in terms of digitization. Finland, for example, is very advanced in digital skills and the digitization of business indicators as well as in digital public services. During the pandemic, numerous digital projects were implemented in Finland to help the population in terms of the correctness of the information related to Covid 19, psychiatric support programs, etc.

During the pandemic, Sweden implemented digital projects to help teachers teach online. Thus, online courses were organized for distance learning, materials and digital resources necessary for online teaching were made available. Another important digital project implemented by Sweden concerns the development of communication channels between those in the field of health and patients or relatives of patients. The physical distance that the pandemic imposed made it difficult for some of the patients diagnosed with chronic diseases to visit a doctor and also for the medical staff and the relatives of the hospitalized patients to communicate.

The pandemic caused by Covid 19 highlighted the digitization issue in the EU. Most employees were forced to work from home, needing high-speed internet; also, certain services, such as paying taxes or bills were digitalized to avoid as much as possible travel and the risk of virus contamination. A large part of companies (especially sales companies) was forced to transfer its business online because of the travel restriction. Thus, it is a good time to analyze what has been achieved at European level in terms of digitalization and what measures are needed to accelerate the process in order to close the gap between EU countries.

To measure the level of digitization achieved in a country, the European Commission has developed the Digital Economy and Society Index (DESI), a composite index that encompasses indicators related to the digital performance and digital competitiveness of EU Member States. Depending on its level, Member States may decide to implement support policies for a particular component (COM, 2020) (Fig. 1).

It is important to note that the best performing states in digital terms do not correspond to the best performing states in terms of the economy, but in the pandemic context, steps have been taken to implement new digital strategies to help improve carrying out activities in crisis situations and not only. These measures will be visible in DESI from the end of 2021. It should be noted that the indicators that make up this Index are influenced and for example, a country that has not made investments in terms of Human Capital will also have low results in terms of the Use of internet indicator. Italy is in this situation, although in terms of connectivity it is at the EU average, very low investments in basic and advanced digital skills have led to a very low rate in the use of online services, including digital public services. For these reasons, Italy, a country with a developed economy, is at the bottom of the Digital Economy and Society Index. The Connectivity component refers to the demand and supply of fixed and mobile broadband. Among the best performing countries in this

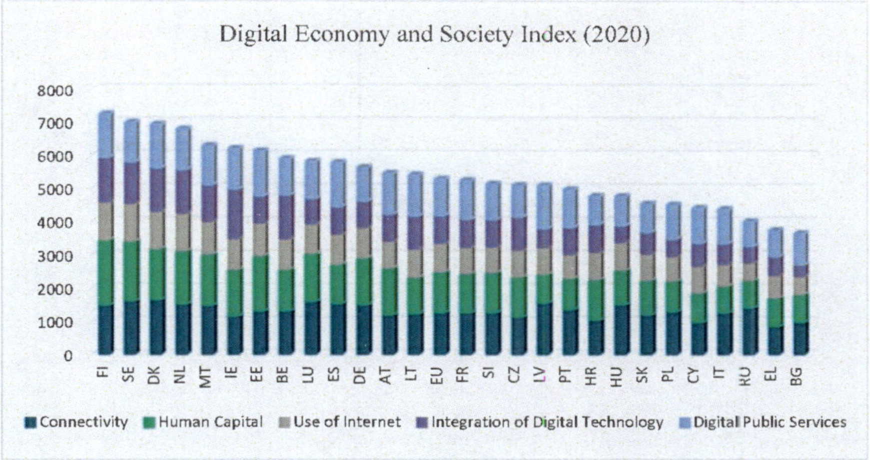

Fig. 1 Digital Economy and Society Index 2020. *Source* Eurostat data

regard are Denmark, Sweden, Luxembourg, and at the opposite pole are Greece, Cyprus, Bulgaria. Although Romania has a good level of connectivity (11th / 27th place), due to the high use of very high speed broadband and the wide availability of high capacity fixed networks, especially in urban areas, it only ranks 26th in the 27 EU Member States in the Digital Economy and Society Index (DESI) for 2020 because areas such as the Use of the Internet or the Integration of Digital Technology are underused. Romania's poor performance in the digital field can be attributed both to the living standards of the population (ensuring basic needs, the need to improve infrastructure in both rural and urban areas), but also to the ever-changing political developments along with development projects (Fig. 2).

In the process of digitization, Connectivity is not enough, it is necessary for the population to have as many skills in using connectivity in order to benefit from the opportunities of digitization. Thus, Human Capital aims at the degree of empowerment of citizens in using the Internet; they can be basic skills, which would make daily activities easier or they can be advanced skills, which have the role of specializing as much labor as possible, therefore, they would contribute to reducing poverty and income inequality (Fig. 3).

Although it is clear that the pandemic can have a positive impact on increasing the number of internet users, the development of digital skills does not automatically come with increased use, it is supported by measures at national level to improve the skills of individuals. There is a significant discrepancy between rural and urban areas in terms of digital skills, and this can be an investment opportunity for policy makers.

With Connectivity and Human Capital, there are several activities that require Use of internet, these activities with online content can be jobs, conferences, entertainment activities. In the context of isolation, much of the time spent outside the

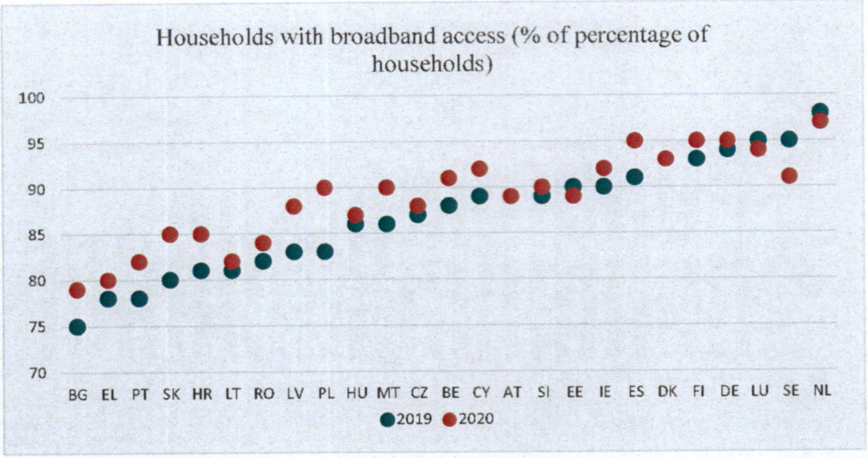

Fig. 2 Households with broadband access (% of percentage of households). *Source* Eurostat data

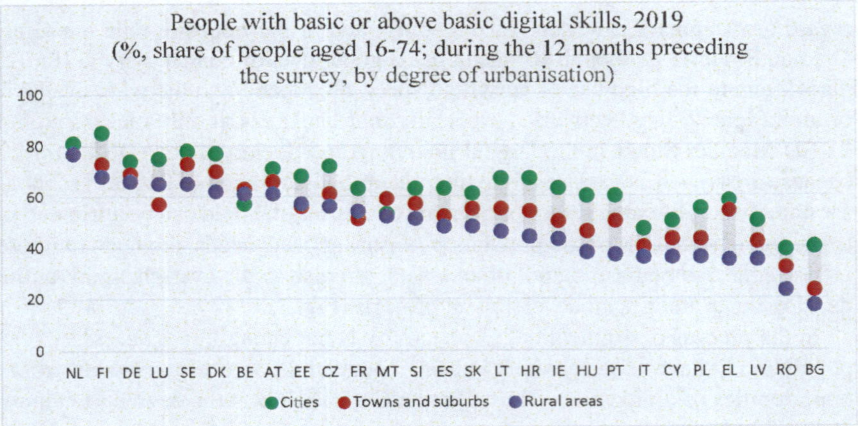

Fig. 3 People with basic or above basic digital skills, 2019. *Source* Eurostat data

home has been replaced by time spent indoors, with online activities being the most convenient choice. This activity is not available for DESI 2020, but can be seen in an increase in the Internet use indicator available for both 2019 and 2020 (Fig. 4).

Most Member States have seen increases in the use of the Internet in 2020 compared to 2019, this increase is due to the "stay at home" measures imposed during the pandemic. The highest increase in the percentage of the population that registered growth is found in SK, RO, CY, SI, and the states that registered decreases in internet use are SE, NL, EE, countries that did not impose restrictions during the pandemic (SE), or who quickly relaxed the restrictions.

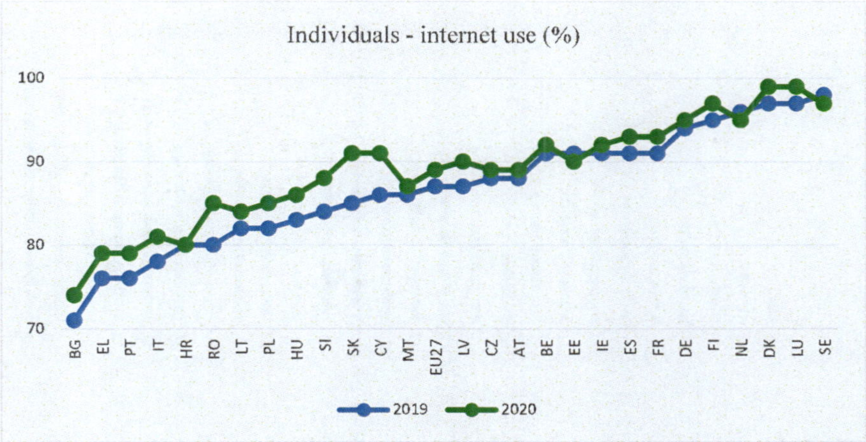

Fig. 4 Individuals - internet use (%). *Source* Eurostat data

Regarding the Integration of digital technology, it measures the degree of digitalization of enterprises and e-commerce. Online commerce has enjoyed a significant rise in some countries such as (RO, HU, HR). Of all these, by far, Romania registered the highest increase among the leading countries (15 pp), and at the opposite pole are EE, SK, SE, FI (between 0 and 3 pp) (Fig. 5).

The fifth component of the DESI index is represented by Digital Public Services, which provides an assessment of the degree of implementation of digitization in public services. The current context highlights the benefits of technology and of the digitization of public services, and the development priorities of all states in the near

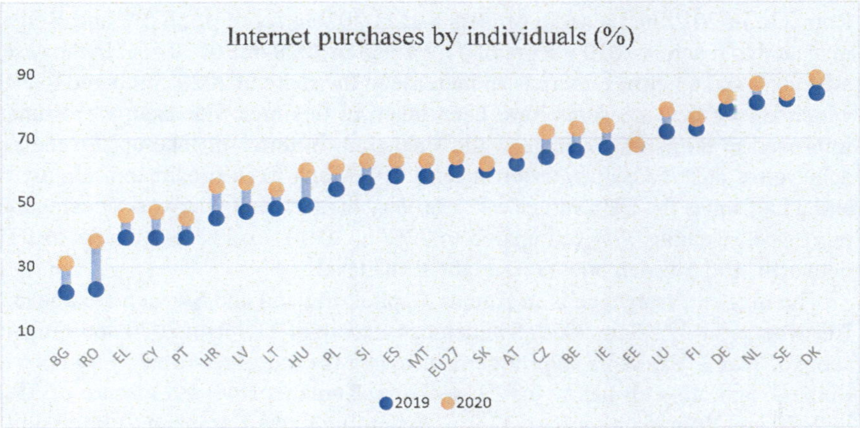

Fig. 5 I Internet purchases by individuals (%). *Source* Eurostat data

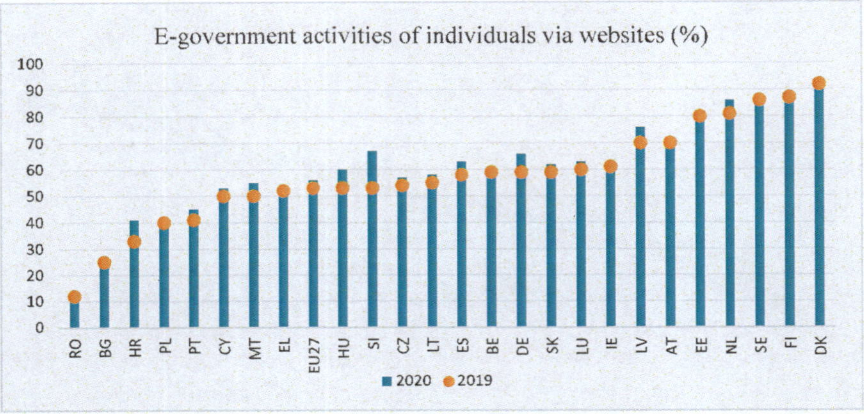

Fig. 6 E-government activities of individuals via websites (%). *Source* Eurostat data

future should include this key issue. For Romania, the digitalization of public services has proved useful during the state of emergency, and the extension of this practice at the level of institutions has already become a priority on Romania's strategic development agenda (Fig. 6).

The graphs below show the data of the 5 indicators that make up the Digital Economy and Society Index for 2019 and 2020 in the case of the best performing countries Finland and Sweden and in the case of the least performing countries Romania and Bulgaria. As we can see, the proximity of the indicator axes means a high score, while the location of the axes further inland indicates a lower score.

There is a gap in terms of the 5 indicators for the four countries analyzed. There is a slight convergence regarding Connectivity, Finland has a score of 54.5% in 2019 and 59.2% in 2020, Sweden recorded a score of 60.1% in 2019 and 64.4% in 2020, Romania in 2019 had a score of 50% and in 2020 a score of 56.2% and Bulgaria managed to reach in 2019 a score of 37.2% and in 2020 one of 38.5%. In the case of all 4 analyzed countries, there is an increase in the score in 2020 compared to 2019, which shows that measures have been taken in this area. For example, Romania improved in terms of coverage while stagnating in terms of take-up. Sweden has achieved a 66% take-up rate for at least 100 Mbps fixed broadband, almost two and a half times the EU average of 26%. Very high capacity network coverage also increased, reaching 77% (compared to 72% in 2018), exclusively thanks to FTTP networks, and Sweden now ranks eight at EU level.

The biggest divergence is in Human Capital, Finland and Sweden are above the EU average of 49.3% in 2020. Sweden has a score of 71.1% in 2020 and Finland a score of 78.4%. Romania and Bulgaria have half the scores recorded by Sweden and Finland, and are well below the EU average. Romania recorded a score of 33.2% in 2020 and Bulgaria a score of 33.9%. According to the European Commission, in Romania less than a third of the people aged 16 to 74 have at least basic digital skills, while 35% have at least basic software skills, compared to the EU average of 61%.

IT specialists represent only 2.2% of the workforce in Romania, while in the EU the average is 3.9%. Romania ranks well among graduates in the IT field, approximately 5.6% of all Romanian graduates are in the IT field, which ranks Romania 5th among Member States.

Regarding digital public services, there is a slight convergence of scores in Finland, Sweden and Bulgaria but Romania remains the last in the EU, with only 48.4% in 2020 well below the EU average of 72%. Regarding the other two indicators, Use of internet and Integration of Digital Technology, the axes indicate major gaps between the countries performing at the level of digitization Finland and Sweden and the less performing countries at the level of digitization Romania and Bulgaria. The scores recorded by Romania and Bulgaria in 2020 are only half of those recorded by Finland and Sweden.

For the Fig. 7, 4 revealing indicators were selected to measure the degree of digitization, noting that these data also capture the effects of the pandemic on digitization. The data presented in the previous graphs represent some scores established based on the information from the year prior to the report.

As we can see, the pandemic has also slowed down the implementation of digitalization policies. The four selected countries made little progress in 2020 compared to 2019, in terms of the 4 indicators. Regarding internet access in 2020, there is a high degree of convergence between the 4 countries, Bulgaria, Romania and Finland registered small increases of the indicator compared to 2019, except for Sweden which registered a small decrease. One possible explanation may be the difficult economic situation of some consumers, caused by the pandemic, who have chosen to give up internet subscriptions. The same situation is encountered in internet use, the only country that has registered a decrease is Sweden, this may be a consequence

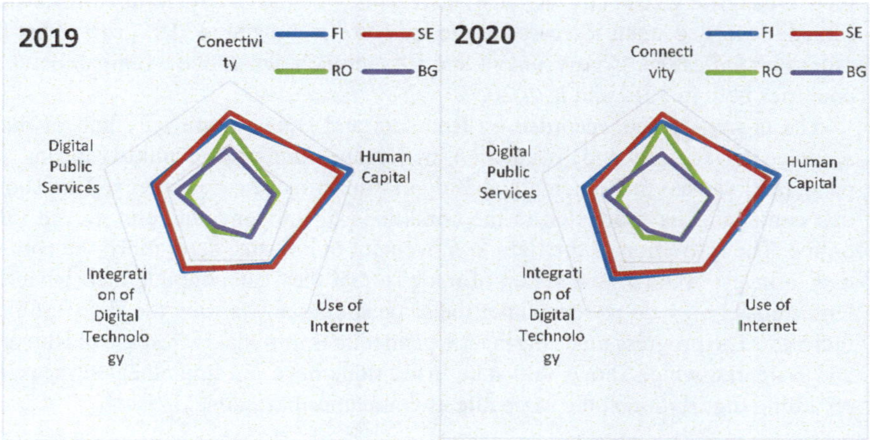

Fig. 7 DESI 2019 – 2020. *Source* Eurostat data

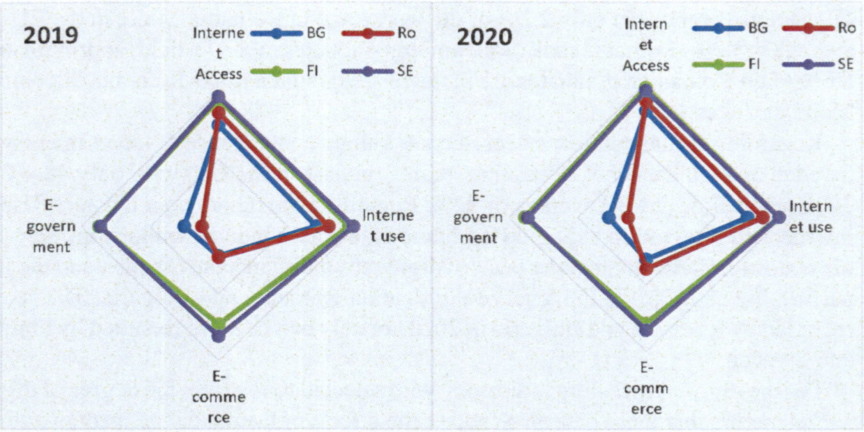

Fig. 8 DESI FI, SE, RO, BG. *Source* Eurostat data

of the fact that this country was the only one in the EU that did not adopt lock-down measures during the pandemic. Since some consumers have given up internet subscriptions, the use of the internet has also decreased.

As we can see, the pandemic has also slowed down the results of implementation of digital policies. To understand better the data and the results we have to take into account the principle of dynamic inconsistency over time. There is a strong need for a coherent digital strategy and an adequate and fast implementation process in terms of digitalization policies. The four selected countries made little progress in 2020 compared to 2019, in terms of the 4 indicators. Regarding internet access in 2020, there is a high degree of convergence between the 4 countries, Bulgaria, Romania and Finland recorded small increases in the indicator compared to 2019, except for the other two indicators, E-government and E-commerce we notice a big gap between countries both in 2019 and in 2020.

The aggregate data recorded by Romania and Bulgaria are only half of those recorded by Finland and Sweden. A major and important evolution during the pandemic, we see in Romania and Bulgaria in terms of e-commerce. This shows that companies have adapted to the conditions of the pandemic and moved sales online. The restriction of the right to movement of citizens, determined the companies to invest in the digitalization of trade so that they can continue their business. Unfortunately, we do not encounter the same situation regarding the E-government indicator. The progress made during the pandemic is minimal in the case of Romania and Bulgaria, which shows that state institutions have not implemented measures regarding digitization at the same rate as companies have done.

5 Conclusions

The transition to a digital economy is a process that has become a priority and stringent for every state in the European Union. The pandemic caused by the covid-19 virus accentuated the need for the digitization process. Most people worked from home, school took place online, the right of movement was restricted so that the internet and online operations became indispensable. From shopping online, paying bills or communicating the results of online medical analysis, EU citizens have felt the benefits of the digitization process.

Unfortunately, not all EU countries are equally developed in terms of digitalization. Through this analysis we managed to prove that the hypothesis of the paper is valid, there is a gap in the digital transition at EU level. The countries that have the best performance in the digital process are the countries in the Nordic sub model and those with the worst results are in the central-eastern countries.

DESI index uses data from 2019, so there is a possibility that the next report will show the progress of many countries that have implemented policies during the pandemic. For this reason, in the second part of the paper we analyzed for Finland, Sweden, Romania and Bulgaria 4 indicators to capture the evolution of the digital process during the pandemic. The data showed that the private sector has adapted to the conditions imposed by the pandemic and has invested in digitalization. At the governmental level, the data indicate little progress, but usually these investments require more time so the results can be seen in the coming years.

The role of public institutions in this process is defining and would allow an accelerated absorption of the digital gaps compared to other European countries. From the E-Government perspective Romania occupies the last place in the ranking of EU countries. The fact that certain components of digital process achieved a very good level, similar to those of the best performing countries (high connectivity and high degree of internet use) clearly shows us that Romania has the capacity to accelerate the digitization process in the next period.

As a recommendation, EU member states need to implement policies to improve citizens' skills in internet uses, which would allow them access to much better paid jobs or keep their current jobs. As mentioned in the literature, people at risk of losing their jobs as a result of digitization are part of the category of people with low internet skills. Member States must take into account public policies that favor these vulnerable groups in order to protect them from unemployment and inequality.

The European Union has made digitalization one of the main keys to post-pandemic recovery and has allocated substantial funding to this end. The responsibility of EU countries is to identify and implement the best public policies that will increase the degree of digitalization and produce added value in the economy and society.

References

Acemoglu, D., Restrepo, P.: The race between man and machine: implications of technology for growth, factor shares and employment, Am. Econ. Rev. **108**(6), 1488–1542 (2018)

Acemoglu, D., Robinson, J.: Why nations fail: The origins of power, prosperity, and poverty. Crown Business (2012).

Arntz, M., Gregory, T., Zierahn U.: The risk of automation for jobs in OECD countries: A comparative analysis. Working Papers No. 189, OECD Social, Employment and Migration (2016)

Banhidi, Z., Dobo, I., Nemeslaki, A.: What the overall digital economy and society index reveals: A statistical analysis of the DESI EU28 dimensions. Reg. Stat. **10**(2), 42–62 (2020)

Deaton, A.: The great escape: health, wealth and the origins of inequality. Litera Publishing House, Bucharest (2017)

De Sousa Jabbour, A., Jabbour, C., Foropon, C., Godinho Filho, M.: When titans meet: Can industry 4.0 revolutionise the environmentally-sustainable manufacturing wave? The role of critical success factors. Technol. Forecast. Sacocial Change **132**, 18–25 (2018)

European Commission: Digital Economy and Society Index 2020–Thematic Chapters (2020)

European Commission: Digital Economy and Society Index 2020–Bulgaria (2020)

European Commission: Digital Economy and Society Index 2020–Finland (2020)

European Commission: Digital Economy and Society Index 2020–Italy (2020)

European Commission: Digital Economy and Society Index 2020–Romania (2020)

European Commission: Digital Economy and Society Index 2020–Sweden (2020)

European Commission: Report of the HLEG on the Impact of the Digital Transformation on EU Labour Markets, Brussel (2019)

Schwab, K.: The fourth industrial revolution. World Econ. Forum (2016)

Solow, R.M.: A contribution to the theory of economic growth. Q. J. Econ. **70**(1), 65–94 (1956)

Stavytskyy, A., Kharlamova, G., Stoica E.: The analysis of the digital economy and society index in the EU. TalTech J. Eur. Stud. **9**(3) (2019)

Stoica, E., Bogoslov, I.: A comprehensive analysis regarding DESI country progress for Romania relative to the european average trend. In Balkan Region Conference on Engineering and Business Education, vol 3(1), pp. 258–266 (2017)

Vasilescu, M., Serban, A., Dimian G., Aceleanu M., Picatoste X.: Digital divide, skills and perceptions on digitalisation in the European Union—Towards a smart labour market. **15**(4), PLoS ONE (2020)

Greening the European Freight Transport

Adriana Grigorescu and Amalia-Elena Ion

Abstract Due to the specificity of the EU economy and the potential for the freight industry, the public policy decision-makers have been actively engaged in determining the sources of unsustainable logistics, with the main drive toward greening the entire trans-European network. The World Bank database allows for the construction of a study based on indicators related to fuel consumption, CO_2 emissions, GHG emissions, for the purpose of estimating the effect they have on the whole EU economy. The public policies on the EU agenda refer to the cleaning of the transport sector for more than a decade now. The current study will estimate the totality of emissions from transport in correlation to the forecasted effects of the coercive legislation for greening the freight network, in three European countries–Cyprus, the Netherlands, and Latvia. The three countries have been selected due to their specificity of freight transport modes that covers their infrastructure. The results are showcasing the possible links to the state of the country's economic sustainability and to the implemented public policies for greening the freight transport industry.

Keywords EU green policy · Freight industry · Freight transport modes · Pollutant emissions · Sustainability

1 Introduction

In line with the information released by the European Commission in 2019 (2019) and the European Parliament in the last two decades is the statement that portraits a world economy in which there is continuous and purposeful commitment towards stabilizing the environment and the climate through actions convergent with sustainability, moderation, resource-efficiency, competitiveness and innovations. There has also been a discourse on the future of the EU transport sector, its business model, and its potential within the economy. The influence of its secondary products, such as CO_2 emissions, is considered to add significantly to the greenhouse gasses that negatively

A. Grigorescu (✉) · A.-E. Ion
National University of Political Studies and Public Administration SNSPA, Bucharest, Romania
e-mail: adriana.grigorescu@snspa.ro

M. Busu (ed.), *Digital Economy and New Value Creation*, Springer Proceedings in Business and Economics, https://doi.org/10.1007/978-3-031-07265-9_4

affect the livelihood of the environment. The greening of the EU freight transport refers to the diminishing quantity released in the breathable air and other negative effects of different types of air pollutants, as a result of a sustainable formula for the usage of multi-modal transport systems. The current paper addresses the idea that no particular way of transport is superior to other in terms of overall environmental output, and in which a multi-modal connectivity inland (rail, road, waterways) could really increase the sustainability indicators, improve the competitivity of economic operators in the sector, and decrease the negative impact on the environment.

This particular research is part of a broader study on the European freight transport industry, and aims at clarifying the proportion of damaging practices in the field compared to the legislative measures that have been enforced accordingly. The decision of transferring the system towards a greener perspective would imply efficiency, effectiveness in financial terms, sustainability, and competitiveness. Generally, the road freight mode is considered to be the best alternative for the transport of goods, especially for its ease of access, for its convenience, for its speed, and even for its cost-effectiveness (Eurostat 2021). Nevertheless, there are many factors influencing the European freight market, and should all be thoroughly assessed before deciding on the next determinants of the industry.

There are patterns to the freight industry in the European Union, consistent with the fact that, as of the Eurostat information (2020), road transport accounted for three quarters of the total inland freight transport, and that one third of the air freight was shipped through German airports. Moreover, the waterway freight transports have been carried through Dutch ports in proportion of 17%. This showcases how the modal split is averaged across the EU countries, although, there are particularities to some of the EU countries, where the proportions are consistently different. The countries with the biggest disparities are Cyprus, the Netherlands and Latvia, from the perspective of modal split of inland freight transport. The current research will, therefore, look into the opposable situations of those three EU member states, and understand the impact the different combinations of transport have on the environment and the economy. By using a regression analysis, the database will forward information on potential correlations between the different freight transport combinations and the overall level of the economy and the public policy compliance within the considered EU member states, on the basis of different variables measuring air pollutants' levels, and pollutants' emissions.

The paper addresses the dilemma of greening the freight transport, and the transport system in Europe, and whether it really transforms the entire model of business or not. The limitations of the study are consistent with the dimension of the researched group of countries, the lack of clustering, the simplicity of presuming precedence in indicators' correlations, but it definitely represents a starting point in a major industry research.

2 Literature Review

Consistent studies have been performed in relation to the topic of freight transport and all the connected elements to it, showcasing a range of different particularities and characteristics of the given industry. The literature comprises studies performed on topics varying from environmental sustainability, intermodal transport mode, tenders and contracts for freight transport, cost theories, business models for freight transport, etc. (Bask and Rajahonka 2017). Across the years, the freight transport market was subjected to a number of forecasting practices, such as policy analysis model, interlinked modules, I/O models for production and distribution, disaggregate model for mode and shipment, assignment module, etc. (Jong et al. 2004) in order to undercover the modal split gravity formulation, the identification of best practices, and the correlation to future possible policy screening, analysis and enactment.

Some authors argue that the intermodal freight transport, although representing a major sector of the industry, is programmed to interfere with problems (Macharis and Bontekoning 2004), ranging from container optimization of transfers during loading phase, joint monitoring between the loading and reloading phase, limitations of storage capacity, waiting times imposed by tactical and technical operations–drayage, terminal, network, intermodal operations, etc. In terms of passenger transport, Buldeo Rai et al. (2017) approach the topic of crowd logistics, identifying models for linking the supply and demand of logistics services, considering the perspective of time and space, and all the interconnectivity with the economy, society and environment, suggesting the adaptation of policy and regulations for sustainability purposes and social development.

The carbon emissions from freight transport have experienced a sharp increase in the last 30 years, as Lv et al. (2019) study explores the case of Chinese urbanization process and its correlations to the freight industry pollutants. Apparently, since 1988, the carbon emissions from freight transport in China have increased more than 30 times, with the road haulage being the largest sector of the market as well as the biggest polluter. Given the positive effects of the urbanization process for the development of the overall economy, it is clear that the former will greatly and positively impact the freight industry. Determined by the grandeur of the market, the different combinations of transport modes become specific imprints of particular areas in the country. Moreover, one of the most significant findings of the study (Lv et al. 2019) portraits the inverted U-shaped correlation between the carbon emissions from different transport modes and the GDP per capita.

Following the theme of sustainability and environment-friendly economies, the work of Muñoz-Villamizar et al. (2020) sheds a light on the methodology for environmental performance assessment of the freight industry with the application of the Overall Greenness Performance instrument, a tool that measures value added-activities, productivity of companies in the transport industry, and the requirements from the latter in order to maintain a standard in environmental impact.

The public policy influencing and regulating the freight transport industry has the capability of altering the context of the business in a positive or negative manner. The function it reproduces as a result of decision-making processes has been observed in the study performed by Marcucci et al. (2017), using simulations on multilayer networks and opinion dynamics for the traffic zone of the Italian capital city. Basically, the research managed to rank plausible policies that might maximize productivity and minimize cost-related attributes, all achievable through the usage of negotiation processes.

Another approach toward combining modes of transport for the growth of the industry, and the dissipation of negative environmental effects of the freight businesses was highlighted in the work of Ambra et al. (2018). Starting from the premise of environmental, economic and societal sustainability achievement, the authors propose a combination of the physical internet and the synchro-modal designs' alignment to the European public policy on the freight industry. The main scope is that of making a breakthrough, and finding the optimum state of capacity utilization.

Since almost 25% of the GHGs in Europe are attributed to the transport sector of the economy, a factor of the fossil-fuel usage especially in the road transport networks, scientists have worked on creating alternative fuels, and on testing other types of fuels with a lower carbon emission; recently, a new type of heavy-duty vehicle for freight transport has been promoted, having an engine that runs on liquefied natural gas. Apparently, LNG has a smaller environmental impact compared to diesel fuels, with drops of 20% of GHG emissions during operations with a heavy-duty vehicle running on a gas fueled engine (Osorio-Tejada et al. 2016). This alternative is backed by regulations and policies enforced through lower road taxes, tolls, and import tariffs.

Some of the most significant findings in the literature (from 2017 to 2020) were retrieved from a research (Winkler and Mocanu 2020) based on the German national transport model (DEMO). The study built on three future scenarios, provisioning the sustainability, performance and function (including modes) of the transport industry in Germany, and, eventually, indicated that road transport continues to occupy the majority of the market, and suggested the regulation to include potential technological developments toward conditioning the industry to perform under sustainable systems and infrastructure.

Considering the growing technological advancements in all sectors of the economy and industry, it is clear that new modes of transportation will arise. This topic was debated in the research of Visser (2018), introducing the concept of automated underground freight transport as an extra mode of transport or even as a replacing mode of transport for the benefit of the environment, and even of the economy.

Subsequently, the policies formulated in regard to the urban freight transport system have been implemented in such a form that is restrictive to objectives and completion targets (Akgün et al. 2019). Therefore, the access to monitored performance on behalf of the UFT is limited from the perspective of public policy, since there is no strategic direction under the form of clear, measurable objectives and milestones. From this approach, it is manageable to assume that not all the decisions transferred in policies are completed, and that the urban freight transport cannot be assessed due to lack of measurable indicators and timeframes. Moreover, other studies (Kin et al. 2017) propose testing the sustainability of the UFT through the perspective of megacities, and urbanization developments. The results of this particular research underline the necessity of implementing complete policies to address the value chain in its entirety. Consistently, a urban telematics system for monitoring, evaluating, controlling and improving the transport within a city was developed and tested by Comi (2018), with results demonstrating optimized planning and delivery operations, as smart tools intuitively upgrade the transport industry.

Scholars have contributed to the field of freight transport by examining the output of different modes of transport, concluding that road transport particularly has a negative effect on the quality of the environment, while rail transport could be the answer to the ever-asked question. Nevertheless, the consistent inequalities between the simulation and the real-case scenario resurface under the form of railroad infrastructure, frequency of trains, delivery reliability, competition, holistic business models (Kumar and Anbanandam 2020).

According to Centobelli et al. (2020), the adoption of green policies and green practices within the freight transport industry relies on four pillars that create the basis for the green index, namely sustainability leaders, information systems adopters, sustainability followers, and green practice adopters. Thus, there is a clear process for the implementation and development of environmental sustainability in the freight transport and logistics firms. The strategic leverage stems from few green practices and information systems' adoption, which determines the growth stage, identified by the organizational and technological leverage (either many green practices and few information systems, or the reverse scenario), and followed by the maturity stage, also called the managerial leverage, where there is an equilibrium between the two debated elements of the perspective. The guideline created by Centobelli et al. (2020) could be proposed for the organizational adoption of greening policies and practices in the freight transport industry.

Another aspect that would create a consistent effect on the emissions of the freight transport industry refers to the modes of propulsion for heavy-duty vehicles for road transport. A study from 2017 (Quiros et al. 2017) introduced the comparison between heavy-duty vehicles fueled by diesel, hybrid diesel, and natural gas, and their impact on the environment, in terms of CO_2, CH_4 and N_2O emissions, on low-speed versus high-speed routes. The natural gas-based engines on the heavy-duty vehicles emitted 15% lower CO_2 levels than conventional diesel heavy-duty vehicles on highway routes, with 12% higher CO_2 level emissions from CNG vehicles compared to diesel engine trucks when the loads weight was higher. Moreover, Ghandriz et al. (2020) worked on a model that employs battery electric heavy-duty vehicles (BEHVs) and automated driving systems (ADS), fueling the industry with the premises for significant cost-reductions, optimization of infrastructure and transport, and competitiveness of ADS and BEHVs render for longer travel ranges, without the necessity of human capital involvement (in driving tasks). Other studies (Mirhedayatian and Yan 2018) experimented on the economic performance of electric vehicles (EVs) as an alternative to traditional road transport modes. Although the policies support the integration of the former in the business model of the freight transport firms, the total logistics cost on the firms increases, while the social welfare improves.

Returning to the problematic of policy making and implementation, Le Pira et al. (2017) address the decision-making process through the perspective of the stakeholders. Therefore, they propose a model for the integration of stakeholders' input in the realization and implementation of urban freight transport policies, by assessing the behavioral knowledge on stakeholders, and the dynamic interaction within the decision-making processes, for the ultimate purpose of general acceptability of policies and consensus on building a stable framework for the UFT (Subanidja and Hadiwidjojo 2017). A study directed toward the impact assessment of freight transport on a particular inland situation in Belgium (Merchan et al. 2019) proposed, under the methodology of the life cycle of negative impacts with effects on climate change, chemical produce on the ozone layer, and matter formation, a comparison between road and rail freight transport. Results have showcased that the environmental impact is maximized when the road freight transport mode is employed, compared to the minimized effect when assessing the rail freight transport mode. There is also the discussion of planning the freight transport according to the city development strategy, by creating a comprehensive transport plan, with projects pertaining to the policy formulation, stakeholders' opinion, ICT inclusion, and to the development of the city through the inclusion of UFT knowledge and practices (Kiba-Janiak 2017). The transport is considered part of the emerging knowledge intensive activities and a potential sector of integrating the newest technologies as well as job creator (Grigorescu et al. 2020).

Across the literature there is a persistent highlight on the effect of the freight transport industry in regards to the quality of the breathable air, and adds on the perspective where supposed practices alongside public policies would determine diminishing negative effects of the former variable on the latter. This particular study builds on the statement extracted from various studies that particular combinations of modes of transport in the freight industry might determine higher or lower negative effects on the air quality. By examining the outliers within the European Union, where there is one particular mode of freight transport that generally covers the market, it is possible to observe the carbon emissions, GHG emissions, methane emissions, and other air pollutants resulted from the burnt fossil fuel due to diesel engines or other types of engines used within this economic sector. This research is meant to perform an analysis on the real negative effects different types of modes of transport in the freight industry have on the quality of breathable air within the European Union.

3 Methodology

Understanding relationships between different variables within the observable universe, in general, represents the underpinning of human civilization. It brings the knowledge one step further, and converges into learning on account of particular phenomena. The manipulation of the near or proxy environment is based on empirical observations or testing possible correlations, all primed by the theoretical background. Although some relationships can be easily observed and tested, some require analytical extraction, with the study of the possible/probable effect of regressors on response variables. This type of analysis has the power of backwards tracking the impact of multiple variables, given the name regression analysis. The variables used in the test were listed in Table 1.

As stated above, the EU countries considered for the study are the Netherlands, Latvia and Cyprus. The main reason for the choice related to this particular decision-making process resides in a statistics' graphic showing the main structures of freight transport modes across EU countries. There were a few countries (the Netherlands, Latvia, Cyprus, Malta and Romania) that stood out in terms of transport modes coverage. Since Malta had presented the same patterns as Cyprus, one of those two countries was randomly selected for the testing phase. Romania is a special case that will be further approached in future studies.

The database was compiled from an annually updated data set downloaded from the World Bank, and was tested for different correlations between variables using STATA 13. The maximum number of observations under a particular variable was 19, for the information tested referred to activities and stats consistent to the 2001–2019 timeline. The method used to test the dependent variables against the GDP per capita and the CO_2 emissions was based on the work of Lv et al. (2019) which demonstrated an inverted U-shaped correlation between the carbon emissions from different transport modes and the GDP per capita (Figs. 1, 2 and 3).

The main hypotheses of the study are unidirectional, and state that for each one of the countries selected the pollutant emissions from freight transport activities is not the main source of air pollution (1), and that the EU public policies for greening the freight transport industry have had a clear impact on the latter (2). Moreover, there is no clear understanding from the tests that a particular combination or lack of transport modes would imply a greener industry and less air pollution (3).

4 Results

Based on the Netherlands' results on the linear regression analysis, it is estimated that the correlation between the CO_2 emissions in metric ton per capita, and GDP per capita at current US value is supported at 95% significance level, showcasing an apparent negative connection, and supporting the statement of Lv et al. (2019) of an inverted U-shaped correlation between the carbon emissions from different transport modes and the GDP per capita. According to the magnitude of the correlation between the two variables, any 1 metric ton drop in the CO_2 emissions per capita will trigger a USD 0.00289 increase in the GDP per capita (current \$). The t-statistic is calculated at −2.45, a value that reads out a typical significant test at 95% level. The p-value registered is 0.02, while the R-squared is 0.3, a high level for social sciences, measuring the goodness of fit of the model (Fig. 4).

Moreover, from the examination of the relationship between the CO_2 emissions (kt) and the quantity of goods transported through rail or air freight modes, there is no apparent influence of those two modes of freight transport and the emissions'

Table 1 Description of variables used in the model

Variable name	Variable description
Air transport, freight (million ton-km)	Air freight is the volume of freight, express, and diplomatic bags carried on each flight stage (operation of an aircraft from takeoff to its next landing), measured in metric tons times kilometers traveled
CO_2 emissions (kt)	Carbon dioxide emissions are those stemming from the burning of fossil fuels and the manufacture of cement. They include carbon dioxide produced during consumption of solid, liquid, and gas fuels and gas flaring
CO_2 emissions (metric tons per capita)	
CO_2 emissions from transport (% of total fuel combustion)	CO_2 emissions from transport contains emissions from the combustion of fuel for all transport activity, regardless of the sector, except for international marine bunkers and international aviation. This includes domestic aviation, domestic navigation, road, rail and pipeline transport, and corresponds to IPCC Source/Sink Category 1 A 3
Fossil fuel energy consumption (% of total)	Fossil fuel comprises coal, oil, petroleum, and natural gas products
GDP (current US$)	GDP at purchaser's prices is the sum of gross value added by all resident producers in the economy plus any product taxes and minus any subsidies not included in the value of the products. It is calculated without making deductions for depreciation of fabricated assets or for depletion and degradation of natural resources. Data are in current U.S. dollars. Dollar figures for GDP are converted from domestic currencies using single year official exchange rates. For a few countries where the official exchange rate does not reflect the rate effectively applied to actual foreign exchange transactions, an alternative conversion factor is used
GDP per capita (current US$)	GDP per capita is gross domestic product divided by midyear population. GDP is the sum of gross value added by all resident producers in the economy plus any product taxes and minus any subsidies not included in the value of the products. It is calculated without making deductions for depreciation of fabricated assets or for depletion and degradation of natural resources. Data are in current U.S. dollars

(continued)

Table 1 (continued)

Variable name	Variable description
GHG net emissions/removals by LUCF (Mt of CO_2 equivalent)	GHG net emissions/removals by LUCF refers to changes in atmospheric levels of all greenhouse gases attributable to forest and land-use change activities, including but not limited to (1) emissions and removals of CO_2 from decreases or increases in biomass stocks due to forest management, logging, fuelwood collection, etc.; (2) conversion of existing forests and natural grasslands to other land uses; (3) removal of CO_2 from the abandonment of formerly managed lands (e.g. croplands and pastures); and (4) emissions and removals of CO_2 in soil associated with land-use change and management. Data are in million metric tons
Methane emissions (kt of CO_2 equivalent)	Methane emissions are those stemming from human activities such as agriculture and from industrial methane production
Nitrous oxide emissions (thousand metric tons of CO_2 equivalent)	Nitrous oxide emissions are emissions from agricultural biomass burning, industrial activities, and livestock management
PM2.5 air pollution, mean annual exposure (micrograms per cubic meter)	Population-weighted exposure to ambient PM2.5 pollution is defined as the average level of exposure of a nation's population to concentrations of suspended particles measuring less than 2.5 microns in aerodynamic diameter, which are capable of penetrating deep into the respiratory tract and causing severe health damage. Exposure is calculated by weighting mean annual concentrations of PM2.5 by population in both urban and rural areas
Pump price for diesel fuel (US$ per liter)	Fuel prices refer to the pump prices of the most widely sold grade of diesel fuel. Prices have been converted from the local currency to U.S. dollars
Railways, goods transported (million ton-km)	Goods transported by railway are the volume of goods transported by railway, measured in metric tons times kilometers traveled
Total greenhouse gas emissions (kt of CO_2 equivalent)	Total greenhouse gas emissions in kt of CO_2 equivalent are composed of CO_2 totals excluding short-cycle biomass burning (such as agricultural waste burning and savanna burning) but including other biomass burning (such as forest fires, post-burn decay, peat fires and decay of drained peatlands), all anthropogenic CH_4 sources, N_2O sources and F-gases (HFCs, PFCs and SF6)

Source World Bank Series Meta Data; (The World Bank 2021)

Variable	Obs	Mean	Std. Dev.	Min	Max
seriesname	19	2010	5.627314	2001	2019
airtranspo~m	19	25.61841	20.79757	.021	49.376
co2emissio~t	16	7249.43	786.3762	5947.874	8577.113
co2emissio~a	16	6.763815	.9539062	5.1998	7.930295
co2emissio~l	14	29.29243	1.383454	26.78063	31.85053
fossilfuel~t	14	96.05741	1.76405	92.90621	97.82739
gdpcurrent~n	19	21529.95	5134.206	10397.9	27844.7
gdppercapi~s	19	26491.77	5323.633	14821.45	35397.36
methaneemi~n	12	618.4977	13.81063	584.682	642.4094
nitrousoxi~t	12	290.2857	18.73317	264.4988	315.1088
pm25airpol~u	9	18.82751	1.137245	16.95802	20.09096
pumppricef~r	8	1.24625	.4205757	.44	1.78
totalgreen~c	10	8445.393	647.0162	7431.193	9165.955

Fig. 1 Descriptive statistics–Cyprus (*Source* STATA compiling)

Variable	Obs	Mean	Std. Dev.	Min	Max
year	19	2010	5.627314	2001	2019
airtranspo~m	19	5.39996	5.322311	.465	18.375
co2emissio~t	16	7380.754	456.7154	6912.295	8316.756
co2emissio~a	16	3.449831	.243069	2.988929	3.849593
co2emissio~l	14	39.90558	3.148993	34.62069	44.85646
fossilfuel~t	14	62.53323	2.807674	56.71743	65.85
gdpcurrent~n	19	24866.48	8642.74	8351.416	35756.23
gdppercapi~s	19	12041.03	4608.182	3573.303	17858.28
ghgnetemis~o	9	-17.9507	3.39629	-22.7434	-13.53093
methaneemi~n	12	3171.659	116.4506	2930.62	3416.83
nitrousoxi~t	12	1376.186	157.86	1226.456	1830.665
pm25airpol~u	9	15.19419	1.431303	13.42646	17.11506
pumppricef~r	8	1.21875	.3591831	.65	1.77
railwaysgo~t	19	17952.68	2189.916	14179	21867
totalgreen~c	12	13662.57	748.2751	12089.66	14670

Fig. 2 Descriptive statistics–Latvia (*Source* STATA compiling)

levels. Nevertheless, the correlation between the response variable CO_2 emissions from transport (% of total fuel combustion) and the regressor – quantity of freight goods transported via railways, estimates a positive effect, with a magnitude of 0.06% increase in the CO_2 emissions from transport for each 1 million ton/km of freight goods shipped via railways.

Variable	Obs	Mean	Std. Dev.	Min	Max
year	19	2010	5.627314	2001	2019
airtranspo~m	19	5233.484	717.0084	4115.852	6444.382
co2emissio~t	16	177335.9	5203.907	168883.7	185355.8
co2emissio~a	16	10.73731	.4723229	10.01385	11.3793
co2emissio~l	14	20.5405	.7339915	19.44127	21.69504
fossilfuel~t	15	92.88754	1.085761	90.92188	94.33435
gdpcurrent~n	19	778002.9	150226.7	431213.4	947997.7
gdppercapi~s	19	46646.16	8488.282	26873.28	57644.48
ghgnetemis~o	9	2.519963	.1663664	2.241624	2.687378
methaneemi~n	12	21150.73	1449.897	19025.82	24013.3
nitrousoxi~t	12	11615.17	2136.786	8924.044	13884.22
pm25airpol~u	9	13.49719	1.213473	12.02524	15.23243
pumppricef~r	8	1.45125	.3639245	.81	1.95
railwaysgo~t	19	6080.105	857.7772	4293	7216
totalgreen~c	12	211193.3	8851.425	195873.8	221355.3

Fig. 3 Descriptive statistics–the Netherlands (*Source* STATA compiling)

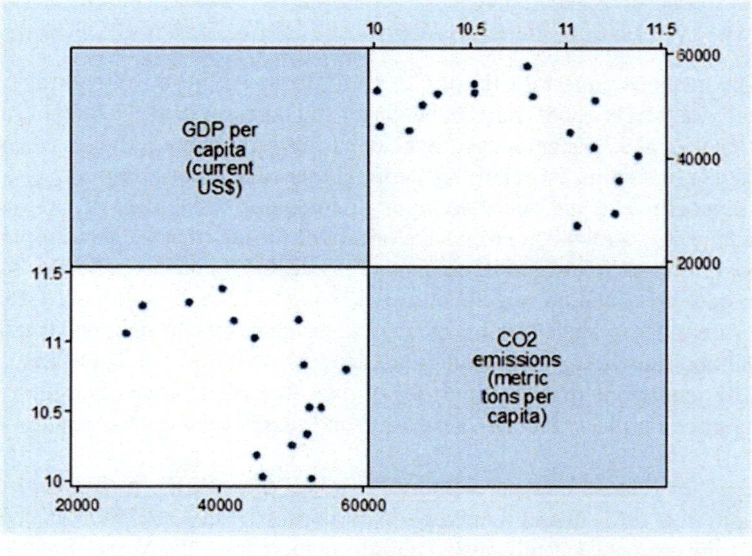

Fig. 4 Scatterplot Matrix presenting the U-shaped distribution of the correlation between the CO_2 emissions (metric tons per capita) and the GDP per capita (current US$)–NL (*Source* STATA compiling)

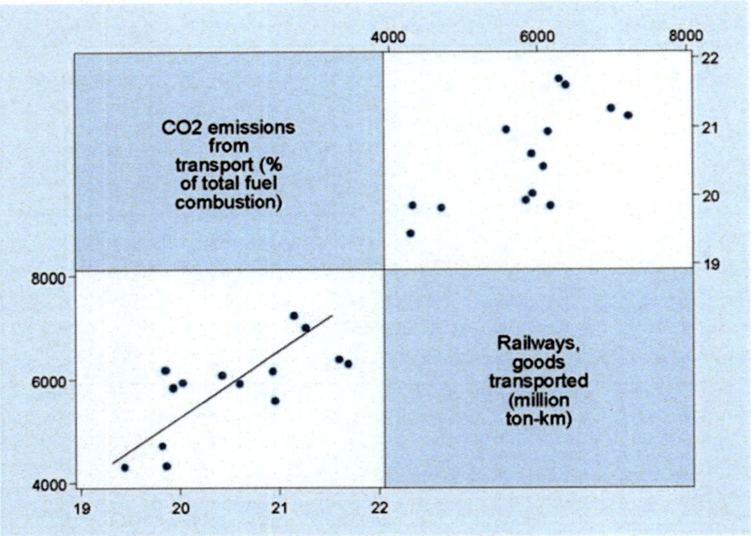

Fig. 5 Scatterplot Matrix presenting the linear correlation between the CO_2 emissions from transport (% of total fuel consumption) and the Railways, goods transported (million ton-km)–NL (*Source* STATA compiling)

The methane emissions (kt of CO_2 equivalent) are negatively correlated to the quantity of freight goods transported via air and railways in the Netherlands, demonstrating that, at 95% significance level, other types of human activities determine the methane production, especially agricultural related activities (Fig. 5).

The coefficients are considerably high compared to the case of CO_2 emissions from transport, allowing for a conclusion around the idea that freight transport related activities are not solely and significantly responsible for the air pollution. Moreover, the model presented no correlation between the PM 2.5 air pollution, mean annual exposure (micrograms per cubic meter) and the quantity of freight goods transported via air and railways. At the same time, the tests showcased no correlation between the CO_2 emissions from transport (% of total fuel combustion), and the regressors pump price for diesel fuel (US$ per liter) and fossil fuel energy consumption (% of total).

The Netherlands transports its freight goods through the road infrastructure in proportion of 50%, inland waterways in proportion of 43%, and more than 6% of the goods are shipped via railways. The information from the World Bank database' analysis does not underline to a correlation between the major freight transport mode–road haulage, and the levels of CO_2 and other pollutants as a result of fossil fuel consumption. The public policy directed toward greening the European freight transport industry has, particularly for the Netherlands, contributed to a diminishing effect of the air pollution. Considering the geographical structure of the country, it is clear that the extensive usage of inland waterways will continue to develop within the freight transport industry, with a support activity from the road haulage for short-distance trips. By testing the correlation between the total GHG emissions (kt of CO_2 equivalent) and the CO_2 emissions from transport (% of total fuel combustion), the model underlines the negative relationship established among the two variables—an increase of 1 kt of GHG emissions determines a 0.00643% drop of CO2 emissions from transport (Fig. 6).

The second country used in the study was Latvia for its freight transport structure, combining a carrier pattern of 76% haulage trips via railroad, and 24% via road. The correlation between the CO_2 emission (metric tons per capita) and the GDP per capita (current US$) is a positive one, tested at 99% significance level, with a t-statistic of 4.54, and a p-value of 0 (Figs. 7 and 8).

The magnitude of the correlation is determined by the following effect: USD 1 increase in the GDP per capita triggers a 0.00429 metric tons per capita increase in CO_2 emissions.

It is, comparably opposite to the case of the Netherlands, a typical formulation for an economy undergoing the development process. The U-shaped structure of the relationship CO_2 emissions—GDP is not formed yet. Generally, the Latvian economy is based on activities that produce increasing amounts of CO_2 emissions, calculated in metric tons per capita.

The freight air transport (million ton-km) has a considerable effect on the CO_2 emissions (kt), as the air transport of 1-million-ton freight goods per km determines an increase of 4631.48 kt of CO_2 emissions. The data was inconsistent with any clear

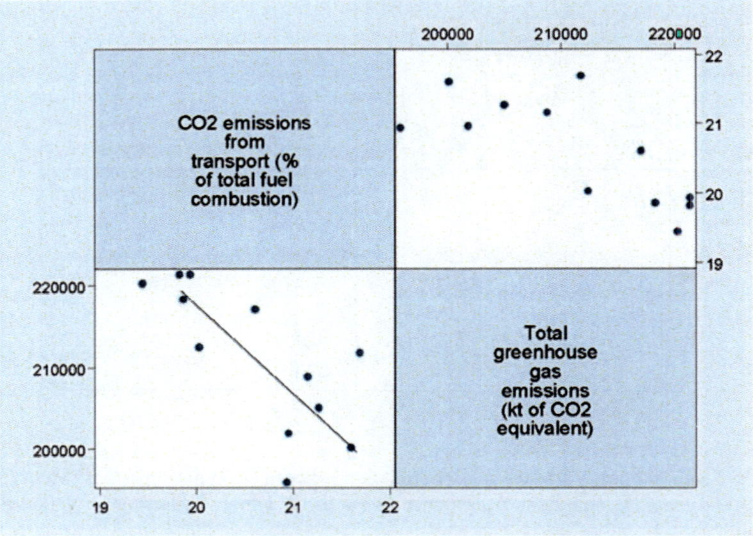

Fig. 6 Scatterplot Matrix showcasing the linear correlation between the CO_2 emissions from transport (% of total fuel combustion) and Total greenhouse gas emissions (kt of CO_2 equivalent)–NL (*Source* STATA compiling

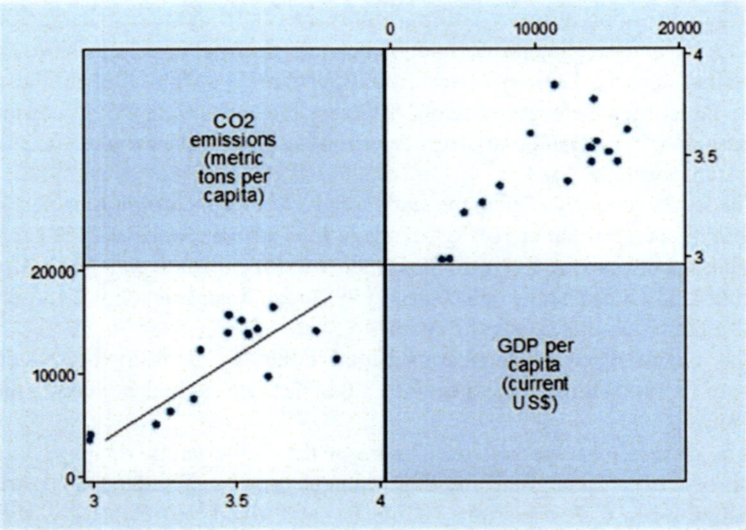

Fig. 7 Scatterplot Matrix of the linear correlation between CO_2 emissions (metric tons per capita) and the GDP per capita (current US$)–LV (*Source* STATA compiling)

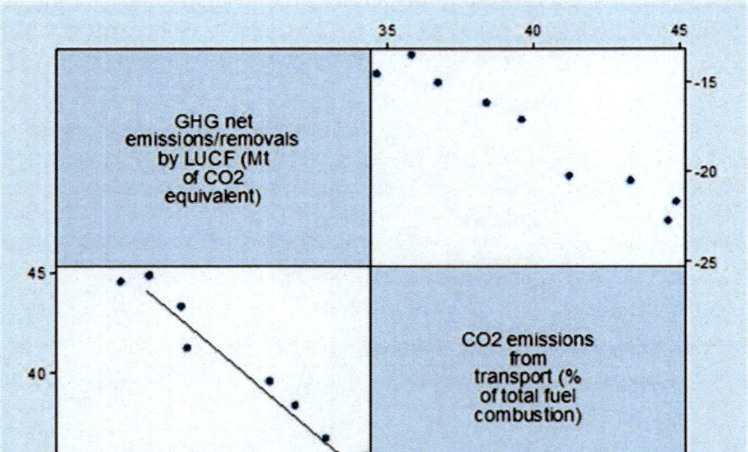

Fig. 8 Scatterplot Matrix of the linear correlation between GHG net emissions (Mt of CO_2 equivalent) and the CO_2 emissions from transport (% of total fuel combustion)–LV (*Source* STATA compiling)

result regarding the existence of any type of correlation between the CO_2 emissions from transport (% of total fuel combustion) and the quantity of freight goods shipped via railroad. Therefore, it is not possible to allocate any type of pollutant observed in the breathable air to the transportation of freight goods via railroad. This finding is consistent with the direction of the public policy in the EU, advocating for a greener performance of the freight transport industry. The results of this study can conclude that an intermodal system, where the freight goods are transported via railroad in their majority on long distances, and via road haulage on short distances, is the best possible solution for a greener EU transport network. Moreover, the correlation between the GHG net emissions/removals by LUCF (Mt of CO_2 equivalent) and the CO_2 emissions from transport (% of total fuel combustion) is negative. The t-statistic is -11.08, and the p-value is 0, with a measurement of the overall fit of the model of 0.94. The magnitude of the effect can be translated into the following: any 1% increase in the CO2 emissions from transport (% of total fuel combustion) triggers a drop of 86.41 mt of GHG net emissions.

Therefore, the main polluter in this case is not the freight transport industry, a probable effect of the European public policy drafted and applied in the direction of more sustainable transport practices.

The third case selected was represented by Cyprus, where the sole mode of freight transport is that performed on road. In terms of correlations between the CO_2 emissions (metric tons per capita) and the GDP per capita (current US$), no consistent pattern resulted from the model. The CO_2 emissions (kt) in Cyprus are influenced by the quantity of freight goods transported by airways, as 1 extra million-ton-km of freight goods shipped via airways triggers an increase of 3218.95 kt of CO_2. The t-statistic is 4.7, significant at 95% level, and the p-value is 0. The overall fit of the model is high, at 0.6119. When testing the correlations between the CO_2 emissions from transport (% of total fuel combustion) and the pump price for diesel fuel (US$ per liter), the results showcased a positive relationship, due to the lack of freight transport modes/alternatives. As the pump price for diesel fuel increases by USD 1 per liter, the CO_2 emissions from transport (% of total fuel combustion) increases by 225.25%. The t-statistic is 3.55, at 95% significance levels, with p-value 0.016, and with an overall fit of the model at 0.71 (Fig. 9).

Moreover, the regression analysis has highlighted also the correlation between CO_2 emissions from transport (% of total fuel combustion) and fossil fuel energy consumption (% of total), where any 1% increase in the latter determines a drop of 68.31% in the former, meaning that the air pollution with CO_2 emissions is not only triggered by the freight transport industry. The effect is even more relevant when observing the influence of Nitrous Oxide emissions (thousand metric tons of CO_2 equivalent) on the overall CO_2 emissions (metric tons per capita). There is a significant amount of pollution relevant to the emissions from agricultural biomass burning, industrial activities, and livestock management (Fig. 10).

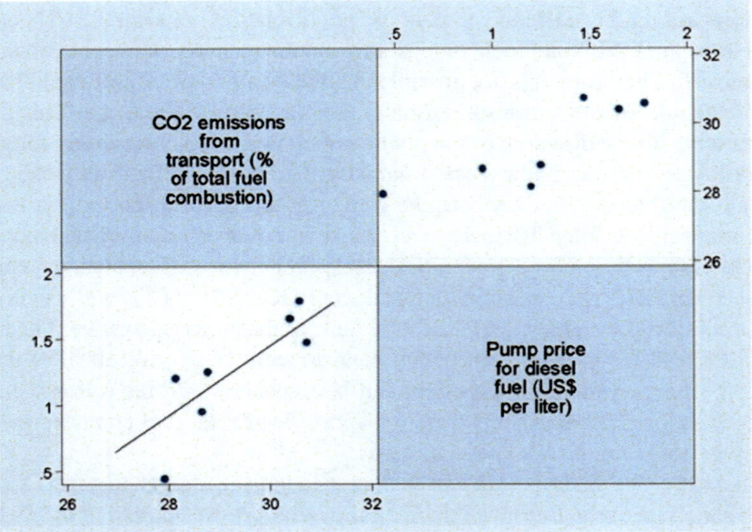

Fig. 9 Scatterplot Matrix of the linear correlation between CO_2 emissions from transport (% of total fuel combustion) and the pump price for diesel fuel (US$ per liter)–CY (*Source* STATA compiling)

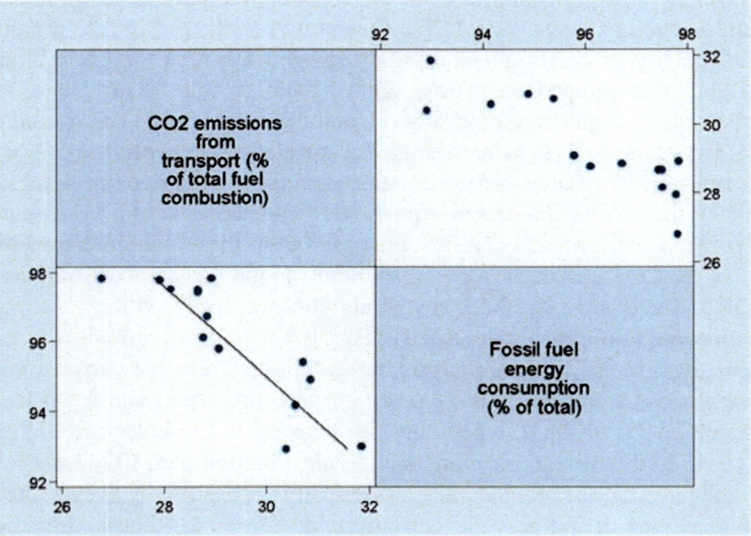

Fig. 10 Scatterplot Matrix of the linear correlation between CO_2 emissions from transport (% of total fuel combustion) and the Fossil fuel energy consumption (% of total)–CY (*Source* STATA compiling)

5 Discussion

Different countries require different solutions for the transport of the freight goods. Moreover, some trips/journeys can be efficient only if the compatible mode(s) of transport is(are) used. In general, road freight transport is the get-to mode of transport, due to its minimal costs, its speed of delivery, its maximum potential for loading/unloading, etc. For decades now, the European Commission, together with the European Parliament have worked on legislation that would ensure the effectiveness, efficiency and sustainability of the freight transport industry within the community. Numerous attempts and actual directives have made their way into the completion of a green body of law, presenting alternatives with lower environmental impact, and positive contribution to the state of climate. Naturally, the highest levels of fossil fuel combustion are released by the diesel engines of heavy-duty road vehicles, while the railroads and inland waterways presumably offer a more sustainable solution to the transport of freight goods.

The current paper was built around the idea of testing out for European countries that do not abide to a general pattern in terms of freight transport structure of modes, and of understanding the impact or effects of the different modes of transport under the form of pollutant emissions. Moreover, it was the case of also putting together the puzzle pieces with regards to the overall economic situation of each country, by employing the usage of GDP in testing the phenomenon.

The countries selected for the model were the Netherlands, Latvia and Cyprus. Each one of those countries has a specificity in terms of combination of modes of transport for freight goods. The Netherlands relies on road and inland waterways for transport, Latvia on railways, and Cyprus solely on road transport. It is also the question of the potential for implementation of the public policies in the EU regarding the greening of the freight transport industry, since there are so many countries with their own particularities—geographical differences, political, social, economic and technological differences, etc. Efficient-based solutions that incorporate the sustainability principle cannot be limited to effects that might be only specific to certain cases.

The results of the regression analysis have disclosed some interesting points. Within the model, a series of variables have been tested, the majority being categorized as pollutants-and emissions-related, such as CO_2 emissions (kt), GHG emissions (mt as CO_2 equivalent), Methane emissions, Fossil fuel energy consumption, Total greenhouse gas emissions, Nitrous oxide emissions, PM 2.5 pollution, etc. Among the regressors there have been variables such as GDP per capita, Pump price for diesel fuel, railways goods (million tons-km), air freight goods (million tons-km), etc.

The main hypotheses of the study are unidirectional, and state that for each one of the countries selected the pollutant emissions from freight transport activities is not the main source of air pollution (1), and that the EU public policies for greening the freight transport industry have had a clear impact on the latter (2). Moreover, there is no clear understanding from the tests that a particular combination or lack of transport modes would imply a greener industry and less air pollution (3).

The main findings of the research have been consistent with the aspects disclosed in the next three paragraphs under the form of an elaborate discussion on the paper's results.

The Netherlands has a competitive and developed economic system, as the results of the testing have clearly underlined, and it is the case of one EU member state that implemented and worked on creating solutions as directed within the greening agenda for the freight transport industry, *validating hypothesis 2*. Here is an example of such a follow-up on the public policy, consistent with a showcase of drop in CO_2 emission levels with the growth and expansion of the GDP per capita indicator. Moreover, the correlation between CO_2 emission levels and the quantity of freight goods transported via railways was not confirmed, *validating hypothesis 1*. There is a slight growth in CO_2 emissions from transport activities, when the quantity of freight goods transported increases, but that is almost insignificant. When looking into the methane emission levels in rapport to the quantity of freight goods transported, it is clear that a significant level of pollution comes from different agricultural activities, livestock management, and other complementary activities, statement backed-up by the next result of the study, where the connection between the PM2.5 air pollution (mean annual exposure) and the quantity of freight goods transported was not confirmed, and *hypothesis 3 was invalidated*. Furthermore, any increase in the level of GHG emissions (coming also from the agricultural activities, livestock management and other complementary activities) determines a drop in the CO_2 emission levels from transport.

The results obtained for Latvia showed a developing economy, focused on using cost-effective modes of freight transport, without bypassing the EU green public policy, and, thus *validating hypothesis 2*. In this particular case, the highest pollution from transport is stemming from the airway mode of freight transport. There was no correlation between CO_2 emissions from transport and the quantity of freight goods shipped via railroad, and, since their infrastructure depends in very high proportion on the railways freight transport, the *first hypothesis was also validated*. To back-up that statement stands the negative correlation between the CO_2 emissions from transport and the GHG emission levels. Thus, the results of the testing *validated hypothesis 3*.

Cyprus was the country in the study that uses only road freight transport due to the state of infrastructure development. There was no particular connection between the economy's indicator and the level of carbon emissions. The air freight goods transport mode is triggering increases in the CO_2 emissions, but they are not significantly high. One interesting aspect disclosed by the results of the testing is based on one fundamental principle of economy. The monopoly of the road haulage market showcases a positive relationship between the CO_2 emissions from transport and the pump price for diesel fuel, as there is no alternative to the freight goods transportation in Cyprus other than the road transport mode. *Hypothesis 1 was validated*, as any growth in the fossil fuel energy consumption determines drops in the CO_2 emissions from transport, underlining that the pollution is not solely coming from the freight transport industry, and that high pollution is stemming from agricultural activities, biomass burning, industrial activities, and livestock management. The other two *hypotheses, 2 and 3, could not be validated or invalidated* in this particular case.

6 Limitations and Conclusion

The transport sector of the economy is a derived demand, and it basically becomes an intrinsic value-adding sector when its activities make that possible. That means, a shipment is continuously performed and provides a consistent return in the economy when it supports a manufacturing or service-based operation. In general, the freight transport sector is not perceived in its totality as a value-adding source for the economy, and mostly is regarded as a secondary system to sustain the three main industries of the global economy. Nevertheless, its revenues in the economy are not the only element to be considered, but also its secondary outputs, such as the carbon emissions and other types of air pollutants that it releases in the atmosphere causing damaging effects on the entire environment.

The current research referred to the cases of Cyprus, the Netherlands and Latvia, and looked into their specificity in terms of freight transport systems. By assessing the outliers of the EU freight transport industry, as each of the three countries presents a significant coverage with a particular mode of transport, the paper disclosed important findings on the degree of air pollutants emissions from each mode of transport and their impact on the overall quality of air and, generally, on the environment. Although road transport haulage is considered the main polluter, when it is tested against air pollution, the results showcase a limited negative contribution on behalf of that mode of transport. This particularity was observed in the regression's results obtained from assessing the case of Cyprus. Moreover, in Cyprus, it has been underlined that the air freight transport has the highest level of air pollutants emissions compared to any other mode of transport. The same results have been obtained in the case of Latvia. While both Cyprus and Latvia have showcased interesting results, the Netherlands' outputs have been the most fascinating. The road haulage's carbon emissions are insignificant within the quantum of GHG levels. Moreover, the emissions of methane as a result of agricultural activities have the highest output of pollutant emissions in the air, underlining, once again, that the freight transport industry, as a support value-adding sector of the economy, does not represent a significant air polluter. Another finding of the research is related to the connection between the PM2.5 air pollution (mean annual exposure) and the quantity of freight goods transported. This is a correlation that was not confirmed, turning the attention from the freight transport industry as a major air polluter to the main sectors of the economy, namely agriculture and manufacturing.

Among the limitations of the paper can be added the fact that the model used a linear regression analysis instead of panel data analysis, that the number of countries in the study was small, the number of observations for each variable in the study were minimal, and also that some information/data could not be added to the study (there was no information on road freight goods, such as quantities, no information on the total number of km of highways in each country, no particular variable for the CO_2 emissions from freight transport specific to each mode of transport, etc.). The study can be considered a first attempt into uncovering the basis of environmental sustainability, efficient freight transport practices, and a combination of the two former elements into a complete and clear business model.

References

Akgün, E.Z., Monios, J., Rye, T., Fonzone, A.: Influences on urban freight transport policy choice by local authorities. Transp. Policy **75**, 88–98 (2019)

Ambra, T., Caris, A., Macharis, C.: Towards freight transport system unification: reviewing and combining the advancements in the physical internet and synchromodal transport research. Int. J. Prod. Res. (2018). https://doi.org/10.1080/00207543.2018.1494392

Bask, A., Rajahonka, M.: The role of environmental sustainability in the freight transport mode choice: a systematic literature review with focus on the EU. Int. J. Phys. Distrib. Logist. Manag. **47**(7), 560–602 (2017)

Buldeo Rai, H., Verlinde, S., Merckx, J., Macharis, C.: Crowd logistics: an opportunity for more sustainable urban freight transport? Eur. Transp. Res. Rev. **9**, 39 (2017). https://doi.org/10.1007/s12544-017-0256-6

Centobelli, P., Cerchione, R., Esposito, E., Shashi.: Evaluating environmental sustainability strategies in freight transport and logistics industry. Bus. Strategy Environ. 1–12 (2020)

Comi, A.: Smart urban freight transport: tools for planning and optimising delivery operations. Simul. Model. Pract. Theory (2018). https://doi.org/10.1016/j.simpat.2018.08.006

European Commission: Communication from the Commission: The European Green Deal. Brussels. https://eur-lex.europa.eu/legal-content/EN/TXT/?qid=1576150542719&uri=COM%3A2019%3A640%3AFIN (2019)

De Jong, G., Gunn, H., Walker, W.: National and international freight transport models: an overview and ideas for future development. Transp. Rev. Transnatl. Transdiscipl. J. **24**(1), 103–124 (2004)

Eurostat: Freight Transport Statistics. Modal split of inland freight transport, 2018. https://ec.europa.eu/eurostat/statistics-explained/index.php?title=Freight_transport_statistics&oldid=492516#Modal_split (2020)

Eurostat: Road freight transport statistics, Statistics Explained. https://ec.europa.eu/eurostat/statistics-explained/pdfscache/9217.pdf (2021)

Ghandriz, T., Jacobson, B., Laine, L., Hellgren, J.: Impact of automated driving systems on road freight transport and electrified propulsion of heavy vehicles. Transp. Res. Part C Emerg. Technol. **115**, 102610 (2020)

Grigorescu, A., Zamfir, A.M., Mocanu, C.: Emerging trends and drivers for knowledge-intensive economy. Manage. Market. **15**(2), 172–185 (2020)

Kiba-Janiak, M.: Urban freight transport in city strategic planning. Res. Transp. Bus. Manag. **24**, 4–16 (2017)

Kin, B., Verlinde, S., Macharis, C.: Sustainable urban freight transport in megacities in emerging markets. Sustain. Cities Soc. (2017). https://doi.org/10.1016/j.scs.2017.03.011

Kumar, A., Anbanandam, R.: Evaluating the interrelationships among inhibitors to intermodal railroad freight transport in emerging economies: a multistakeholder perspective. Transp. Res. Part A **132**, 559–581 (2020)

Le Pira, M., Marcucci, E., Gatta, V., Ignaccolo, M., Inturri, G., Pluchino, A.: Towards a decision-support procedure to foster stakeholder involvement and acceptability of urban freight transport policies. Eur. Transp. Res. Rev. **9**, 54 (2017)

Lv, Q., Liu, H., Yang, D., Liu, H.: Effects of urbanization on freight transport carbon emissions in China: common characteristics and regional disparity. J. Clean. Prod. **211**, 481–489 (2019)
Macharis, C., Bontekoning, Y.M.: Opportunities for OR in intermodal freight transport research: a review. Eur. J. Oper. Res. **153**, 400–416 (2004)
Marcucci, E., Le Pira, M., Gatta, V., Inturri, G., Ignaccolo, M., Pluchino, A.: Simulating participatory urban freight transport policy-making: accounting for heterogeneous stakeholders' preferences and interaction effects. Transp. Res. Part E **103**, 69–86 (2017)
Merchan, A.L., Léonard, A., Limbourg, S., Mostert, M.: Life cycle externalities versus external costs: the case of inland freight transport in Belgium. Transp. Res. Part D Transp. Environ. **67**, 576–595 (2019)
Mirhedayatian, S.M., Yan, S.: A framework to evaluate policy options for supporting electric vehicles in urban freight transport. Transp. Res. Part D Transp. Environ. **58**, 22–38 (2018)
Muñoz-Villamizara, A., Santos, J., Montoya-Torres, J., Velázquez-Martínez, J.: Measuring environmental performance of urban freight transport systems: a case study. Sustain. Cities Soc. **52**, 101844 (2020)
Osorio-Tejada, J.L., Llera-Sastresa, E., Scarpellini, S.: Liquefied natural gas: Could it be reliable option for road freight transport in the EU? Renew. Sustain. Energy Rev. (2016). https://doi.org/10.1016/j.rser.2016.12.104
Quiros, D.C., Smith, J., Thiruvengadam, A., Huai, T., Hu, S.: Greenhouse gas emissions from heavy-duty natural gas, hybrid, and conventional diesel on-road trucks during freight transport. Atmos. Environ. **168**, 36–45 (2017)
Subanidja, S., Hadiwidjojo, D.: The influence of knowledge management "bottleneck" on company's performance. Manage. Market. **12**(3), 402 (2017)
The World Bank: Data Catalog. Data Access and Licensing. https://datacatalog.worldbank.org/public-licenses#cc-by (2021)
Visser, J.G.S.N.: The development of underground freight transport: an overview. Tunn. Undergr. Space Technol. **80**, 123–127 (2018)
Winkler, C., Mocanu, T.: Impact of political measures on passenger and freight transport demand in Germany. Transp. Res. Part D **87**, 102476 (2020)

Work Ability Index-Evidence from the Human Resource Field in Romania

Ramona Ştefania Igreţ, Cristian Virgil Marinaş, Monica Roman, and Simona Irina Goia Agoston

Abstract Nowadays the talent shortage is at its highest level in many emerging economies from Europe, many employers reporting difficulties in filling their vacancies with high skilled and competent human resources. The main causes are related to talent migration, education-labor market mismatch, demographic ageing or disrupting technologies that are changing the nature of jobs. In this context, keeping high skilled employees in a good working condition should be a priority for employers. The purpose of this study is to investigate the individual work ability in the case of human resource (HR) professionals from Romania in order to detect at an early stage, work – related health problems and early retirement risks. We use the Work Ability Index (WAI), a well-established instrument for the assessment of the work capacity. The instrument contains seven dimensions and it was applied on a sample of 81 respondents, from various economic sectors. The results show that in the case of human resource professionals, the overall score of the Work Ability Index is more influenced by the mental rather than physical abilities and it is not significantly affected by respondent's age, gender or education.

Keywords Work Ability Index (WAI) · HR professionals · Romania · Talent shortage · Sickness absence · Early retirement

R. Ş. Igreţ (✉) · C. V. Marinaş · M. Roman · S. I. Goia Agoston
The Bucharest University of Economic Studies, Bucharest, Romania
e-mail: ramona.igret@man.ase.ro

C. V. Marinaş
e-mail: cristian.marinas@man.ase.ro

M. Roman
e-mail: monica.roman@csie.ase.ro

S. I. Goia Agoston
e-mail: simona.goia@ase.ro

M. Busu (ed.), *Digital Economy and New Value Creation*, Springer Proceedings in Business and Economics, https://doi.org/10.1007/978-3-031-07265-9_5

1 Introduction

Employers around the world are facing important labor force challenges: a talent shortage, a high proportion of applicants that lack either hard or soft skills (Manpower 2018), ageing population especially in the developed countries (Eurostat 2019). Besides these common issues, organizations doing business in Romania have to face even more workforce related problems like talent migration or a high rate of functional illiteracy among young people.

The talent shortage is at its highest level. According to Manpower 2018 study, 45% of the employers around the world are reporting difficulties in filling their vacancies, as they cannot find the necessary soft or hard skills on the labor market. The same study states that Romania occupies the second place in the world regarding the difficulty in filling positions, as 81% of the employers encounter difficulties in reaching candidates or even cannot find the right ones. We can find a lot of reasons why talent shortage is so high in our country. However, the most important ones are the ageing population, the high rate of migration, and the functional illiteracy rate among youths.

Eurostat 2018 data report that almost one fifth of the total population of Europe (19.7%) are individuals aged 65 years or more and their share is projected to reach 28.5% in 2050 or almost 150 million inhabitants. Besides that, the life expectancy at birth increased in many European countries, exceeding 80 years. In Romania, 3.5 million persons were aged 65 years or more in 2018, which represents more than 18% of the total population of the country. The life expectancy also has registered a positive trend from 73.7 years in 2009 to 75.3 years in 2017. Moreover, by 2050, individuals aged 55 years or older are projected to account for almost 45% of the total population in Romania.

Besides the ageing population, the emigration rate in Romania increased dramatically over the last 10 years. According to UN International Migration Reports, it is estimated that more than 3.5 million Romanians have emigrated in more developed countries, especially in Italy, Spain, Germany and United Kingdom. A recent OECD report (OECD 2019) states that in 2016, Romania was the second country in the world after China with respect to the magnitude of immigration flows to OECD countries. Moreover, the OECD data show that almost two-thirds of Romanian emigrants between 2004 and 2016 were under 30 years old and more than a half had a high or medium level of education.

Another very important concern regarding the Romanian labor market is the high proportion of functional illiteracy among youth. According to PISA 2015 tests, more than a third of young people in our country are unable to understand core school subjects of science, reading or mathematics (OECD 2018). This fact has long term consequences. The young generation will enter soon the labor market being unable to perform jobs that require high soft or hard skills. The challenges are even higher in the context of SARS-CoV-2 (COVID–19) pandemic and the necessity of teleworking.

This paper aims at investigating the individual work ability of a sample of human resource professionals from Romania using the tool Work Ability Index, which was initially developed in Finland and afterwards applied in several other countries (Bridger 2018). In the next sections of the paper it is presented a brief literature review, followed by the methodology of the research and the results of the survey, presented according to various categories of respondents and variables. The paper ends with a section of conclusions, which summarizes the main outcomes of the study and provides future directions for research.

2 Literature Review

As mentioned in the Introduction, some of the major challenges related to the labor market refer to ageing population in Europe, lack of skills of young graduates, mismatch education-labor market. On the top of that Romania faces also high migration figures and youth illiteracy. All these workforce features and challenges create the need to think about solutions to persuade people work longer, be active and willing to perform. Increasing workers' sustainable employment brings advantages for both companies and individuals: companies benefit from the experienced and high skilled workers and individuals obtain a higher income, as most of the retirement pensions in Romania cannot cover current expenses.

Worker demographics can hardly change (Rothmore and Gray 2019), so besides national strategies to encourage late retirement, companies need to define actions to maintain the work ability of their employees as long as possible. Working for a longer period means also prolonging the exposure to work hazards and risks as the work capacity is also declining with age. Safety at work should be a high priority of our country as Romania registered the highest number of fatal accidents per 100 000 employees among the EU Member States (5.56 fatal accidents per 100 000 persons employed) (Eurostat 2018).

Work ability is defined as a relationship between an individual's physical and mental functioning capacity and the work requirements (Juszczyk et al. 2019), a person's potential to manage the working tasks, considering her/his health, working conditions and mental capability (Ilmarinen and Lehtinen 2004). The balance model of work ability considers the person's work demands and the individual's resources to cope with these requirements (Freyer et al. 2018). A good work ability is related to the ability of an individual to perform any type of work that can be valued (Belevska et al. 2019). Work ability index (WAI) is a tool universally used to record the work ability of an individual (Ilmarinen 2009; van den Berg et al. 2009) and a lot of research has been conducted using this instrument. Its validity and reliability has been researched within many studies with positive results (Freyer et al. 2018; Kaewboonchoo and Ratanasiripong 2015; Geissler et al. 2005; Kujala et al. 2005).

Several recent studies reported relevant associations between a range of workplace risks and hazards and the work ability scores of the employees. Rothmore and Gray (2019) state a direct association between reduced WAI scores and physical and psychosocial risk factors like: working with arms above shoulders, pushing and pulling with force, equipment, workstations, work–life balance, managers' attitude to health and safety, facilities for breaks, learning opportunities, consultation, flexibility, stress or unnecessary work. Furthermore, studies reviewing the effects of work-related and individual factors on WAI reported a decreased work ability index with older age, especially in the case of women (Monteiro et al. 2006) and a lower WAI associated with hard life situation outside work, raising children or low self-confidence (Van den Berg et al. 2009). The characteristics of work such as possibilities to control one's work and workload, work ergonomics and management also influence WAI scores (Pohjonen, 2001). WAI is also a practical tool in predicting sickness absence among young employees (Ohta et al. 2017; Reeuwijk et al. 2015). Research conducted on Dutch financial office workers conclude that WAI can fairly well identify officers at risk for prolonged sickness absence from workers with low or without absence (Reeuwijk et al. 2015). A high work ability is associated with higher performance at work, improved well–being and motivation among employees and decreasing the likelihood of early retirement (Juszczyk et al. 2019). Thus, we consider it important for the HR professional bodies and companies doing business in Romania and worldwide to evaluate the HR professionals' WAI levels in order to develop measures to improve the work ability and identify health and early retirement risks.

The purpose of this study is to use the WAI (Work Ability Index) instrument to investigate the individual work ability of a sample of young human resource professionals (HR professionals) from Romania in order to detect at an early stage, work–related health problems and early retirement risks. Although the HR profession requires mostly mental work and it is, most of the time, a pleasant and rewarding activity, many tasks and responsibilities might determine physical and psychological diseases. The profession challenges are very high, imposed by new realities in the labour market: upheavals in jobs and organizations, the scarcity of talented people, very mobile workers, the remote working or disrupting technology. However, we assumed from the very beginning that the work ability of young professionals is high enough. What is important in the study is to discover the main factors that are influencing a decrease in work ability and propose solutions to counteract them.

3 Methodology

In this research we used a cross-sectional study design and descriptive statistics in order to investigate the work ability of a sample of Romanian HR professionals. The respondent subjects were young full time employees between 20 and 49 years old, participating in a human resource management master programme and HR professionals actively involved in HR social media hubs from Romania.

The tool of research was the English version of WAI (Work Ability Index) instrument, developed by the Finland Institute of Occupational Health Research (Ilmarinen 2009; Tuomi et al. 1998) and routinely applied for the assessment of work ability of employees (Freyer et al. 2018; Ohta et al. 2017; El Fassi et al. 2013; Karttunen

Table 1 WAI dimensions and range of scores

Dimension	Range of scores
i. Current work ability compared with lifetime best	1–10 points
ii. Work ability in relation to the demands of the job[(a)]	2–10 points
iii. Number of current diseases diagnosed by a physician[(b)]	1–5 or 7 points
iv. Estimated work impairment due to diseases	1–6 points
v. Sick leave during the past year (12 months)	1–5 points
vi. Own prognosis of work ability 2 years from now	1, 4 or 7 points
vii. Mental resources	1–4 points

(a) work ability is assessed in relation with both mental and physical demands of the job
(b) a list of 51 diseases
Source Tuomi et al. 1998

and Rautiainen 2009; van den Berg et al. 2009). The questionnaire also contains additional questions about demographic characteristics, the field of activity and size of the company.

WAI tool involves a questionnaire to conduct a self-assessment and is based on the assumption of a single factor underlying the construct of work ability (Freyer et al. 2018). WAI consists in a seven dimensions self-assessment related to (i) current work ability compared with lifetime best, (ii) work ability in relation with job demands, (iii) reported diseases diagnosed by a physician, (iv) estimated work impairment due to diseases, (v) sick absence over the past 12 months, (vi) self-prognosis of work ability 2 years from now and (vii) mental resources. The individual's work ability is calculated by summing up the scores of these seven dimensions and ranges from 7 to 49 points. According to the manual, the higher the value, the better work ability has the evaluated employee. The results are classified into four categories or levels: poor (7–27), moderate (28–36), good (37–43) and excellent (44–49) (Table 1).

The questionnaire, developed in Google Docs, was administered via e-mail to HR master students and shared on 3 h hubs. The participation was voluntary and the data strictly confidential. After 3 months, a total number of 81 valid questionnaires was received and analyzed. About 65% (53 respondents) were HR master students and practitioners and 35% (28 respondents) were other professionals from the field. The data obtained was processed using R statistical software. Descriptive statistics were used to analyze the demographic characteristics of the sample and the WAI final score and dimensions.

4 Results and Discussions

In this section there are presented the main results of the research.

Table 2 Characteristics of the sample (N = 81)

Variables	n (%)
Gender	
Male	18 (22.2%)
Female	63(77.8%)
Prefer not to say	0 (0%)
Age	
20–29	52 (64.2%)
30–39	21 (26%)
40–49	8 (9.8%)
Education	
University degree	41 (50,6%)
Master degree	34 (42%)
Phd. Degree	6 (7.4%)
Type of organisation (number of employees)	
Under 9 employees	6 (7.4%)
Between 10 and 49 employee	25 (30.9%)
Between 50 and 249 employees	9 (11.1%)
More than 250 employees	41 (50.6%)

Source Authors' own illustration

4.1 Demographics

A total number of 81 h professionals completed the WAI questionnaire. Most of the participants were female (77.8%) and 22.2% male. The high percent of female respondents is explained by the high number of women involved in HR activities. About 65% of the participants were 20–29 years of age, 26% were 30–39 years old and 9.8% were aged between 40–49 years. All of the persons involved in the study had a university degree. Regarding the size of the organizations the respondents were working in, more than a half were big size companies, with more than 250 employees (50.6%), the rest of them being SMEs (small and medium sized enterprises). A summary of the characteristics of the sample is provided in Table 2.

4.2 Work Ability Index Results

The work ability index (WAI) mean score was 40.8, higher for female HR specialists (41.2) (SD = 4.62) than for males (39.6) (SD = 6.45). We used Shapiro–Wilk's test to verify the normality of the data. From the output, the p-value > 0.05 implies that the distribution of the data are not significantly different from normal distribution.

A high percent of the respondents had a good (48%) and excellent (33.3%) work ability. We expected a high rate of good work ability, as the respondents were young people. Moreover, these results are in line with other research studies conducted on work ability (Kaewboonchoo and Ratanasiripong 2015; Karttunen and Rautiainen 2009). However, more than 18% of the HR professionals had a moderate and poor work ability. The percent is not very high, still, a moderate work ability means that the employee is struggling to cope with daily tasks and in order to restore work ability, immediate measures need to be taken (Kloimuller et al. 2000). One might consider that employees performing mostly mental work (as HR professionals do) should have higher WAI scores. Still, research conducted in the work ability index states that five out of seven studies analyzing WAI scores reported a positive association between high mental work demands and poor WAI scores (van den Berg et al. 2009). Moreover, we must state that the pandemic (COVID-19) brought a lot of challenges for the HR departments and a high amount of mental stress. Employees with poor work ability might perform less, be demotivated and disengaged or even take time off sick (Fig. 1 and Table 3).

Taking into consideration the fact that the respondents are young employees, the age group does not significantly correlate with WAI scores. However, analyzing the descriptive statistics we can observe a slight decrease in the WAI mean for older female respondents. Further research needs to be done for more accurate results (Table 4).

Examining the results of the seven items of the work ability index, we can observe several characteristics of our respondents.

Analysing the first item of WAI index, current work ability compared to the lifetime best, we can see that respondents consider, on a scale from 0 to 10 (0 = unable to work, 10 = best work ability ever) to have an average of 8.46 work ability. The average is higher for women respondents (8.71) compared to male respondents (7.56). Moreover, there is a positive correlation between this item and the work ability index score ($r = 0.62$), higher for males ($r = 0.77$) than for females ($r = 0.53$).

The second dimension of WAI refers to the work ability in relation to the demands of the job. This dimension comprises 2 items, first referring to the physical demands and the second referring to the mental demands of the job. In this study we assumed that the HR profession is both physically and mentally demanding, so we didn't

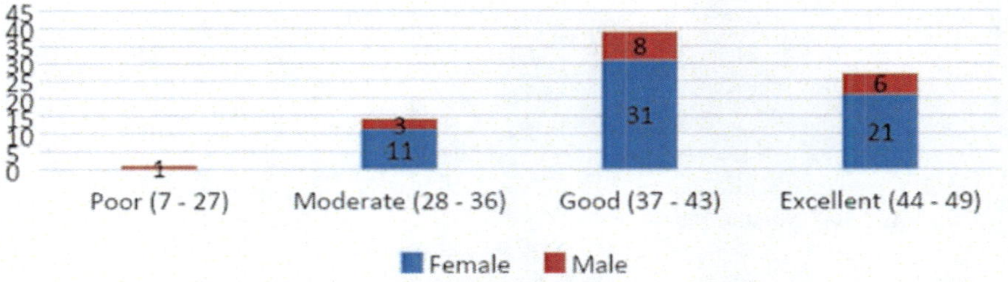

Fig. 1 Work ability index by levels and gender. *Source* Authors' own illustration

Table 3 Work ability index (WAI) by gender

		Female	Male	Total
WAI	Mean	41.2	39.6	40.8
	Median	41	40.5	41
	Standard deviation	4.62	6.45	5.08
	Min–Max	30–49	25–49	25–49
Work ability index (WAI) levels	Poor (7–27)			
	n	0	1	1
	%	0	5.6	1.3
	Moderate (28–36)			
	n	11	3	14
	%	17.5	16.7	17.3
	Good (37–43)			
	n	31	8	39
	%	49.2	44.4	48.1
	Excellent (44–49)			
	n	21	6	27
	%	33.3	33.3	33.3
	Total			
	n	63	18	81
	%	100	100	100

Source Authors' own illustration

Table 4 Work ability index (WAI) by gender and age group

		Age group		
		20–29	30–39	40–49
Females	$n = 63$	40	16	7
	Mean	41.8	40.6	39.1
	SD	4.43	5.42	3.34
	Min	33	30	36
	Max	49	49	45
Males	$n = 18$	12	5	1
	Mean	39.8	42.2	25
	SD	5.72	4.97	NA
	Min	30	37	25
	Max	48	49	25

Source Authors' own illustration

weigh the responses. The average of work ability respondents in relation with their job demands is 8.38, with a 1.48 standard deviation. This dimension has a positive correlation with the total WAI index (r = 0.73) (Table 5).

A very relevant dimension related to work ability is the third item contained in the index, respective, the reported diseases or injuries diagnosed by a physician. The questionnaire addresses 51 types of diseases/injuries organized through 14 main categories (e.g. injuries from accidents, musculoskeletal diseases, respiratory diseases, etc.). The scoring of the responses is done according to a very simple methodology (for at least 5 diagnosed diseases declared = 1 point, 4 diseases = 2 points, 3 diseases = 3 points, 2 diseases = 4 points, 1 disease = 5 points, no disease = 7 points). The average of responses was 4.73 with a 2.15 standard deviation, with a slight difference between women (mean = 4.76) and men (mean = 4.61). According to the results, 60.5% of the respondents declare they have at least one disease/injury diagnosed by a physician. Taking into account the low age of the respondents, it is a significant result in the context of work ability, with possible future negative consequences over the ability to work. Most of the diseases are related to musculoskeletal disorders (37), respiratory diseases (20), digestive diseases (17), injuries form accidents (12) or skin disease. There is a positive correlation between this item and the final work ability index score (r = 0.53), higher for female respondents (r = 0.61) than for males (r = 0.33).

Although it is very important for an employer to know at least some relevant information about the general health of their employees (information that is shared by the employee and is not considered too personal), even more important is to understand how a disease/injury might impair their employees' work. On a scale from 1 (entirely unable to work) to 6 (no hindrance to work), the respondents registered an average of 5.11 with a 1.04 standard deviation, higher for females than for men. It is an important result as it states that, although more than 60% of the employees declare they have a disease, they don't consider it to be an impairment to their current work. The correlation between this item and the final work ability score is positive (r = 0.67) higher for males (r = 0.80) than for females (r = 0.60).

The 5th item of the index deals with the number of whole days an employee has been off work because of health problems during the past 12 months. 69% of the respondents did not take days off work because of health problems, which is a positive thing.

Regarding the self-prognosis of work ability 2 years from now, the respondents were asked to declare if they believed in their ability to do their current job two years from the moment they had responded. They received three options to score: 1 (unlikely), 4 (not certain) and 7 (relatively certain). More than 80% of the respondents were relatively certain that they can continue their work, 12.4% were not certain and only one respondent declared that it was unlikely to be able to do his current job.

The last item named mental resources addresses the general opinion of the employees regarding their life in general, both during work and leisure time. They were asked if they had enjoyed regular daily activities, if they had felt hope for the future and if they had been active and alert. The score ranges from 1 to 4. The average of responses was 3.41 with a 0.75 standard deviation, meaning that most of them felt

Table 5 Work ability index single item distribution of scores

WAI item	Score n (%)											Total	Range of scores[a]
	0	1	2	3	4	5	6	7	8	9	10		
Current work ability	0 (0)	1 (1.2)	0 (0)	0 (0)	0 (0)	0 (0)	7 (8.6)	11 (13.6)	20 (24.7)	15 (18.5)	27 (33.3)	81 (100.0)	0–10
Work ability in relation to the demands of the job			1 (1.2)	0 (0)	0 (0)	1 (1.2)	6 (7.4)	10 (12.3)	24 (29.6)	16 (19.8)	23 (28.3)	81 (100.0)	2–10
Reported diseases diagnosed by a physician		9 (11.1)	5 (6.2)	12 (14.8)	11 (13.6)	12 (14.8)		32 (39.5)				81 (100.0)	1–5, 7
Estimated work impairment due to diseases		0 (0)	1 (1.2)	7 (8.6)	12 (14.8)	23 (28.4)	38 (46.9)					81 (100.0)	1–5
Sick leave		1 (1.2)	0 (0)	8 (9.9)	16 (19.8)	56 (69.1)						81(100.0)	1–5
Self-prognosis of work ability 2 years from now		6 (7.4)			10 (12.4)			65 (80.2)				81 (100.0)	1, 4 or 7
Mental Resources		1 (1.2)	10 (12.4)	25 (30.9)	45 (55.6)							81 (100.0)	1–4

[a] Tuomi et al. 1998

Source Authors' own illustration based on Tuomi et al. 1998

Table 6 Work ability index items. Descriptive statistics

WAI seven items	Mean (SD)	Female mean (SD)	Male mean (SD)	Range
Current work ability	8.46 (1.56)	8.71 (1.30)	7.56 (2.04)	0–10
Work ability in relation to the demands of the job	8.38 (1.48)	8.38 (1.31)	8.39 (2.00)	2–10
Reported diseases diagnosed by a physician	4.73 (2.15)	4.76 (2.20)	4.61 (2.00)	1–5, 7
Estimated work impairment due to diseases	5.11(1.04)	5.21 (0.97)	4.78 (1.22)	1–6
Sick leave	4.56 (0.77)	4.57 (0.66)	4.5 (1.10)	1–5
Self-prognosis of work ability 2 years from now	6.19 (1.78)	6.19 (1.72)	6.17 (2.01)	1, 4 or 7
Mental resources	3.41 (0.75)	3.35 (0.72)	3.61 (0.85)	1–4

Source Authors' own illustration

they had enough mental resources to continue their daily activities. Still, 11 out of 81 respondents registered scores below 3, which means that they were not especially happy about their recent activities and the hope for a better future was low (Table 6).

5 Conclusion

Considering the challenges existing on the European labor market such as aging population, mismatch between skills and competences developed in schools and universities and the ones required on the labor market and talent shortage, complemented by the issues existing at national level in Romania, such as a high emigration rate and functional illiteracy as shown by PISA studies, we consider that it is highly important for companies as well as for authorities in charge with policies related to demography and labor market to assess the work ability of different occupational groups. The assessment of the future ability to perform a certain type of work and the chances of being unable to perform the current job should also have a preventive role. The results obtained could be used in order to develop and implement policies and measures both at macroeconomic but also at organizational level in order to diminish the chances of prolonged sickness absence from work, early retirement due to health problems and reduced performance at the job.

The present paper aims at assessing the work ability of professionals from the HR field using a well-established instrument, WAI, initially developed by Finnish researchers and subsequently successfully tested in other countries. The results reveal a mean score of 40.8 which falls in the category Good (37–43 points). However, considering the individual scores, almost 20% of the respondents are in the category poor and moderate. This prompts for immediate action from the employers in order to restore the work ability of the employees and avoid law performance, medical leaves and even early retirement. Analyzing the descriptive statistics resulting from the present study we can observe a slight decrease in the WAI mean for older female respondents, which is in line with the results obtained in other comparative studies on international level (Monteiro et al. 2006) using the same tool, WAI.

As far as we know by researching the academic literature, no further studies deploying the WAI have been conducted so far in Romania. Considering the fragile situation on the Romanian labor market as well as the challenges faced by both health and pension systems, further studies deploying WAI, as a reliable and well-established tool, can be conducted. One of the main limitations of the present research consists in the fact that it was performed on a single category of employees, namely HR professionals. Another limitation refers to the relative homogeneity of the respondents related to the age distribution (most of them are young persons, below 40). In a next stage, authors aim at extending the research to other professional groups, including persons which perform physical work. Also, a more heterogeneous group of respondents from the perspective of their demographics is targeted. A further questionnaire which contains questions related to the private/personal situation/status of the respondents could be distributed together with the WAI form and some relevant correlations between WAI and these variables could be inferred.

References

Belevska, M., Jovanovic, J., Dastevska, E.G., Velkovski, Z.: The importance of individual predictors and psychosocial working conditions in assessing the work ability index of people with low vision. Int. J. Occup. Saf. Ergon. (2019). https://doi.org/10.1080/10803548.2019.1613811

Bridger, R.S.: Introduction to Human Factors and Ergonomics, 4th edn. CRC Press, Taylor and Francis Group (2018)

El Fassi, M., Bocquet, V., Majery, N., Lair, M.L., Couffignal, S., Mairiaux, P.: Work ability assessment in a worker population: comparison and determinants of Work Ability Index and Work Ability score. BMC Public Health **13**, 305 (2013). https://doi.org/10.1186/1471-2458-13-305

Eurostat: Accidents at work statistics. https://ec.europa.eu/eurostat/statistics-explained/index.php/Accidents_at_work_statistics#Number_of_accidents (2018). Accessed 7 Sept 2019

Eurostat: Ageing Europe. Looking at the lives of older people in the EU. https://ec.europa.eu/eurostat/documents/3217494/10166544/KS-02-19%E2%80%91681-EN-N.pdf/c701972f-6b4e-b432-57d2-91898ca94893 (2019). Accessed 3 Oct 2019

Freyer, M., Formazin, M., Rose, U.: Factorial validity of the Work Ability Index among employees in Germany. J. Occup. Rehabil. **29**(2), 433–442 (2018)

Geissler, H., Tempel, J., Geissler–Gruber, B.: Can the Work Ability Index also be used by non-medical professionals? A comparative study. In: International Congress Series 1280, pp. 281–285 (2005). https://doi.org/10.1016/j.ics.2005.02.050

Ilmarinen, J.: Work ability – a comprehensive concept for occupational health research and prevention. Scand. J. Work Environ. Health **35**, 1–5 (2009). https://doi.org/10.5271/sjweh.1304

Ilmarinen, J., Lehtinen, S. (Eds.): Past, present and future of work ability. Finnish Institute of Occupational Health, Helsinky (2004).

Juszczyk, G., Czerw, A.I., Religioni, U., Olejniczak, D., Walusiak-Skorupa, J., Banas, T., Mikos, M., Staniszewska, A.: Work Ability Index (WAI) values in a sample of the working population in Poland. Ann. Agric. Environ. Med. **26**(1), 78–84 (2019). https://doi.org/10.26444/aaem/91471

Kaewboonchoo, O., Ratanasiripong, P.: Psychometric properties of the Thai version of the work ability index (Thai WAI). J. Occup. Health **57**, 371–377 (2015). https://doi.org/10.1539/joh.14-.0137-oa

Karttunen, J., Rautiainen, R.H.: Work Ability Index among Finnish Dairy Farmers. J. Agric. Saf. Health **15**(4), 353–364 (2009). https://doi.org/10.13031/2013.28889

Kloimuller, I., Karazman, R., Geissler, H., Karazman-Morawetz, I., Haupt, H.: The relation of age, work ability index and stress-inducing factors among bus drivers. Int. J. Ind. Ergon. **25**, 497–502 (2000)

Kujala, V., Remes, J., Ek, E., Tammelin, T., Laitinen, J.: Classification of Work Ability Index among young employees. Occup. Med. **55**, 399–401 (2005)

Manpower: Solving the talent shortage. Build, buy, borrow and bridge. https://go.manpowergroup.com/hubfs/TalentShortage%202018%20(Global)%20As (2018). sets/PDFs/MG_TalentShortage2018_lo%206_25_18_FINAL.pdf?hsLang=en. Accessed 5 Sept 2019

Monteiro, M.S., Ilmarinen, J., Filho H.R.C.: Work ability of workers in different age groups in a public health institution in Brazil. Int. J. Occup. Saf. Ergonomics **12**(4), 417–427 (2006). https://doi.org/10.1080/10803548.2006.11076703.

OECD: PISA 2015. Results in focus. https://www.oecd.org/pisa/pisa-2015-results-in-focus.pdf (2018). Accessed 15 Aug 2019

OECD: Recent trends in emigration from Romania, in talent Abroad: A review of romanian emigrants, OECD Publishing, Paris, https://doi.org/10.1787/00fb26e2-en (2019).

Ohta, M., Higuchi, Y., Kumashiro, M., Yamato, H., Sugimura, H.: Decrease in Work Ability Index and sickness absence during the following year: a two-year follow-up study. Int. Arch. Occup. Environ. Health **90**, 883–894 (2017). https://doi.org/10.1007/s00420-017-1251-x

Pohjonen, T.: Perceived work ability of home care workers in relation to individual and work-related factors in different age groups. Occup. Med. **51**(3), 209–217 (2001)

Reeuwijk, K.G., Robroek, S.J., Niessen, M.A., Kraaijenhagen, R.A., Vergouwe, Y., Burdorf, A.: The prognostic value of the Work Ability Index for sickness absence among office workers. PLoS ONE **10**, e0126969 (2015). https://doi.org/10.1371/journal.pone.0126969

Rothmore, P., Gray, J.: Using the work ability index to identify workplace hazards. Work **62**, 251–259 (2019)

Tuomi, K., Ilmarinen, J., Jahkola, A., Katajarinne, L., Tulkki, A.: Work Ability Index, 2nd edn. Finnish Institute of Occupational Health, Helsinki (1998)

Van den Berg, T.I.J., Elders, L.A.M., de Zwart, B.C.H., Burdorf, A.: The effects of work-related and individual factors on the Work Ability Index: A systematic review. Occup. Environ. Med. **66**(4), 211–220 (2009)

The Impact of Digitalization in Romanian Automotive Industry

Mihaela Cornelia Sandu, Irina-Virginia Drăgulănescu, and Ane-Mari Androniceanu

Abstract Nowadays, digitalization is as important as innovation for the good well of the economy. In different industries, robots of different categories have taken over almost entirely the production process from humans. In this way the production was increased and the final cost decreased. In all day practice we can find difference between digitization, digitalization and digital transformation, so we will explore this area in the theoretical part of the paper. Also, we refer to industry 4.0 that describes the trend toward automatization and data exchange in technologies and manufacturing process. In the practical part of the paper, the purpose was to determine the impact of digitalization on a Romanian company, Automotive Dacia. Because of the lack of information about the company, we will consider fixed assets as the investment in modernization and digitalization. We will develop a linear regression model having as dependent variable the profit and independent variable the fixed assets and employees. We saw the direct impact of fixed assets on profit and the negative impact of employees. As we expected, the production increased using the new technology even if the initial cost to implement these new technologies for the company was significant. The limitation of our study is given by the lack of transparency of the company but in the future, we intend to find more data and expand the study in another paper.

Keywords Industry 4.0 · Digitization · Digitalization · Digital transformation · Automotive industry

M. C. Sandu · I.-V. Drăgulănescu (✉)
University of Bucharest, Bucharest, Romania
e-mail: irina.dragulanescu@faa.unibuc.ro

M. C. Sandu
e-mail: mihaela.sandu@faa.unibuc.ro

A.-M. Androniceanu
Bucharest University of Economic Studies, Bucharest, Romania

M. Busu (ed.), *Digital Economy and New Value Creation*, Springer Proceedings in Business and Economics, https://doi.org/10.1007/978-3-031-07265-9_6

1 Introduction

Digitalization is today almost as popular as innovation for companies. Digitalization promises many opportunities and benefits: from process optimizations to productivity increases and cost reductions. Also, people and them behaviors change from a day to another. In this case digital natives prefer and demand a different kind of experience. In the context of pandemic, companies need to meet their customers' needs as easily as possible, and also the physical interaction to be limited. As we expected, there are companies that does not succeed in digitize their own processes and services. It seems that about 70% of digital transformation processes fail. So, the manager has to work with experts in the field, allocate the necessary budget and to be patient if he wants to succeed with digitization of the company.

2 Literature Review

According to Digital Economy and Society Index (DESI), in 2020 Romania was on 26th place from all European Union Member States. Especially in urban areas, Romania has a good connectivity dimension because of high-speed broadband and high-capacity fixed networks. Our country was on the 5th place in European Union for high-speed broadband services, 49% of homes have this kind of subscription. The digitalization of the Romanian economy lagged behind due to the lack of people's digital skills. In terms of ICT graduates, Romania occupies the 5th place, with 5.6% of all graduates (in European union the average was 3.6%). In terms of using internet services and digitalization of public services, Romania occupies the last place among European Union Member States (European Commission 2019).

In 2020, for Connectivity Romania ranked 11th in EU with a score of 56.2 (EU average was 50.1). Another element of DESI is human capital; for this one Romania ranked 27th in EU with a score of 33.2 (EU average was 49.3). In terms of using internet services, Romania ranked 28th in EU with a score of 35.9 (EU average was 58). For integration of digital technology Romania ranked 27 in EU with a score of 24.9 (EU average was 41.4). the last indicator of DESI, public digital services ranked Romania 28th in EU with a score of 48.4 (EU score was 72), (European Commission 2019).

New technologies are making more and more difference in the global economy, so it is very important for Romania to use new economic models, robots, digitalization and mechanization. Without a substantial investment in technology, Romania we will not cope the fourth industrial revolution and so, the Romanian economy will not be competitive at European and also global level (Curierul National 2019).

In term of competitiveness, the elements that will make more and more difference will be determine by using robotics, automating processes, optimizing production flows with dedicated software and AI. A good example for our country is the automotive industry. While big automotive companies are launching electric cars and uses robots to build them, Romanian companies build cars with internal combustion engines and use the classic production process. If Romanian investors doesn't want to miss the global race for competitiveness, they should focus on new technologies (Curierul National 2019).

In this paper we will discuss about the digitalization within the company Automotive Dacia, one of the most important automotive companies from Romania. We will see in the paper the differences between digitization, digitalization and digital transformation. Also, we will talk about industrial revolution 4.0 and its effect on global economy. The paper end with the analysis of data about Automotive Dacia, a Romanian automotive factory which is in full development and digital modernization. Developing an econometrical analysis by constructing a linear regression model, we will see which is the impact of fixed assets on digitalization of Automotive Dacia. Data used in this model correspond only to the factory from Mioveni of Automotive Dacia.

The first industrial revolution took place in period 1760–1840 and is called Industry 1.0. In this period manufacturing evolved from work executed by people and animals to work performed by people that use mechanisms put in place by the power of water. Period 1870–1914 took place the second industrial revolution, Industry 2.0, also known as technological revolution. The appearance and expansion of railway and telegraph permitted people and thought to move faster. Another emblematic element for this period was the electricity that permitted people to build modern factories. The end of twentieth century was marked by the appearance of and so the digital revolution; we deal with another industrial revolution called Industry 3.0. Up to this moment, the last industrial revolution is Industry 4.0. new technologies and manufacturing processes that include artificial intelligence will be developed (Greensoft 2019).

European Commission said in 2016 that industry 4.0 and digitalization are the way to be followed. Those two affect business models and the entire context in which business operate today. The existence of a common goal among all stakeholders are issues extremely important: in addition to enterprises, they concern the social partners at all levels, academia, research, regional and local public actors, the education sector and consumers (Van Iersel et al. 2016).

When we talk about digital technologies, we think about various disciplines including those relating to management, technology, information systems studies and so on.

The wide variety of technologies and tools that can be used that mainly concern the internet, ecommerce, social media, big data, cloud computing, the internet of things (IoT), advanced automation, artificial intelligence and machine learning determine different definitions for digitization, digitalization and digital transformation. Also, there are cases when the terms digitization and digital transformation are used as synonyms. (OECD 2016; Kotarba 2017; Rachinger et al. 2019; Ulas 2019).

Digitization indicates the process of converting or encoding information from analog into digital form so that it can be managed by computers (Dougherty and Dunne 2012; Loebbecke and Picot 2015; Li et al. 2016; Sebastian et al. 2017). Also, by digitization can be understood the integration of information technology (IT) in existing activities of the company or the development of resources through IT (Lai et al. 2010; Vendrell-Herrero et al. 2017; Rubino et al. 2020).

Digitalization refers to the use of IT or digital technologies with the aim to change existing business processes (Li et al. 2016). Digitalization and digitization have a point where they have a boarder concept, the implementation phase of digital technologies within the companies in order to optimize existing processes or improve their coordination (Pagani and Pardo 2017; Verhoef et al. 2019).

Digital transformation concerns the implementation of the process of technological transformation. It identifies a new way of rethinking the company, a profound change necessary to take full advantage of the opportunities that arise from the mix of technologies to layout. Digital transformation impacts the entire company and its management methods (Amit and Zott 2001) by going beyond the simple digitization that is configured through the modification of processes and organizational activities. Therefore, companies are required to implement a digital transformation process, that is a change in the business development logic (Li et al. 2018) or in the value creation process (Gölzer and Fritzsche 2017; Rubino et al. 2020).

At International Atomic Energy Agency, we can find the following classification and definitions: Digitization means to create a digital representation for physical objects or characteristics; Digitalization means to establish or to improve the process by leveraging digital technologies and digitized data; Digital transformation represents the business transformation enabled by digitalization. If we construct a pyramid using these three concepts, at the basis will be digitization, above digitalization and, on the top, digital transformation. (Gupta 2020).

3 Short History of Automotive Dacia Company

Automotive Dacia is a Romanian company founded in 1966 at Mioveni. From 1999 is part of Renault Group. In 2011 Dacia opened a new branch in Samoca, Maroc. In 2012 a new branch is opened in Maroc at Tangier. Also, Dacia has branches in Russia, Iran, Irak, Algeria and India. Through its attractive, qualitative and good prices models Dacia gained customers from 44 countries (www.grouprenault.ro). Dacia is the most important economic and commercial ambassador, the percent from GDP determined by the company being around 3% (Munteanu 2018).

The modernization-automation process of Dacia company is a long one. In 2014 Dacia had only one robot used in the manufacturing process and in 2019 had over 800 robots. So, if in 2015 only 5% of the production line was robotized in the next years it reached at 20%. These automations have targeted and continue to target demanding stations where operators are forced to handle larger parts involving awkward, difficult and ergonomically tiring movements. The modernization will impact the number of workers. As part of the production processes will be taken over by the robots, there will be a reduction of staff in the respective sections. As in the past, this reduction will be based on volunteering by offering compensatory packages, tempting especially for those that are close to retirement age. At the same time, there will be a redistribution of staff to other departments (Autotestmagazin 2015).

Dacia brand sales have doubled in the last eight years and reached a total of 736,570 vehicles in 2019, up 5% from the previous year and rising faster to the threshold of 1 million units, which should exceed in the following years, with the increase of production capacities in Mioveni, Tangier and Casablanca. The Romanian plant produced almost 350,000 units in 2019. At European level, taking into account the EU and the EFTA states (Iceland, Norway, Liechtenstein, Switzerland), Dacia occupies the 13th place in the ranking of best-selling brands with 581,543 cars sold. France remains the market where Dacia sells the most cars, with over 140,500 vehicles delivered, but down 1% compared to 2018. In second place rose strongly Italy, with almost 88,500 units, up 33% year on year previous. Turkey scored the largest decline, at 33% to 20,000 units. On a Romanian market, Dacia ended 2019 with sales increasing by 1.5% compared to the previous year to almost 55,463 vehicles. On the other hand, Dacia car registrations rose by 36% last year to 50,311 units (Group Renault 2019).

4 Digitalization in Automotive Industry

With 28% of all industrial robots, we can say that the automotive industry is the most important consumer in this area. In 2017 and 2018 there were registered an ascendent trend and the highest value of 125,581 units in 2018. The demand for robots from automotive industry decreased in 2019 by 16% reaching at 105,379 units. Globally the production of vehicle declined in 2018 and 2019. The new trend in the automotive industry is given by electric engines which will take place to the internal combustion engines. The decreasing demand will determine some problems in reaching the need for capacity expansion to reach this new goal. From 2014 to 2019, the robot installations in the automotive industry increased with an average of 2% every year (CAGR). The economic crisis from 2008–2009 determined a restructuration in car manufacturers business. The automotive industry has recovered after 2010 and the investments in production modernization have determined the request for robots (Guerry et al. 2020).

European automotive production was in a continuous growing in period 2008–2019. Passengers car production (Fig. 1a) and commercial car production (Fig. 1b) registered a similar trend; in Fig. 1a negative fluctuation in 2009, 2012 and 2019.

European countries that have the highest density of robots in manufacturing industry in 2019, according to International Federation of Robotics (Guerry et al. 2020), are: Germany (346), Sweden (274), Denmark (243), Belgium (214), Italy (212), Netherlands (194), Spain (191), Austria (189), France (177), Switzerland (161) and Finland (149). The operational stock of robots registered an ascendent trend in period 2017–2019 (almost 762,000 in 2017, 847,000 In 2018 and 923,000 in 2019).

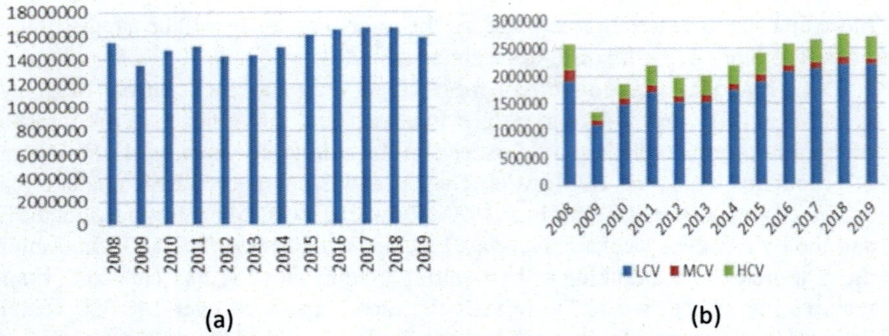

Fig. 1 Passenger (**a**) and commercial (**b**) car production in EU, 2008–2019. *Source* European Automobile Manufacturers' Association

After the annus horribilis in which there was the sharpest decline ever in car sales in the European Union, due to the pandemic, the European Automobile Manufacturers' Association (ACEA) predicts that the 2021 will mark a first step on the road to recovery, with an increase in sales of around 10% compared to 2020 (Quinfinanza 2021). The Covid-19 crisis is expected to continue in the first quarter of 2021, but the auto market expects to recover in the second half of the year, thanks to the progress of vaccination programs. The EU car market fell by 23.7%, dropping to 9.9 million units in 2020. A drop of three million cars compared to 2019.

Right now, it is more important than ever to work hand in hand with EU policy makers to ensure the European automotive industry's competitiveness on the world stage. Producers fear that the future Euro 7 standard, in preparation by the European Commission, will mark the end of diesel and gasoline engines. So, decarbonization, with the digitalization, is also changing the nature of the technologies and components of future vehicles. Moreover, the market for electric and hybrid plug-in vehicles exploded in 2020, "driven by industry investments and government support measures" according to ACEA, with a market share expected by 10.5%, from 3% in 2019. With the right public policies, including the massive development of alternative energy charging infrastructures in all EU states, this favorable trend can and must continue.

Directly or indirectly, European automotive workers number 14.6 million, or 6.7% of all EU jobs. The approximately 3.7 million of the 11.5% EU manufacturing jobs are in the automotive sector. Motor vehicles represent € 440.4 billion in taxes in major European markets. The automotive industry generates a trade surplus of 74 billion euros for the EU. The revenue generated by the automotive industry accounts for more than 7% of the EU GDP. By investing 60.9 billion Euros per year in research and development, the automotive industry is Europe's largest private contributor to innovation, representing 29% of total EU expenditure.

5 Methodology

Our aim was to determine the profit impact due to investment in digitization as robots, machinery and software. For this goal we develop a quantitative analysis using linear regression model. Therefore, we consider as dependent variable the profit (million lei) while the independent variables the employees and fixed assets. We conducted the analysis for 2005–2019 period, (www.listafirme.ro). The data we used correspond only to the factory from Mioveni of Automotive Dacia.

We know that fixed assets are a long-term tangible equipment or piece of property owned by a company and that is used to generate income. Fixed assets can include constructions, digital and other kind of equipment, property, technology and cars, (www.investopedia.com). We will assume in the paper that fixed assets represent the investment in modernization and digitalization of the factory.

6 Results and Discussions: Digitalization of Automotive Dacia

In Fig. 2 we present the evolution of profit, employees and fixed assets for Dacia in period 2005–2019.

It can be observed an ascendent trend for these variables. As we said before Dacia sales growth significant; also, to more labor was need to cover the demand and so the number of employees' growth; the modernization of Dacia involved buying robots, machines and software so the fixed assets also growth.

In Table 1 we summarize the results of the multiple linear regression model.

The results of F test show that there is a significant relation between profit, fixed assets and employees. The coefficients of independent variables indicate that a growth with one unit of fixed assets determine growth of profit by 0.21 units and we can base on this result because p-value for fixed assets is less than 0.05 so fixed assets has a significant statistical influence on profit; a growth of employees by one unit

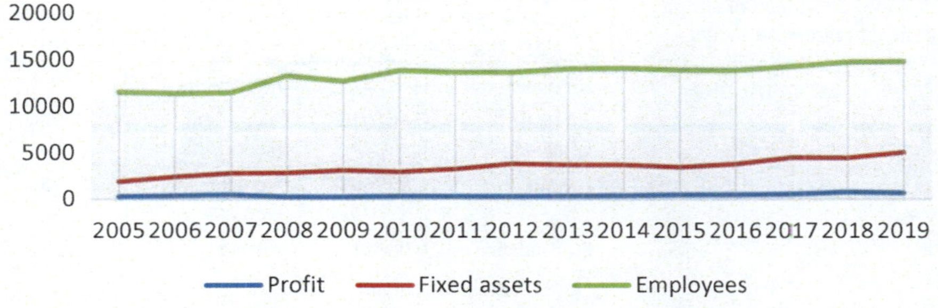

Fig. 2 Financial data for Dacia, 2005–2019. *Source* Authors' own research

Table 1 Results of multiple linear regression model

Indicator	Value of the indicator	Indicator	Value of the indicator
R-square	0.570275	Intercept	632.9159
R-square adjusted	0.498654	Coefficient for fixed assets	0.215456
F value	7.962413	Coefficient for employees	-0.07252
Significance F	0.006297		
t-statistic for fixed assets	3.119281	t-statistic for employees	-1.40514
p-value	0.008867	p-value	0.185337

Source Authors' own research

determine a decrease of profit by 0.07 units but we cannot base on this result because p-value for employees is bigger than 0.05 so employees does not have an significant statistical influence on profit. According to R-square we can say that 57.03% from profit variation is determined by the variation of fixed assets and employees. Looking at R-square adjusted we can observe a big difference from R-square so the information from this linear regression is not reliable. In this case the next step is to eliminate employees and to construct another linear regression model. The results for this new model are presented in the Table 2.

For this new model, the results of F test show that there is a significant relation between profit and fixed assets. The coefficients of independent variables indicate that a growth with one unit of fixed assets determine growth of profit by 0.13 units and we can base on this result because p-value for fixed assets is less than 0.05 so fixed assets has a significant statistical influence on profit. So, 1,000,000 lei invested in fixed assets will determine a grow with 130,000 lei of profit. According to R-square we can say that 49.95% from profit variation is determined by the variation of fixed assets. Looking at R-square adjusted we observe a small difference from R-square so the information from this linear regression is reliable.

Table 2 Results of simple linear regression model

Indicator	Value of the indicator	Indicator	Value of the indicator
R-square	0.49957	Intercept	-53.6764
R-square adjusted	0.461076	Coefficient for fixed assets	0.132086
F value	12.97768	t-statistic for fixed assets	3.602455
Significance F	0.003217	p-value	0.003217

Source Authors' own research

As we expected, the result of this simple linear regression model indicate that the digitization of Dacia has a positive impact.

7 Conclusion

In our days, together with innovation, digitalization is very important for the good running of the company on a competitive market, in full development. Digitalization can determine many opportunities and benefits, from process optimization to productivity increases and cost reduction.

In the practical part of our paper, we discussed about a Romanian company, Automotive Dacia. We wanted to determine what is the impact of digitalization on production. For this, we developed a linear regression model using profit as dependent variable and employees and fixed assets as independent variables. The period analyzed was 2005–2019.

By using our linear multiple model we determined that employees have a negative impact but the variable was not statistically significant. From the linear simple model, we determined that digitalization represented by fixed assets of the company, have a positive impact, as 1,000,000 lei invested in fixed assets determine a grow with 130,000 lei of profit.

Limitation of the study is represented first of all by the lack of information. We could get a better result if we had access to the company's financial data, to see the real impact on production of the number of robots used in the study period, from 2005 to 2019. We intend to explore this idea in a future paper.

References

Amit, R., e Zott, C.: Value creation in e-business. Strat. Manag. J. **22**(6-7), 493–520 (2001)

Aurotestmagazin, Dacia si robotii. http://www.autotestmagazin.ro, last accessed 2021/01/5

Curierul National, România continuă să se claseze printre ultimele locuri în regiune la digitalizarea economiei http://www.curierulnational.ro, last accessed 2021/01/6

Dougherty, D., Dunne, D.: Digital science and knowledge boundaries in complex innovation. Organ. Sci. **23**(5), 1467–1484 (2012)

European Comission, Indicele economiei și societății digitale (DESI) Raportul de țară din 2019 România. http://www.europa.eu. Last accessed 2020/12/5

European Automobile Manufacturers' Association. http://www.acea.com. Accessed 2020/11/7

Gölzer, P., Fritzsche, A.: Data-driven operations management: Organisational implications of the digital transformation in industrial practice. Product. Plan. Control **28**(16), 1332–1343 (2017)

Greensoft, Industria 4.0 nu este o revolutie, este o evolutie!, http://www.greensoft.ro, last accessed 2020/09/16

Groupe Renault, Groupe Renault Romania, o poveste de success. http://www.grouprenault.ro. Last accessed 2020/12/5

Guerry, M., Bieller, S., Muller, C., Kraus, W.: World Robotics 2020 Industrial Robots, International Federation of Robotics. http://www.ifrg.org. Accessed 2020/11/15

Gupta, S.: What is digitization, digitalization, and digital transformation? https://www.arcweb.com/blog/what-digitization-digitalization-digital-transformation. Accessed 2020/12/13

Kotarba, M.: Measuring digitalization-key metrics. Found. Manag. **9**(1), 123–138 (2017)

Lai, K.H., Wong, C.W.Y., Cheng, T.C.E.: Bundling digitized logistics activities and its performance implications. Ind. Mark. Manage. **39**(2), 273–286 (2010)

Li, F., Nucciarelli, A., Roden, S., Graham, G.: How smart cities transform operations models: A new research agenda for operations management in the digital economy. Prod. Plan. Control **27**(6), 514–528 (2016)

Loebbecke, C., Picot, A.: Reflections on societal and business model transformation arising from digitization and big data analytics: A research agenda. J. Strat. Inf. Syst. **24**(3), 149–157 (2015)

Munteanu, M.: Uzina Dacia produce 3% din PIB, AutoExpert Industry. http://www.autoexpertindustry.ro. Accessed 2020/11/7

OECD: Stimulating digital innovation for growth and inclusiveness the role of policies for the successful diffusion of ICT, OECD Digital economy papers No. 256, OECD publishing (2016)

Pagani, M., Pardo, C.: The impact of digital technology on relationships in a business network. Ind. Mark. Manage. **67**, 185–192 (2017)

Quinfinanza: Mercato auto, ACEA: ripartirà nel 2021, previsto un +10% rispetto al 2020. http://www.quinfinanza.it. Accessed 2020/10/3

Rachinger, M., Rauter, R., Müller, C., Vorraber, W., Schirgi, E.: Digitalization and its influence on business model innovation. J. Manuf. Technol. Manag. **30**(8), 1143–1160 (2019)

Rubino, M., Vitolla, F., Raimo, N.: Il Processo di digitalizzazione aziendale e la digital transformation. In: Smart Technologies. Digitalizzazione E Capitale Intellettuale, pp. 54–74. FrancoAngeli Publishing, Milano, Italy (2020)

Sebastian, I.M., Ross, J.W., Beath, C., Mocker, M., Moloney, K.G., Fonstad, N.O.: How big old companies navigate digital transformation. MIS Q. Exec. **16**(3), 197–213 (2017)

Ulas, D.: Digital transformation process and SMEs. Proc. Comp. Sci. **158**, 662–667 (2019)

Van Iersel, J., Konstantinou, N.: Avizul Comitetului Economic și Social European pe tema „Industria 4.0 și transformarea digitală: calea de urmat" in Jurnalul official al Uniunii Europene. http://www.europa.eu. Accessed 2020/12/3

Vendrell-Herrero, F., Bustinza, O.F., Parry, G., Georgantzis, N.: Servitization, digitization and supply chain interdependency. Ind. Mark. Manage. **60**(1), 69–81 (2017)

Verhoef, P.C., Broekhuizen, T., Bart, Y., Bhattacharya, A., Dong, J.Q., Fabian, N., Haenlein, M.: Digital transformation: A multidisciplinary reflection and research agenda. J. Bus. Res. in press, pp. 1–13 (2019)

www.acea.com. Accessed 2020/10/13

www.grouprenault.ro. Accessed 2020/10/27

www.listafirme.ro/automobile-dacia-sa-160796/, last accessed 2020/11/25

www.investopedia.com. Accessed 2020/11/9

The Climate Refugees—A Different Effect of Climate Change

Florina Bran, Carmen Valentina Radulescu, Alexandru Dumitru Bodislav, and Iulian Gole

Abstract Climate change has become a major issue in the international community from many points of view: political, urgency, sustainable development for future generations, etc. The results of the last elections in some European countries (Switzerland, Germany), as well as some other areas around the world (Australia, Canada) are clear: the Greens have conquered an unprecedented victory in the political arena. This great wave of environmentalism, described as 'historic', now seems inevitable in the political field, as well as in the level of reforms that the latter will bring. Indeed, those in power cannot voluntarily deny the environmental awareness conveyed by the general public, because it is undeniable—there is continuous pressure coming from the simple people. However, despite green political discourses, issues related to "climate refugees" are proving to be minimised. In this paper, we will try to provide data to understand what the size of the climate factor (or its consequences) is, as a cause of migration compared with economic or political reasons. We will also try to suggest some solutions in order to ameliorate the situation of people forced to leave their homes because of weather-related events. To better understand this issue, it is interesting to point out at first that the subject of climate refugees is not something recent but rather ancient. Indeed, this question has existed for a long time. We will quickly trace the historical evolution of this debate mixing migration and environment.

Keywords Climate change · Migration · Environment policies · Natural disasters

F. Bran (✉) · C. V. Radulescu · A. D. Bodislav · I. Gole
Bucharest University of Economic Studies, Bucharest, Romania
e-mail: florinabran@yahoo.com

C. V. Radulescu
e-mail: sburlacu@amp.ase.ro

A. D. Bodislav
e-mail: alex.bodislav@ase.ro

I. Gole
e-mail: iuliangole@yahoo.com

M. Busu (ed.), *Digital Economy and New Value Creation*, Springer Proceedings in Business and Economics, https://doi.org/10.1007/978-3-031-07265-9_7

1 Introduction

According to one of the latest reports from the Intergovernmental Panel on Climate Change (2018), more than 280 million people worldwide will be forced into exile in the next 80 years due to climate change—its impact manifesting itself through floods, storms, earthquakes, droughts, and other natural disasters. In this paper, we will try to provide data in order to understand what the size of the climate factor (or its consequences) is, as a cause of migration compared with economic or political reasons.

1.1 Literature Review and Background

Only two levels of headings should be numbered. Lower level headings remain unnumbered; they are formatted as run-in headings.

A research group from the University of Neuchâtel (Piguet et al. 2011), started from the idea that environmental migration is a recent phenomenon, as is often illustrated. Nevertheless, according to their research, the debate over migration and the environment already existed in the nineteenth century. Indeed, we have historical evidence and testimony that confirms the existence of major migratory flows following natural disasters. However, in the twentieth century, the ecological and/or environmental aspect, as an explanatory factor of migratory flow, was minimised for several reasons (Rădulescu et al. 2020).

Historically speaking, it proved that it is the responsibility of states to push people to become refugees (Bran et al. 2019); it was also the case when refugees created states. In other words, the debate on migration was centred on political issues (Bodislav et al. 2019). Towards the end of the 20th and the beginning of the twenty-first century, alarmist predictions about the number of people who would be forced to migrate for climate causes raised awareness among all government actors, in addition to civil society (Burlacu et al. 2018a, b, 2020a, b). It is in this context that we see in 1990 the first UN intergovernmental report on climate change and its impact on migration. At the same time, the International Conference on Population and Development, held in Cairo in 1994, stressed the need to consider immigration applications from countries whose existence is imminently threatened by global warming and consequently by climate change, to be judged by the available scientific data (Bran et al. 2018). These discussions had two effects: first is the awareness of the general public regarding the stakes of the situation, and second, the upgrading of the agenda of these issues from the side of state governments concerned (Profiroiu et al. 2020). Today, a huge number of international conferences are being held to find a solution to the degradation of our planet and its impact on migratory flows (Bran et al. 2020). The majority of states recognize this reality, but no government seems willing to grant (legal) refugee status to these displaced persons due to climate change reasons (Negescu Oancea et al. 2020).

Regarding this type of migrations, up until 2020, there was a debate within the international discussion, as it opens a politico-legal reflection (Alpopi et al. 2018). At present, the 1951 Geneva Convention Relating to the International Status of Refugees (signed by all western states) defines a refugee as a "person who owing to a well-founded fear of being persecuted for reasons of race, religion, nationality, membership of a particular social group, or political opinion, is outside the country of his nationality, and is unable to or, owing to such fear, is unwilling to avail himself of the protection of that country". Thus, this very narrow definition completely denies the climatic factor â. Therefore, an asylum application for environmental reasons cannot be legally recognized by western states. This legal, or even 'empty', gap contradicts human rights principles, including universal protection. Based on this finding, in 1985, the United Nations Environment Program (UNEP) provided a first definition for what an environmental refugee denotes: "any person forced to leave his or her traditional home temporarily or permanently due to a clear degradation of its environment that upsets its living environment and seriously unbalances its quality of life". Despite this clarification, no tangible reforms have been implemented in favour of these environmental refugees (Rădulescu et al. 2018a, b).

However, since 2015, the Nansen Initiative, adopted by 109 states, (which is a consultative process intended to build consensus on the development of a protection agenda addressing the needs of people displaced across international borders in the context of disasters and the effects of climate change) has been trying to bring institutional solutions to address the need for these refugees upstream. In other words, this process involves reaching a regional or even international consensus on the principles and fundamentals of refugee protection.

If we look into details it means that states define which the "acceptable" and "unacceptable" migrants are, according to specific criteria (Burlacu et al. 2010). Currently, the majority of rich states legitimise political refugees and deny the right to asylum to economic refugees. Yet this clear distinction is generally inconsistent, even without an empirical basis. On the ground, as in some African states, political causes are closely linked to the economic sphere. Nevertheless, this structuring interdependence is (un)voluntarily ignored by European states, as it allows them to reject certain asylum seekers.

Nevertheless, the situation of climate refugees seems to suffer a 180 degrees turnaround. The issue of climate refugees made headlines in 24th of October 2019, when the UN Human Rights Committee made a historic ruling on the issue: it is considered illegal to deport a migrant to his country of origin when his life is endangered by climate change.

The judges relied on the case of Ioane Teitiota, a Pacific Islander living in Kiribati. This archipelago is particularly threatened by rising waters. Ioane Teitiota filed an asylum application in New Zealand, but authorities refused to grant him protection under the reasoning that the fisherman-farmer life was not directly threatened. The UN Committee has just condemned this interpretation, which could lead states around the world to now consider climate change in the study of asylum applications.

The decision of the UN Human Rights Committee to recognize climate change as a ground for asylum is described as "historic" by experts, but its immediate effects are very limited, according to others, since this judgement does not create refugee status, it does not mean that many individuals will now meet the conditions set by the UN Committee which remain though very strict.

Nevertheless, there is almost always a mix between different initial reasons for migrations (political, human rights, economic, religious, climate, etc.), therefore we should analyse credible data to understand the complexity of the situation.

2 Methodology

By the means of descriptive and comparative analysis we exemplify with real data what the roots of climate refugee's situation are. We started our analysis with an international status perspective of the climate refugee position, from the point when international organisations and individual countries considered there is no ground to give the status of refuge to displaced people due to climate change until the last advancement when UNHRC accepted for the first time that climate change could be considered as a real reason for an asylum application.

We analysed data from different sources to see what the causes for displacements are, how many people are yearly affected in different geographical areas, and we schematically represented why the people from Tuvalu, Kiribati, and the Maldives had already suffered important losses and why their situations should be attentively considered.

3 Results and Discussions

Calamities have caused around 265 million deracination since 2008 (IDMC 2019), which represents more than three times the number of refugees who left their houses because of conflict and violence. The size of the phenomena pushed the international agenda to create the International Displacement Monitoring Centre to create data and reports.

The situation is likely to get worse and more difficult in the future. Weather hazards account for more than 87% of calamity deracination. In the table below we can see the size in total refugees by each event, weather related or geophysical Table 1.

If we take into consideration the stronger impact of climate change and the growing concentration of people in areas exposed to storms and floods we can predict easily that more and more people are at risk of being displaced. People that left their houses because of natural disasters are in the same situation, at the end of the day, as those who left because of violence or conflict. Many have lost their families, houses, assets, and revenues and face difficult times having low access to basic needs such as water supply, food, basic health care services, and education (Fig. 1).

In this graphic we can see that the numbers of refugees varies from one year to another because of big meteorological events. For example, in 2018, the departure of

Table 1 Displacement by hazard categories (2008–2018)

1. Geophysical	**12,73%**
1.1 Earthquake (incl. tsunami)	12,23%
1.2 Volcanic eruptions	0,46
1.3 Dry mass movements	0,03%
2. Weather related	**87,27%**
2.1 Floods	50,62%
2.2 Storms	35,54%
2.3 Drought	0,77%
2.4 Wildfires	0,61%
2.5 Extreme temperatures	0,40%
2.6 Wet mass movements	0,34%

Source Disaster Displacement, A global review, 2008–2018, IDMC

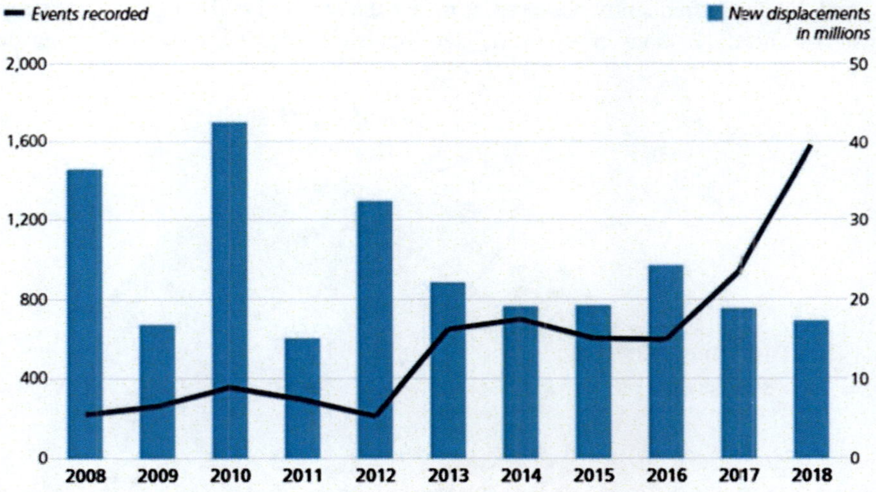

Fig. 1 Calamities events and people displaced 2008–2018. *Source* Disaster Displacement, A global review, 2008–2018, IDMC

9.4 million people was due to 10 events. It is clear that some years are outliers, 2010 being one of the most complex, with floods in China and Pakistan that made millions of houses disappear, massive rains associated with El Niño in Latin America and the Caribbean, on top of that being the earthquake in Haiti.

In the table below is presented the distribution of displacements for 2008–2018, on different geographical zones. Data is provided by "A global review report, 2008–2018, IDMC" (Table 2).

Table 2 The distribution of displacements for 2008–2018, in different geographical zones

East Asia and Pacific	141 311 000	53,21%
South Asia	69 419 000	26,14%
The Americas	30 788 000	11,59%
Sub-Saharan Africa	21 919 000	8,25%
Europe and Central Asia	1 200 000	0,45%
Middle East and North Africa	927 900	0,35%

Sources A global review report, 2008–2018, IDMC

Natural calamity displacement has happened in almost 190 countries in the period of 11 years. A bit more than 80% of all new displacements registered during 2008–2018, which means around 210 million people, are originally situated in the Asia–Pacific regions (Fig. 2).

Dangerous situations related to meteorological events were accountable for 87% of all refugees. As long as we see an increase in the populations in areas where storms and floods are frequently observed, it means that we will register even more people at risk shortly. In many cases, the displacements were organised by local authorities,

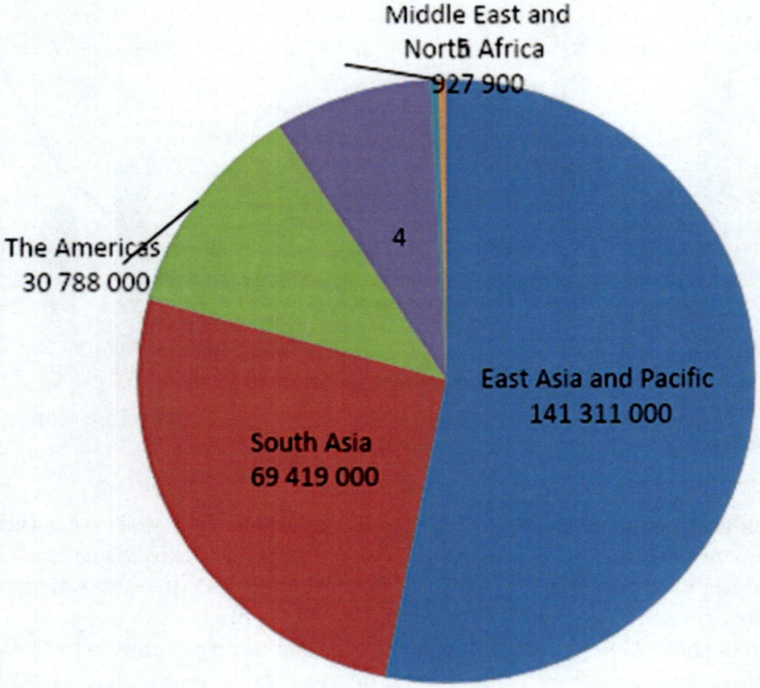

Fig. 2 Regional distribution of displacements 2008–2018. *Source* Own representation with data from Disaster displacements, A global review report, 2008–2018, IDMC

as a preventive measure to avoid victims or collateral damages but since there are no clear standards in order to classify when was the exact moment when people moved (before, during, or after calamity) it is very complicated to have a clear situation, about the moment in time and the number of people affected or about how disaster risks are managed.

Of course, there are events like earthquakes or tsunamis that are very disproportionate in terms of consequences in the sense that it may rarely occur but might cause millions of displacements. Usually, the refugees resulting from these events usually end up in more complicated situations, because they don't have the time to prepare for the departure and their periods of deracination are often extended.

For example, in 2018, from 1.5 million refugees from the 2010 earthquake in Haiti, 38 000 people are still living in temporary shelters, compared with 62 600 in 2016. Even though it is not a frequent phenomenon, during the interval 2008–2018, the earthquake is responsible for 12% of natural calamities disaster. Since 2017, credible data about drought, erosions of river banks started to be officially registered, so maybe the new numbers will show another perspective in a few years. Anyway, the real situation is far from being perfectly truly expressed in figures of reports concerning the refugees. Despite many countries that suffered severe droughts in 2018, only 9 countries reported this event (Afghanistan, Brazil, Burundi, Ethiopia, Iraq, Madagascar, Mongolia, Senegal, and Somalia), a total 765,000 refugees being registered. Another 49,000 displacements were coming from Afghanistan, Bangladesh, Myanmar, and Vietnam due to riverbank erosion.

Countries with large populations like China and India will register a higher number of refugees but we have to consider the overall situation, as a percentage of all inhabitants, or as a percentage of the area community. The situation is different when we are talking about the affected small islands because usually many if not all, the people living in the respective areas are exposed to the same risk (for example–Pam cyclone in Tuvalu 2015; Winston cyclone in Fiji 2016; Hurricane Maria in the Dominican Republic 2017; Super typhoon Yutu in Maria Island 2018).

Many specialists, as those grouped under The Intergovernmental Panel on Climate Change, confirmed that global warming is the cause of rising sea levels, ocean acidification, increasing sea surface temperature, and reduction of oceanic oxygen, which will affect some specific islands and their communities severely. It is obvious that there is interconnectivity between different effects which will finally make the entire economical and biologic systems of these islands to be badly and irremediably damaged. Their inhabitants will remain without viable solutions to survive so they will be forced to leave their homes (Fig. 3).

Due to human industrial activities, the planet faces climate change having a direct consequence of the increasing temperatures everywhere in the world. Therefore the temperature of the ocean increases and finally the oxygen level is reduced.

As the first effect of this situation, the coral reefs are dying which will reduce the fish stocks and the affluence of tourists, making life complicated for island habitants. The increasing level of temperature will melt the glaciers which will contribute to the sea level rise. This will bring beach destruction so again a low level of tourists, but also it will damage the quality of soil for agriculture, which finally will make

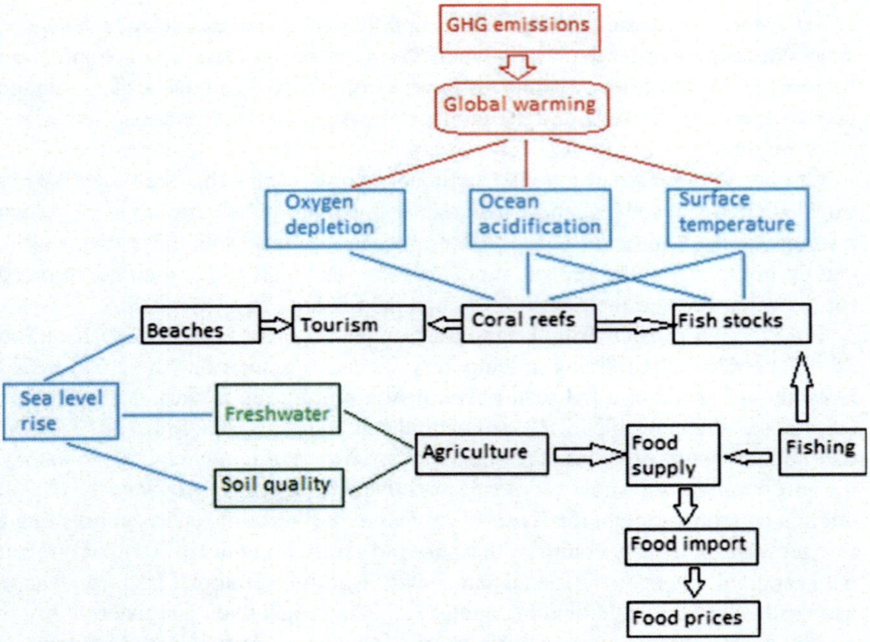

Fig. 3 Climate change impacts for under sea level islands. *Source* Own representation, adapted from IDMC, 2018

their existence on the island. Confronted with all these troubles the islanders will see their lives getting more and more difficult and ultimately will be forced to leave their lands.

Tuvalu, Kiribati, and the Maldives had already suffered important losses and some experts demonstrated that these territories will be underwater till 2050 (Mcadam 2011). Despite the pieces of evidence presented by researchers, there is an evident lack of leadership, interests, and empathy coming from developed countries that have the potential and the political power to impose solutions to protect these vulnerable populations. An immediate answer is expected rather from New Zealand, Australia, and the United States since they are geographically closer but all the big GHG contributors should be taken responsible for that.

Starting with 2019 we have a much broader view about what is happening in the refugee camps. There are 33.4 million new internal displacements from 145 countries—it is very important to pay attention to this data because, even though some people are not leaving their countries of origin, since they are leaving their homes it means that they became refugees as well. Their losses are also due, in part, to environment exchange.

While conflicts and violence gave rise to 8.5 million new displacements in 2019, we had 24.9 million new displacements coming from 140 countries—a record since

Table 3 Number of people displaced in 2019

Conflict and violence		Disaster	
Syria	1 847 000	India	5 018 000
Democratic Republic of Congo	1 672 000	Philippines	4 094 000
Ethiopia	1 052 000	Bangladesh	4 086 000
Burkina Faso	513 000	China	4 034 000
Afghanistan	461 000	United States	916 000

Source Global Report on Internal Displacement 2020, IDMC

2012 and 3 times higher than refugees of war and violence. The table below reflects the most relevant data by countries affected (Table 3).

All data presented express the importance of the issue, climate refugees (internal displacements or cross border) being one of the indirect global warming consequences.

Although emergency plans are created and relocation solutions are searched, we don't see an international collaboration aiming to find solutions for these people. Even though some countries developed good practices, the international community didn't agree yet to work on a common framework. If globally speaking, there is a visible combat against global warming and the issue attracts the attention over time, the large majority of the countries didn't have the motivation or political power to impose solutions to protect climate refugees.

Their inaction probably is because they have to deal already with a high number of other events like war, terrorism, and conflict refugees. After all, the issue doesn't directly affect their countries. Still, if we look at the amplitude of the climate change phenomena, the number of this type of people in need will grow exponentially in years after, and consequently asylum demands will explode. Needless to say, developed countries should have a closer look at the issue and invest in finding viable solutions to avoid a wave of unexpected immigration.

4 Conclusion

The answer for this problem should start with the recognition of the climate refugee status as being an individual in need of protection—as a matter of fact, we are talking about a person who doesn't have access to resources, lost his home, found himself in a situation where can no longer live (more or less these are the same circumstances for a war refugee). The second step, the most important and the most difficult, is to fight effectively against climate change progression in order to let time to affected countries and those who want to accept refugees to prepare, to create emergency plans, to find the optimal and sustainable solutions.

While the situation of a war refugee takes time to assess and it is individual, the climate refugee affair is general and could be applied to an entire country or territory.

Of course, the pressure is not at the highest limit so there will be a long time to take action, since there is no imminent danger, as in a war case. A third step, also difficult since there is always a crisis somewhere, should be the international community to find necessary funds to be used to relocate climate refugees but when there is common willingness, there is always a way to find solutions.

Releasing huge GHG emissions into the atmosphere due to our industrial activities lead the world in a situation of unprecedented rates of change. If we continue emitting greenhouse gases at our current rate, we will most probably face an increasing temperature period (certainly after 2050) but still, there is a little time left to react. Some of the global warming effects are already visible: extreme storms, floods, extended droughts and heatwaves, vegetation fires, rising sea levels. Vulnerable communities are already in danger; many people are suffering from losing their homes, assets, sources of income, their cultures—their lives are at risk because of an unstable climate. Even if we name them climate refugees or not, these people's tragedy is a consequence of climate change so the international community, especially the biggest GHG emitters, should assume the responsibility and make greater efforts to help them.

References

Alpopi, C., Burlacu, S., Iovițu, M.: Procesul de globalizare şi politicile ecologice. In: Competitivitatea şi Inovarea în Economia Cunoaşterii. Vol.2, 28–29 septembrie 2018, Chișinău, Republica Moldova: Departamentul Editorial-Poligrafic al ASEM, pp. 317–324 (2018). ISBN 978-9975-75-931-1

Bodislav, A.D., Rădulescu, C.V., Moise, D., Burlacu, S.: Environmental Policy in the Romanian Public Sector. The Bucharest University of Economic Studies Publishing House, 312 (2019)

Bran, F., Alpopi, C., Burlacu, S.: Territorial development-disparities between the developed and the least developed areas of Romania. LUMEN Proc. **6**(1), 146–155 (2018)

Bran, F., Rădulescu, C.V., Bodislav, D.A., Burlacu, S.: THE ANTHROPIC PRESSURE ON THE FOREST SPACE. DYSFUNCTIONS AND RISKS IN ROMANIA. Quality-Access to Success, 20 (2019)

Bran, F., Rădulescu, C.V., Bodislav, D.A., Burlacu, S.: Environmental risks in the context of globalization. Economic Convergence in European Union, 350 (2020)

Burlacu, S.: Evaluation of training process by fuzzy techniques. Administratie Si Management Public **15**, 162 (2010)

Burlacu, S., Gavrilă, A., Popescu, I.M., Gombos, S.P., Vasilache, P.C.: Theories and models of functional zoning in Urban space. Revista De Manag. Comparat Int. **21**(1), 44–53 (2020a)

Burlacu, S., Bodislav, D.A., Rădulescu, C.V.: E-Commerce and global food resources. Manag. Chall. Contemp. Soc. Proc. **11**(2), 48 (2018a)

Burlacu, S., Alpopi, C., Popescu, M.L.: Omul și degradarea mediului natural. Efecte distructive. In: Competitivitatea şi Inovarea în Economia Cunoaşterii. Vol.1, 28–29 septembrie 2018b, Chisinau, Republica Moldova: Departamentul Editorial-Poligrafic al ASEM, pp. 159–165 (2018b). ISBN 978-9975-75-932-8

Burlacu, S., Vasilache, P.C., Velicu, E.R., Curea, Ş.C., Margina, O.: Management of water resources at global level. In Proceedings of the International Conference on Economics and Social Sciences (pp. 998–1009). Sciendo (2020b)

IOM Haiti: 2010 Earthquake. Report package—EQ2010", 1 May 2018 (2018). Retrieved from https://haiti.iom.int/event/2010-earthquake

IDMC: Disaster displacement: A global review, 2008–2018 (2019). Retrieved from https://reliefweb.int/report/world/disaster-displacement-global-review-2008-2018

Intergovernmental Panel on climate Change: Global warming of 1.5°C. An IPCC Special Report on the impacts of global warming of 1.5°C above pre-industrial levels and related global greenhouse gas emission pathways, in the context of strengthening the global response to the threat of climate change, sustainable development, and efforts to eradicate poverty (2018). Retrieved from https://report.ipcc.ch/sr15/pdf/sr15_spm_final.pdf

International Covenant on Civil and Political Rights, UNHCR (2019). Retrieved from https://tbinternet.ohchr.org/_layouts/15/treatybodyexternal/Download.aspx?symbolno=CCPR%2fC%2f127%2fD%2f2728%2f2016&Lang=en

IPCC, Climate Change 2014: Synthesis Report. Contribution of Working Groups I, II and III to the Fifth Assessment Report of the Intergovernmental Panel on Climate Change. IPCC, Geneva, Switzerland, 151 (2014). Retrieved from https://www.ipcc.ch/report/ar5/syr/

Jane Mcadam: Swimming against the tide: Why a climate change displacement treaty is not the answer. Int. J. Refugee Law **23**(1), March 2011, 2–27 (2011)

Nansen initiative (2015). Retrieved from https://environmentalmigration.iom.int/projects/nansen-initiative

Negescu, M.D., Burlacu, S., Mitriţă, M., Buzoianu, O.C.A.: Managerial analysis of factoring at the international level challenges of the contemporary society. Proceedings; Cluj-Napoca **13**(1), 99–102. Cluj-Napoca: Babes Bolyai University (2020)

Piguet, E., Pecoud, A., De Gucheteneire, P.: Changements climatique et migrations: Quels risques, quelles politiques ? L'information Géographique **2011**, 86–109 (2011)

Profiroiu, C.M., Bodislav, D.A., Burlacu, S., Rădulescu, C.V.: Challenges of sustainable urban development in the context of population growth. Eur. J. Sustain. Devel. **9**(3), 51–51 (2020)

Rădulescu, C.V., Bodislav, D.A., Burlacu, S.: Demographic explosion and it governance in public institutions. Manag. Chall. Contemp. Soc. Proc. **11**(1), 18 (2018a)

Rădulescu, C.V., Dobrea, R.C., Burlacu, S.: The business management of distress situations. The 12th international management conference "Management Perspectives in the Digital Era". Novembre 1st-2nd, 2018b, BUCHAREST, ROMANIA, 1, 741-747 (2018b)

Rădulescu, C.V., Bran, F., Burlacu, S., Dobrea, C.R., Diaconu, S.: Challenges regarding food resources in the context of globalization and population growth. In Proceedings of the International Conference on Economics and Social Sciences (pp. 1041–1052). Sciendo (2020)

Resolution 429 of the United Nations General Assembly (1951). Retrieved from http://www.cas/com/discoveryguides/refugee/review2.php

The Impact of the Culture and the Social Norms on the Characteristics of Entrepreneurs: The Case of East Asian Countries

Khanh Hung Doan

Abstract Culture and social norms play an important role in the development of each country, not only its economic development but also its human development. A country with a positive culture and social norms will contribute to economic development by supporting entrepreneurship, educating and training the personality, and promoting productive businesses. A country with a good culture will benefit all stakeholders. This paper focuses on researching and evaluating the differences in the impact of culture and social norms on the characteristics of entrepreneurs in some East Asian countries: China, South Korea, Japan, and Vietnam. These are countries with similar cultural characteristics but their developments of economy and entrepreneurship are different. The research is based on 214 samples from Global Entrepreneurship Monitor—National Expert Survey—Individual Level Data in 2015. In the case of these four countries, in spite of similar culture and social norms, the results show that there are differences in the impact of culture and social norms on the characteristics of entrepreneurs. Therefore, it can be said that the culture and social norms of a country are not able to fully explain the characteristics and behaviors of entrepreneurs. The reason is due to the heterogeneity of the business environment of entrepreneurs in different countries. The results promote research about the impact of culture and social norms on entrepreneurship and the characteristics of the entrepreneurs.

Keywords Characteristics of entrepreneur · Culture · Social norms · Entrepreneurship · East Asia

1 Introduction

Nowadays, with the economic development of countries and the world, more and more enterprises are established and strongly developed. Besides the advantages, there are also many risks and challenges such as the economic, social, and business environment affecting enterprises (Păunescu et al. 2018; Priede-Bergamini

K. H. Doan (✉)
The Bucharest University of Economic Studies, 010374 Bucharest, Romania
e-mail: doankhanhhung.hat@gmail.com

M. Busu (ed.), *Digital Economy and New Value Creation*, Springer Proceedings in Business and Economics, https://doi.org/10.1007/978-3-031-07265-9_8

et al. 2019; Păunescu and Mátyus 2020; Păunescu and Molnar 2020). Therefore, enterprises need to have strategies to change or respond to the increasing requirements as well as the changing external environment. From there, it can contribute to helping enterprises to survive and to develop. The main factor for the development of an enterprise is the entrepreneur (Mas-Tur and Soriano 2014; Tu et al. 2014). Today, entrepreneurs are recognized as a force contributing to economic growth in the world. Developing countries that want to develop rapidly must pay special attention to the entrepreneur force. The entrepreneur is the main driving force for business development as well as socio-economic development.

Many researchers pointed out that the factors related to the business environment (such as infrastructure, information technology, and social environment) have a direct impact on entrepreneurship behavior, as well as the characteristics of the entrepreneur (Ozaralli and Rivenburgh 2016; Păunescu and Molnar 2020). However, research methods on business backgrounds as well as the business environment affecting entrepreneurship are still limited (Kallas 2019; Păunescu and Molnar 2020). Among the factors of the business environment that affect entrepreneurs, culture and social norms are among the most important factors (Giannetti and Simonov 2004; Hong and Kacperczyk 2009; Porfíri et al. 2018). The development of the entrepreneur or people cannot be separated from culture and social norms. Culture and social norms influence and form human personality. It is the necessary factor to ensure the quality of humans. The development of the enterprise and the characteristics of the entrepreneur are influenced by the cultural and social norms characteristics of each country or each region. Besides, culture and social norms also contribute to national economic development and enterprise development (Fritsch and Wyrwich 2017; Harutyunyan and Özak 2017; Porfíri et al. 2018).

Therefore, it is necessary to consider the role of culture and social norms for entrepreneurs. Countries need to consider, in particular, the influence of culture and social norms on entrepreneurs. From there, countries can choose the right cultural and social norms development models, enhance the characteristics of entrepreneurs, contribute to accelerating economic growth, promote entrepreneurship, as well as contribute to the existence and the development of the enterprises. However, there are still questions regarding how context factors such as culture and social norms influence the behavior of entrepreneurs when making comparisons across countries, including countries with similar cultures and social norms.

This paper provides a summary of relationships and the influence of culture and social norms on entrepreneurial characteristics. Moreover, the identification of key trends and gaps between culture and social norms for traits of the entrepreneur is also presented. After that, research was further evaluated on the distinct influence of culture and norms on entrepreneurial characteristics in some East Asian countries. In recent years, East Asia is a region with the highest GDP growth rate in the world. This contributes to a large proportion of the general development of the world. Along with the development of the world economy in general and the development of East Asia in particular, the contribution of East Asia to the world economy is more and more important. More precisely, China and Japan are the second biggest and third biggest economies in the world (WorldBank 2020). Furthermore, these East Asian countries are also countries with similar cultures (Sadao 1993).

The research is based on data collected from Global Entrepreneurship Monitor (GEM)—National Expert Survey—Individual Level Data in 2015. Finally, there are some conclusions and suggestions to improve research on the impact of the culture and social norms on the characteristics of entrepreneurs in different countries. The paper is organized as follows: the first part refers to a brief literature review of the studies, the second part explains the methodology used by the authors in this paper and the third part includes the research results, while the last part concludes and discusses the research results.

2 Literature Review

2.1 Characteristics of Entrepreneur

Nowadays, along with the process of globalization and economic development in the world, enterprises encounter a risky and challenging business environment (Păunescu et al. 2018; Priede-Bergamini et al. 2019; Păunescu and Mátyus 2020; Păunescu and Molnar 2020). Therefore, entrepreneurs need to implement business strategies appropriately in order to survive and to develop. The research by Amiri et al. (2009) on Iran's SMEs showed that only 10% of entrepreneurs could do business on their own while others fail even before starting a business. At the international level, entrepreneurship has high failure rates. It is happening because 20% of enterprises will be removed during the first year of operation (Franco and Haase 2010). Meanwhile, it will reach 66% by the end of the sixth year (Franco and Haase 2010). Driessen and Zwart (2007) stated that 50% of enterprises will disappear in the first five years of establishment. Therefore, if there is not enough business capacity, an enterprise will not exist, especially in the current fierce competition environment and many risk factors. In fact, there are many factors that can affect the business efficiency of the enterprise. On the entrepreneur side, to improve their business capabilities, they often pay attention to financial and non-financial factors as external factors. In addition, most enterprises are not aware of the important role of the entrepreneur's capacity in entrepreneurial activities. Some studies show that there is an indispensable relationship among characteristics of the entrepreneur, entrepreneurial competency, business efficiency, and business success (Hastuti 2020; Machmud and Hidayat 2020). Therefore, entrepreneurship and the characteristics of the entrepreneur are related to the business efficiency of the enterprises (Mitchelmore and Rowley 2010; Al Mamun and Fazal 2018). Entrepreneurs must pay special attention to improving their capabilities to promote their business efficiency and business capacity.

According to Mojab et al. (2011), competencies are classified into knowledge, characteristics, and skills. On the other hand, competencies are divided into natural and artificial capacities (Ismail 2012). Competencies are naturally formed within an entrepreneur as personality traits, attitudes, self-image, and social roles, while artificial competencies are contingent such as skills, knowledge, and experience (Ismail 2012). Moreover, Driessen and Zwart (2007) also identified four types of entrepreneurial competencies, each consisting of a group of features: knowledge (market, people, finance, production), competencies (autonomy, performance, power), competencies (management, motivations, organizational planning, financial governance) and characteristics (risk tolerance, association, tolerance of ambiguity, etc.). These characteristics of the entrepreneur are related to the entrepreneur's business efficiency (Van Auken et al. 2006; Rauch and Frese 2007; Deniz et al. 2011). Besides, researchers have argued that entrepreneurship is based on the presence of personal characteristics as well as the external environment (Arenius and Minniti 2005; Păunescu et al. 2018; Păunescu and Molnar 2020). Therefore, it can be said that the characteristics of entrepreneurs are influenced by many different factors.

2.2 *The Influence of Culture and Social Norms on Entrepreneurs*

The business environment has a strong impact on business efficiency (Ahmad et al. 2010; Krause et al. 2010; Eling and Schaper 2017). In addition, local norms, beliefs, laws, rules, etc. can shape the ability of entrepreneurs (Lyons et al. 2012; Păunescu and Molnar 2020). When the business environment changes (for example, institutions, policies, security, natural disasters, incidents, epidemics, etc.), it will lead to risks to the business operations of enterprises (Păunescu and Argatu 2020; Păunescu and Mátyus 2020). One of the most important factors of the business environment is culture and social norms (Hong and Kacperczyk 2009; Porfíri et al. 2018). Culture and social norms affect an individual's institutional environment (Davari et al. 2018). However, the link among culture, social norms, and entrepreneurship has not been fully established or explored (Krueger et al. 2013; Wach 2015; Porfíri et al. 2018).

Social norms are an explanation of different mandatory, permissible, or forbidden actions in different cultures and situations (Ostrom 2000; Liñán and Chen 2009). Moreover, social norms have an origin in a country's cultural context and have differences between countries. Another study on the definition of social norms is that social norms are the human perception of the rules in which people behave (Etzioni 2000). These rules can be improved by social norms (Etzioni 2000). Elster (1989) believes that if norms are social, they are relevant to people. And in this way, social norms are always to obey the approval or disapproval of people (Elster 1989; Festre 2010). While social norms influence social behavior, they can easily shape individual behavior (Wenzel 2004; Davari et al. 2018). Social norms include an individual's perception and encouragement for an important opinion on a particular topic (Oh and Hsu 2001) and will be divided into subjective norms and beliefs (Herold et al. 1998).

Levie and Autio (2008) considered social norms such as social image and national attitudes towards entrepreneurship. Social image is defined as the social status of entrepreneurs and the media's attention to entrepreneurship (Levie and Autio 2008). The national attitude means prioritizing entrepreneurship as a desirable career (Zali et al. 2013). Research by Barazandeh et al. (2015) analyzed the relationship between the capabilities of the entrepreneurs and social norms. The results showed that there is a positive effect between the social norms and the capabilities of the entrepreneurs (Barazandeh et al. 2015). In addition, Giannetti and Simonov (2004), Emami and Khajeheian (2019) found evidence that social norms have an influence on entrepreneurship. Thus, social norms positively impact and improve entrepreneurship. This is based on how it affects the characteristics of the entrepreneurs.

Besides, culture is one of the important factors influencing entrepreneurial activities (Krueger et al. 2013; Wach 2015; Porfíri et al. 2018). Although there are no laws to enforce social norms, culture can play an important role (Ajzen 2002; Porfíri et al. 2018). Chrisman et al. (2002) argued that culture influences the way entrepreneurs perceive the environment, so that culture influences how entrepreneurs choose strategies and the success rate of projects. Or it can be said that national culture shapes individual behavior by directly affecting the development of certain personalities and promoting individual participation in society (Mueller and Thomas 2001). The research of Levie and Autio (2008) also distinguishes cultures and social norms in the analysis of entrepreneurship and entrepreneurial intention in the world. Considering skills of the entrepreneurs is entrepreneurial competency (Zali et al. 2013), many studies demonstrating the influence of culture on using skill (Connell et al. 2001), development skills (Dickson and Riegel 2009), and acquiring skills (Dickson and Riegel 2009) of entrepreneurs.

Culture can affect entrepreneurship through two main mechanisms. Firstly, a supporting culture leads to social legitimization, making business more valued and socially recognized within that culture. Moreover, it could create a favorable environment (Davidsson 1995; Porfíri et al. 2018). Therefore, many people will try to initiate their adventurous activities, regardless of their personal beliefs and attitudes (Etzioni 1987). Secondly, a culture that shares more entrepreneurial values and knowledge will lead to more individuals displaying psychological traits and attitudes consistent with entrepreneurship (Krueger 2003). Thus, many people will try to become entrepreneurs (Mueller and Thomas 2001). In this sense, appreciation of entrepreneurship in society leads to more positive attitudes and intentions of individuals (Linan et al. 2011). Thus, policy-makers can use cultural traits as a tool to stimulate entrepreneurship (Huggins and Thompson 2012; Wach 2015) and to influence creativity activities and entrepreneurship (Ulijn and Brown 2004).

There are many different cultural backgrounds that affect many aspects of an individual's social environment. Entrepreneurs share a set of cultural and social values, some of which are entirely based on nationality (Muzychenko 2008). The cultural community characteristics of regions can commonly affect the entrepreneurial culture in these places, for example, entrepreneurial capacity. Many studies show a significant relationship between community culture and entrepreneurial attitudes at the

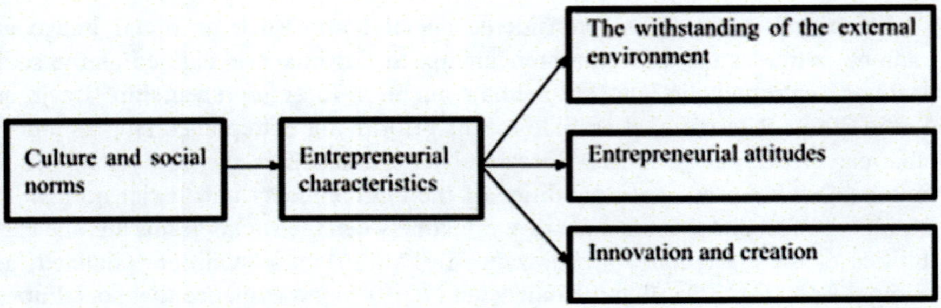

Fig. 1 Research model. *Source* Author's own research (2021)

local level (Huggins and Thompson 2012; Wach 2015). Moreover, compliance with social norms, social cohesion, and work and education is positively related to the entrepreneurial culture. A cohesive community increases business confidence from individuals. In terms of country level, the entrepreneurship level between countries is always different. This difference has its origin in national cultural differences (Harms and Groen 2017). And the fact is that different countries may have different cultures and social norms. Thus, the degree of entrepreneurship is influenced by culture, and social norms can also be different. Besides, the cultural characteristics of different regions can affect entrepreneurial culture, especially the capacity of the entrepreneur (Huggins and Thompson 2012). The research of Castaño et al. (2015) showed that cultural, social, and economic factors influence entrepreneurship in European and Latin American countries. The degree of influence of these factors depends on different countries (Castaño et al. 2015).

Therefore, it can be said that culture and social norms affect the characteristics of the entrepreneur. These include the ability of the entrepreneur to be influenced by the external environment, as well as the characteristics, the attitude, the creativity, and the innovation of the entrepreneur (see Fig. 1).

3 Methodology

3.1 Research Hypotheses

Based on the results of the literature review, the synthesis, and the analysis of the author, the following research hypotheses are given to measure the differences in the level of national culture and social norms that influence the characteristics of entrepreneurs. These research hypotheses are based on the measuring of the culture that affects entrepreneurship (Huggins and Thompson 2012).

H1: There are differences in the influence of national culture in terms of the withstanding of the external environment of the entrepreneurs. (External environment).

Hypothesis H1 is based on the research of Chrisman et al. (2002). Chrisman et al. (2002) pointed out that culture will influence how entrepreneurs perceive the environment, which affects entrepreneurship.

H2: There are differences in the influence of national culture in terms of the attitude of the entrepreneurs. (The entrepreneur's attitude).

Hypothesis H2 is based on the evaluation by Lyons et al. (2012). Lyons et al. (2012) argued that "*local norms, beliefs, laws, rules, and history may not only shape the ability of entrepreneurs to thrive, but may dramatically affect which individuals are able to become entrepreneurs, why they become entrepreneurs, and what goals they seek to achieve through their business ventures*". In addition, Barazandeh et al. (2015) found that there is a positive impact between social norms on entrepreneurial competencies. Therefore, the national culture influences the attitude of the entrepreneur when starting a business.

H3: There are differences in the influence of national culture in terms of the innovation and creativity of the entrepreneurs. (Innovation and creation).

The cultural background and regulations are becoming more important for entrepreneurial innovation. With the development of globalization in recent years, there are more and more new opportunities for entrepreneurs to pursue innovative and creative activities (Levie and Autio 2011; Autio et al. 2014). Therefore, when culture influences the characteristics of the entrepreneur, it will affect the creativity and innovativeness of that entrepreneur. Moreover, Ulijn and Brown (2004) found that culture affects creative activities and entrepreneurship.

3.2 Data Collection

Data of this research are collected based on GEM—National Expert Survey—Individual Level Data in 2015. The GEM program has begun to explore and to evaluate the role of entrepreneurship in national economic growth. This project has developed a worldwide survey of entrepreneurship. Since the GEM project's inception in 1999, GEM has recently investigated entrepreneurship in more than 80 countries. The main purpose of the GEM project is to measure entrepreneurship in each country. All GEM data is published and collected from the website www.gemconsortium.org. This database is widely used in entrepreneurship research by many different researchers (Arenius and Minniti 2005; Huggins and Thompson 2012; Barazandeh et al. 2015; Castaño et al. 2015; Honjo 2015; Wach 2015; Filculescu 2016; Mrożewski and Kratzer 2016; Harms and Groen 2017) because this data allows for evaluation comparison between different countries in the world as well as in-depth analysis of factors affecting the entrepreneurship of each country.

To test the research hypotheses, the author collects data from four countries in East Asian cultural regions, namely Vietnam, China, Japan, and South Korea. Sadao (1993) was the first person to define the concept "東亜文化圏" (East Asian cultural region). He conceived of a culture of China and other East Asian countries that were different from the West. According to Sadao (1993), this region is influenced by

Table 1 Economic overview of East Asian countries in 2019

Country	GDP (Billion USD)	GDP Growth (%)	GDP per capita (USD)
China (PRC)	14,342.903	6.1	10,261
Japan	5,081.769	0.7	40,246
South Korea	1,646.739	2.0	31,846
Vietnam	261.921	7.0	2,715

Source WorldBank (2020)

Confucian and Buddhist ideology and shares the same political and social structure. This region includes China, Japan, the Korean peninsula (North Korea and South Korea), Vietnam, extending to the lands between Mongolia and the Himalayas (Sadao 1993). However, in this paper, we only study countries such as Vietnam, China, Japan, and South Korea because these are the countries with data collected by GEM and these are the core countries in the East Asian cultural region. Besides, these are countries that are geographically close to each other and have similar cultures and social norms. However, they have different levels of economic development (see Table 1). Japan and South Korea are developed countries with high GDP per capita value compared to developed countries. China and Vietnam are developing countries but have relatively high average annual GDP growth rates. China and Vietnam are some of the countries with high growth rates in the world. In addition, China and Japan are the second-biggest and third-biggest economies in the world, respectively (WorldBank 2020).

The statistical descriptive results of the observed samples collected from GEM data of four countries, Japan, South Korea, Vietnam, and China, are shown in Table 2. Total 214 samples are used for calculation. In particular, Japan, China, and Vietnam each country consists of 36 observed samples, accounting for 16.8% of the total observed samples. South Korea has 106 observed samples, accounting for 49.5% of the total observed samples.

Table 2 Description of data from GEM

Country	Frequency	Percent	Valid percent	Cumulative percent
Japan	36	16.8	16.8	16.8
South Korea	106	49.5	49.5	66.4
Vietnam	36	16.8	16.8	83.2
China (PRC)	36	16.8	16.8	100.0
Total	214	100.0	100.0	

Source Global Entrepreneurship Monitor (2015)

3.3 Data Analysis

To analyze the influence of national culture and social norms on the characteristics of the entrepreneur, research uses three questions in GEM—National Expert Survey—Individual Level Data in 2015. These questions respectively include:

- External environment—*In my country, the national culture encourages entrepreneurial risk-taking.*
- Entrepreneurial attitudes—*In my country, the national culture emphasizes the responsibility that the individual (rather than the collective) has in managing his or her own life.*
- Creation and innovation—*In my country, the national culture encourages creativity and innovativeness.*

The scale used in the study is Likert's scale with five levels (1: "Completely False", 2: "Somewhat False", 3: "Neither True nor False", 4: "Somewhat True", 5: "Completely True").

The analysis method used in this paper is One-way ANOVA analysis method by SPSS 23.0 software. One-way ANOVA is used to consider whether there is a significant difference in the mean in the population for the observed variables, here is the difference in terms of countries for the criteria. If the results of One-way ANOVA are appropriate (Sig. value < 0.05), then the author continues to use the post-hoc analysis of Turkey with the assumption that it has equal variance between groups of criteria. These results show in more detail the difference (with Sig. value < 0.05) or no difference for the mean in the population (with Sig. value > 0.05) of countries for different criteria. Finally, with the statistical results depicting the data on the mean of the data according to the criteria and countries, we can see the differences across the countries.

4 Results

4.1 Culture and Social Norms, and Entrepreneurship of the East Asian Countries

To evaluate the entrepreneurship environment, the study uses Global Entrepreneurship Index (GEI) for analysis. The results of the entrepreneurship environment in the East Asia countries are presented in Table 3. The results provide that, in 2019, South Korea was the highest mark of entrepreneurship (58.1), then Japan (53.3), China (45.9), and finally Vietnam (26.0).

To evaluate the influence of culture and social norms, Global Entrepreneurship Monitor (GEM) uses the Cultural and Social Norms Index. This index is an important criterion to evaluate the role of culture and social norms on the entrepreneurship of each country. The assessment results of the influence of culture and social norms on

Table 3 Global entrepreneurship index of the East Asian countries over the years

Country	Global entrepreneurship index		
	2015	2017	2019
China (PRC)	36.4	36.3	45.9
Japan	49.5	51.7	53.3
South Korea	54.1	50.5	58.1
Vietnam	28.8	22.0	26.0

Source Global Entrepreneurship Index (2020)

Table 4 Cultural and social norms index of the East Asian countries over the years

Country	Cultural and social norms index		
	2015	2017	2019
China (PRC)	2.98	3.23	3.81
Japan	2.35	2.26	2.67
South Korea	2.93	2.99	2.92
Vietnam	3.23	3.62	3.62

Source Global Entrepreneurship Monitor (2020)

the entrepreneurship of East Asian countries (including China, Japan, South Korea, and Vietnam) are shown in Table 4.

The results show that there is a change in the influence of culture and social norms on the entrepreneurship of East Asian countries. In the period from 2015 to 2019, China is a country with a significant increase in the Cultural and Social Norms Index, from 2.98 points in 2015 to 3.81 points in 2019 (an increase by 0.83 points). This is the highest increase among East Asian countries. Meanwhile, the Cultural and Social Norms Index of Vietnam and Japan had a slight increase of 0.39 and 0.32 points, respectively. In addition, during this period, there is little change in the index for South Korea. We can see that culture and social norms are increasingly important to the characteristics of Vietnamese, Japanese, and especially Chinese entrepreneurs.

4.2 *Analysis of the Differences of Cultures and Social Norms of Countries for Entrepreneurs*

Table 5 presents the results of Testing the differences of the external environment across East Asian countries for entrepreneurs. One-way ANOVA test results show that there are significant differences between the countries being compared (Sig. value < 0.05). Therefore, the results of post-hoc by Tukey's test method are used for evaluation. The analysis results are presented in Table 6.

In terms of the acceptability of the external environment among the countries, post-hoc analysis Tukey test identifies the presence of differences between the sample of

Table 5 Testing the differences of the external environment across East Asian countries for entrepreneurs – One-way ANOVA results

	Sum of squares	*df*	Mean square	F	Sig
Between Groups	30.952	3	10.317	7.451	*0.000**
Within Groups	288.029	208	1.385		
Total	318.981	211			

* The statistical value is significant at the 0.05 level. *Source* Author's own research results (2021)

Table 6 Testing the differences of the external environment across East Asian countries for entrepreneurs—Multiple comparisons results

(I) Country	(J) Country	Mean difference (I-J)	Std. Error	Sig
Japan	*South Korea*	*-0.79048**	*0.22968*	*0.004**
	Vietnam	*-1.23016**	*0.27934*	*0.000**
	China (PRC)	*-1.03571**	*0.27934*	*0.002**
South Korea	*Japan*	*0.79048**	*0.22968*	*0.004**
	Vietnam	-0.43968	0.22727	0.217
	China (PRC)	-0.24524	0.22727	0.703
Vietnam	*Japan*	*1.23016**	*0.27934*	*0.000**
	South Korea	0.43968	0.22727	0.217
	China (PRC)	0.19444	0.27736	0.897
China (PRC)	*Japan*	*1.03571**	*0.27934*	*0.002**
	South Korea	0.24524	0.22727	0.703
	Vietnam	-0.19444	0.27736	0.897

* The mean difference is significant at the 0.05 level. *Source* Author's own research results (2021)

entrepreneurs from Japan, South Korea, China, and Vietnam (Table 6). On average, entrepreneurs in Japan reported the acceptability of external environments lower than that of entrepreneurs in other countries such as South Korea, China, and Vietnam (Table 7).

The results of comparing differences in entrepreneur's attitudes among East Asian countries are provided in Table 8. The results show a statistically significant difference

Table 7 Description of evaluation about the external environment of entrepreneurs

Country	N	Mean	Std. Deviation	Std. Error	Minimum	Maximum
Japan	35	1.7143	1.04520	0.17667	1.00	4.00
South Korea	105	2.5048	1.15295	0.11252	1.00	5.00
Vietnam	36	2.9444	1.26366	0.21061	1.00	5.00
China (PRC)	36	2.7500	1.27335	0.21223	1.00	5.00

Source Author's own research results (2021)

Table 8 Testing the differences of the entrepreneur's attitudes across East Asian countries for entrepreneurs – One-way ANOVA results

	Sum of squares	*df*	Mean square	F	Sig
Between groups	17.231	3	5.744	3.487	*0.017**
Within groups	342.656	208	1.647		
Total	359.887	211			

* The statistical value is significant at the 0.05 level. *Source* Author's own research results (2021)

(Sig. value < 0.05) in the impact of culture on the attitudes of the entrepreneurs in East Asia.

Post-hoc analysis by the Tukey method is presented in Table 9. The result shows that the only difference is significant between the sample of entrepreneurs in Japan and the sample of entrepreneurs from South Korea. Meanwhile, entrepreneurs from Japan reported, on average, a lower level of the influence of national culture and social norms on entrepreneur's attitudes than entrepreneurs from South Korea (Table 10).

Table 9 Testing the differences of the entrepreneur's attitudes across East Asian countries for entrepreneurs—Multiple comparisons results

(I) Country	(J) Country	Mean difference (I-J)	Std. Error	Sig
Japan	*South Korea*	*-0.8000**	*0.25051*	*0.009**
	Vietnam	-0.66111	0.30468	0.135
	China (PRC)	-0.71667	0.30468	0.090
South Korea	*Japan*	*0.8000**	*0.25051*	*0.009**
	Vietnam	0.13889	0.24789	0.944
	China (PRC)	0.08333	0.24789	0.987
Vietnam	Japan	0.66111	0.30468	0.135
	South Korea	-0.13889	0.24789	0.944
	China (PRC)	-0.05556	0.30252	0.998
China (PRC)	Japan	0.71667	0.30468	0.090
	South Korea	-0.08333	0.24789	0.987
	Vietnam	0.05556	0.30252	0.998

* The mean difference is significant at the 0.05 level. *Source* Author's own research results (2021)

Table 10 Description of evaluation about the entrepreneur's attitudes of entrepreneurs

Country	N	Mean	Std. Deviation	Std. Error	Minimum	Maximum
Japan	35	2.2000	1.25558	0.21223	1.00	5.00
South Korea	105	3.0000	1.25576	0.12255	1.00	5.00

Source Author's own research results (2021)

Table 11 Testing the differences of the innovation and creation across East Asian countries for entrepreneurs—One-way ANOVA results

	Sum of Squares	*df*	Mean Square	F	Sig
Between Groups	26.434	3	8.811	5.792	*0.001**
Within Groups	316.448	208	1.521		
Total	342.882	211			

* The statistical value is significant at the 0.05 level. *Source* Author's own research results (2021)

Finally, the study will test the differences among entrepreneurs on innovation and creation between countries. One-way ANOVA test results are presented in Table 11. The results show that there are significant differences (Sig. value < 0.05) between countries for innovation and creation, and this result is the condition for performing post-hoc Tukey analysis.

The post-hoc Tukey analysis results are shown in Table 12. Thus, there is a significant difference between the report of the entrepreneurs from Japan and entrepreneurs from South Korea and Vietnam. Entrepreneurs from Japan reported lower than entrepreneurs from Vietnam and South Korea about the impacts of national culture and social norms on innovation and creation of the entrepreneurs (see Table 13).

With the results comparing the differences in the influence of national culture and social norms on entrepreneurial characteristics in the East Asian countries above, it can be said that there is practical evidence to support the hypotheses such as after:

H1: There are differences in the influence of national culture in terms of the withstanding of the external environment of the entrepreneurs (External environment).

Table 12 Testing the differences of the innovation and creation across East Asian countries for entrepreneurs—Multiple comparisons results

(I) Country	(J) Country	Mean difference (I-J)	Std. Error	Sig
Japan	*South Korea*	*-0.78312**	*0.23851*	*0.007**
	Vietnam	*-1.13889**	*0.29073*	*0.001**
	China (PRC)	-0.50000	0.29073	0.316
South Korea	*Japan*	*0.78312**	*0.23851*	*0.007**
	Vietnam	-0.35577	0.23851	0.444
	China (PRC)	0.28312	0.23851	0.636
Vietnam	*Japan*	*1.13889**	*0.29073*	*0.001**
	South Korea	0.35577	0.23851	0.444
	China (PRC)	0.63889	0.29073	0.127
China (PRC)	Japan	0.50000	0.29073	0.316
	South Korea	-0.28312	0.23851	0.636
	Vietnam	-0.63889	0.29073	0.127

* The mean difference is significant at the 0.05 level. *Source*: Author's own research results (2021)

Table 13 Description of evaluation about the innovation and creation of entrepreneurs

Country	N	Mean	Std. Deviation	Std. Error	Minimum	Maximum
Japan	36	2.3611	1.31264	0.21877	1.00	5.00
South Korea	104	3.1442	1.17761	0.11547	1.00	5.00
Vietnam	36	3.5000	1.25357	0.20893	1.00	5.00

Source Author's own research results (2021)

The results show that there are differences between entrepreneurs in Japan and entrepreneurs in the other three countries about the impact of the national culture on the ability of entrepreneurs to withstand the external environment. On average, the report on cultural exposure to external environmental exposure shows that the evaluation of entrepreneurs from Japan is lower than the evaluation of entrepreneurs from South Korea, China, and Vietnam. The entrepreneurs from Vietnam had the highest reported levels. Therefore, we conclude that hypothesis H1 is accepted.

H2: There are differences in the influence of national culture in terms of the attitude of the entrepreneurs (The entrepreneur's attitude).

The test shows that there is a significant difference between Japanese entrepreneurs and Korean entrepreneurs. On average, entrepreneurs from Japan reported lower than entrepreneurs from South Korea about the influence of national culture on attitudes of entrepreneurs. Thus, the supporting evidence of H2 is weak, and we only partially accept the hypothesis H2.

H3: There are differences in the influence of national culture in terms of the innovation and creativity of the entrepreneurs (Innovation and Creation).

The results of the statistical test show significant differences between entrepreneurs from Japan and entrepreneurs from South Korea and Vietnam. The mean value of the report from Japanese entrepreneurs is lower than the mean values of the reports from entrepreneurs in all the other two countries. Therefore, there is evidence to support hypothesis H3.

5 Discussions and Conclusions

Although the East Asian countries have similar culture and social norms, the analysis above shows that there are differences in the influence of national culture on the characteristics of entrepreneurs coming from Japan and other countries. In addition, the results also show that the level of cultural contribution to entrepreneurship, innovation, and creation in these countries is quite low (the mean values are assessed low). The degree of contribution to the influence of national culture and social norms is not related to the size of the economy, the growth of the economy, and the entrepreneurship index in the countries studied. Entrepreneurs from Japan reported a lower level of characteristics of entrepreneurs than their counterparts in all the other three countries. Moreover, out of the four countries included in the sample, Vietnam, on average, has high levels of creation and innovation.

Out of the four countries studied, Japan is the most developed economy with the highest GDP per capita than other countries, as well as the second-highest entrepreneurship index (lower than South Korea). However, Japan has the lowest rank of cultural influence among the countries studied. This result shows that the support of national culture and social norms in Japan for entrepreneurship and the characteristics of the entrepreneur are less than in other countries. Japanese entrepreneurs are influenced by other factors more than the factor of national culture and social norms. In contrast, for Vietnam and China, which are two developing countries, the influence of culture and social norms on entrepreneurship is more than that of the other two developed countries, Japan and South Korea. This shows that economic development affects the impact of national culture and social norms on entrepreneurship and the characteristics of the entrepreneur. Therefore, we can conclude that national culture and social norms influence entrepreneurship and the characteristics of the entrepreneurs. However, the economic development index, such as the size of the economy, and the entrepreneurship index of a country are not good factors for making forecasts about the impact of culture and social norms on entrepreneurship in those countries. So, further studies on the factors affecting entrepreneurship and characteristics of the entrepreneur should try to incorporate more aspects relevant to the external environment. In addition, future research needs analysis to identify the supporting factors for entrepreneurship, creation, and innovation in different countries or in individual sectors of different industries.

This study also has certain limits related to the nature and characteristics of the data used for analysis in this paper. The samples are based on a global structured survey conducted by GEM. However, the number of samples used in the study is limited. The number of samples surveyed in each country of Vietnam, Japan, and China is only 36 samples and the total number of samples is 214 samples. In addition, based on the data that were available, the research is also unable to analyze in-depth other factors that influence entrepreneurial characteristics.

References

Ahmad, N.H., Ramayah, T., Wilson, C., Kummerow, L.: Is entrepreneurial competency and business success relationship contingent upon business environment? Int. J. Entrep. Behav. Res. **16**(3), 182–203 (2010)

Al Mamun, A., Fazal, S.A.: Effect of entrepreneurial orientation on competency and micro-enterprise performance. Asia Pacific J. Innov. Entrepr. **12**(3), 379–398 (2018)

Ajzen, I.: Residual effects of past on later behavior: habituation and reasoned action perspectives. Pers. Soc. Psychol. Rev. **6**(2), 107–122 (2002)

Amiri, M., Zali, M.R., Majd, M.: Limitation on emergent businesses. J. Entrepreneurship Dev. **2**(1), 81–102 (2009)

Autio, E., Kenney, M., Mustar, P., Siegel, D., Wright, M.: Entrepreneurial innovation: the importance of context. Res. Policy **43**(7), 1097–1108 (2014)

Arenius, P., Minniti, M.: Perceptual variables and nascent entrepreneurship. Small Busin. Econ. Entrepr. J. **24**, 233–247 (2005)

Barazandeh, M., Parvizian, K., Alizadeh, M., Khosravi, S.: Investigating the effect of entrepreneurial competencies on business performance among early-stage entrepreneurs Global Entrepreneurship Monitor (GEM 2010 survey data). J. Glob. Entrep. Res. **5**(1), 1–12 (2015)

Castaño, M.S., Méndez, M.T., Galindo, M.Á.: The effect of social, cultural, and economic factors on entrepreneurship. J. Bus. Res. **68**(7), 1496–1500 (2015)

Chrisman, J.J., Chua, J.H., Steier, L.P.: The influence of national culture and family involvement on entrepreneurial perceptions and performance at the state level. Entrep. Theory Pract. **26**(4), 113–130 (2002)

Connell, J., Lynch, C., Waring, P.: Constraints, compromises and choice: comparing three qualitative research studies. Qualit. Report **6**(4), 1–15 (2001)

Davari, A., Emami, A., Ramadani, V., Taherkhani, S.: Factors influencing academic entrepreneurship: a case-based study. J. Sci. Technol. Policy Manag. **9**(3), 284–295 (2018)

Davidsson, P.: Culture, structure and regional levels of entrepreneurship. Entrep. Reg. Dev. **7**(1), 41–62 (1995)

Deniz, N., Boz, İT., Ertosun, Ö.G.: The relationship between entrepreneur's level of perceived business-related fear and business performance. Procedia Soc. Behav. Sci. **24**, 579–600 (2011)

Dickson, V.V., Riegel, B.: Are we teaching what patients need to know? Building skills in heart failure self-care. Heart & Lung J. Acute Critical Care **38**(3), 253–261 (2009)

Driessen, M.P., Zwart, P.S.: The entrepreneur scan measuring characteristics and traits of entrepreneurs (2007). http://www.necarbo.eu/files/E-scan%20MAB%20Article.pdf. Accessed 2021/03/01

Eling, M., Schaper, P.: Under pressure: how the business environment affects productivity and efficiency of European life insurance companies. Eur. J. Oper. Res. **258**(3), 1082–1094 (2017)

Elster, J.: Social norms and economic theory. J. Econ. Persp. **3**(4), 99–117 (1989)

Emami, A., Khajeheian, D.: Social norms and entrepreneurial action: the mediating role of opportunity confidence. Sustainability **11**(1), 158–175 (2019)

Etzioni, A.: Entrepreneurship, adaptation and legitimation: a macro-behavioral perspective. J. Econ. Behav. Organ. **8**(2), 175–189 (1987)

Etzioni, A.: Social norms: internalization, persuasion, and history. Law Soc. Rev. **34**(1), 157–178 (2000)

Festre, A.: Incentives and social norms: a motivation-based economic analysis of social norms. J. Econ. Surv. **24**(3), 511–538 (2010)

Filculescu, A.: The heterogeneous landscape of innovation in female-led businesses—cross-country comparisons, Management & Marketing. Chall. Knowl. Soc. **11**(4), 610–623 (2016)

Franco, M., Haase, H.: Failure factors in small and medium-sized enterprises: qualitative study from an attributional perspective. Int. Entrep. Manag. J. **6**(4), 503–521 (2010)

Fritsch, M., Wyrwich, M.: The effect of entrepreneurship on economic development—an empirical analysis using regional entrepreneurship culture. J. Econ. Geogr. **17**(1), 157–189 (2017)

Giannetti, M., Simonov, A.: On the determinants of entrepreneurial activity: social norms, economic environment and individual characteristics. Swedish Econ. Policy Rev. **11**(2), 269–313 (2004)

Global Entrepreneurship Monitor. https://www.gemconsortium.org/. Accessed 2021/01/18

Global Entrepreneurship Index. https://thegedi.org/. Accessed 2021/01/18

Harms, R., Groen, A.: Loosen up? Cultural tightness and national entrepreneurial activity. Technol. Forecast. Soc. Chang. **121**, 196–204 (2017)

Harutyunyan, A., Özak, Ö.: Culture, diffusion, and economic development: the problem of observational equivalence. Econ. Lett. **158**, 94–100 (2017)

Hastuti, S.W.M.: Characteristics of entrepreneurship: business performance of small entrepreneurs. In: 2020 International Proceedings on International Conference of Interdisciplinary Sciences, pp. 54–64. Prosiding Seminar, Indonesia (2020)

Herold, E.S., Maticka-Tyndale, E., Mewhinney, D.: Predicting intentions to engage in casual sex. J. Soc. Pers. Relat. **15**(4), 502–516 (1998)
Hong, H., Kacperczyk, M.: The price of sin: the effects of social norms on markets. J. Financ. Econ. **93**(1), 15–36 (2009)
Honjo, Y.: Why are entrepreneurship levels so low in Japan? Jpn. World Econ. **36**, 88–101 (2015)
Huggins, R., Thompson, P.: Entrepreneurship and community culture: a place-based study of their interdependency. Entrep. Res. J. **2**(1), 1–34 (2012)
Ismail, T.: The development of entrepreneurial social competence and business network to improve competitive advantage and business performance of small medium sized enterprises: a case study of batik industry in Indonesia. Procedia Soc. Behav. Sci. **65**, 46–51 (2012)
Kallas, E.: Environment-readiness entrepreneurship intention model: the case of Estonians and the Russian-speaking minority in Estonia. SAGE Open **9**(1), 1–15 (2019)
Krause, M., Ackermann, M., Gayoso, L., Hirtbach, C., Koppa, M., Brêtas, L.: Formalisation and business development in Mozambique: How Important are Regulations? Studies (53), Deutsches Institut für Entwicklungspolitik (DIE), Bonn, Germany (2010)
Krueger, N.F.: The cognitive psychology of entrepreneurship. In Handbook of entrepreneurship research: An Interdisciplinary Survey and Introduction, pp. 105–140. Springer, Boston, USA (2003)
Krueger, N., Liñán, F., Nabi, G.: Cultural values and entrepreneurship. Entrep. Reg. Dev. **25**(9–10), 703–707 (2013)
Levie, J., Autio, E.: A theoretical grounding and test of the GEM model. Small Busi. Econ. Entrepr. J. **31**(3), 235–263 (2008)
Levie, J., Autio, E.: Regulatory burden, rule of law, and entry of strategic entrepreneurs: an international panel study. J. Manage. Stud. **48**(6), 1392–1419 (2011)
Linan, F., Urbano, D., Guerrero, M.: Regional variations in entrepreneurial cognitions: start-up intentions of university students in Spain. Entrep. Reg. Dev. **23**(3–4), 187–215 (2011)
Liñán, F., Chen, Y.W.: Development and cross-cultural application of a specific instrument to measure entrepreneurial intentions. Entrep. Theory Pract. **33**(3), 593–617 (2009)
Lyons, T.S., Alter, T.R., Audretsch, D., Augustine, D.: Entrepreneurship and community: The next frontier of entrepreneurship inquiry. Entrep. Res. J. **2**(1), 1–24 (2012)
Machmud, A., Hidayat, Y.M.: Characteristics of Islamic entrepreneurship and the business success of SMEs in Indonesia. J. Entrep. Educ. **23**(2), 1–16 (2020)
Mas-Tur, A., Soriano, D.R.: The level of innovation among young innovative companies: the impacts of knowledge-intensive services use, firm characteristics and the entrepreneur attributes. In: Service Business 8(1), pp. 51–63. Springer, Berlin Heidelberg, Germany (2014)
Mitchelmore, S., Rowley, J.: Entrepreneurial competencies: a literature review and development agenda. Int. J. Entrep. Behav. Res. **16**(2), 92–111 (2010)
Mojab, F., Zaefarian, R., Azizi, A.H.D.: Applying competency-based approach for entrepreneurship education. Procedia Soc. Behav. Sci. **12**, 436–447 (2011)
Mrożewski, M., Kratzer, J.: Entrepreneurship and country-level innovation: Investigating the role of entrepreneurial opportunities. J. Technol. Transf. **42**(5), 1125–1142 (2017)
Mueller, S.L., Thomas, A.S.: Culture and entrepreneurial potential: A nine country study of locus of control and innovativeness. J. Bus. Ventur. **16**(1), 51–75 (2001)
Muzychenko, O.: Cross-cultural entrepreneurial competence in identifying international business opportunities. Eur. Manag. J. **26**(6), 366–377 (2008)
Oh, H., Hsu, C.H.C.: Volitional degrees of gambling behaviors. Ann. Tour. Res. **28**(3), 618–637 (2001)
Ostrom, E.: Collective action and the evolution of social norms. J. Econ. Persp. **14**(3), 137–158 (2000)
Ozaralli, N., Rivenburgh, N.K.: Entrepreneurial intention: antecedents to entrepreneurial behavior in the USA and Turkey. J. Glob. Entrep. Res. **6**(1), 1–32 (2016)
Păunescu, C., Argatu, R.: Critical functions in ensuring effective business continuity management. Evidence from Romanian companies. J. Busin. Econom. Manag. **21**(2), 497–520 (2020)

Păunescu, C., Mátyus, E.: Resilience measures to dealing with the COVID-19 pandemic Evidence from Romanian micro and small enterprises. Manag. Mark. Chall. Knowl. Soc. **15**(1), 439–457 (2020)

Păunescu, C., Molnar, E.: Country's entrepreneurial environment predictors for starting a new venture—evidence for Romania. Sustainability **12**(18), 7794 (2020)

Păunescu, C., Popescu, M.C., Duennweber, M.: Factors determining desirability of entrepreneurship in Romania. Sustainability **10**(11), 3893 (2018)

Porfírio, J.A., Mendes, T.C., Felício, J.A.: From entrepreneurship potential in culture and creative industries to economic development: the situation of UK and southern European countries. Int. Entrepren. Manag. J. **14**(2), 329–343 (2018)

Priede-Bergamini, T., López-Cózar-Navarro, C., Benito-Hernández, S., Rodríguez-Duarte, A., Platero, M.: The dual effect of the age of the entrepreneur on the innovation performance of the micro-enterprises. Int. J. Entrep. Ventur. **11**(1), 81–102 (2019)

Rauch, A., Frese, M.: Let's put the person back into entrepreneurship research: A meta-analysis on the relationship between business owners' personality traits, business creation, and success. Eur. J. Work Organ. Psy. **16**(4), 353–385 (2007)

Sadao, N.: The formation of the East Asian world. Selected Translations of Japanese Scholars' Research on Chinese History, Vol. 2. Beijing: Zhonghua shuju (1993)

Tu, C., Hwang, S.N., Wong, J.Y.: How does cooperation affect innovation in micro-enterprises? Manag. Decis. **52**(8), 1390–1409 (2014)

Ulijn, J., Brown, T.E.: Innovation, entrepreneurship and culture, a matter of interaction between technology, progress and economic growth? An introduction. Innovation, entrepreneurship and culture: The interaction between technology, progress and economic growth. Edward Elgar Publishing, USA (2004)

Van Auken, H., Stephens, P., Fry, F.L., Silva, J.: Role model influences on entrepreneurial intentions: A comparison between USA and Mexico. Int. Entrep. Manag. J. **2**(3), 325–336 (2006)

Wach, K.: Impact of cultural and social norms on entrepreneurship in the EU: cross-country evidence based on GEM survey results. Zarządzanie w Kulturze **16**(1), 15–29 (2015)

Wenzel, M.: The social side of sanctions: personal and social norms as moderators of deterrence. Law Hum Behav. **28**(5), 547–567 (2004)

WorldBank: (2020). https://data.worldbank.org/. Accessed 2021/03/03

Zali, M.R., Bastian, B.L., Qureshi, S.: Promoting innovation in the MENA region: the role of social norms and individual factors in entrepreneurial networks. Int. J. Busin. Global. **11**(4), 413–426 (2013)

Evolution and Challenges in the Digital Markets. The Case of Facebook in the Framework of EU Law

Maria Iulia Ariciu and Mihail Busu

Abstract In this paper the authors are making an overview of the Digital Markets evolution. The legal framework for e-commerce at the European level is introduced and the challenges and the evolution of one of the biggest digital platforms are discussed. The authors are focusing their work on Facebook as one of the well-known multisided digital platforms and the biggest social platform. Facebook acquisition campaign is analyzed in the context of competition on the digital market and several antitrust cases are enumerated. Final discussions are drawn in the Conclusions section.

Keywords Digital markets · Competition · Acquisitions · Facebook · GAFAM · EU

1 Introduction

Online platforms (in general, no matter if we are referring to search engines, social media platforms, e-commerce platforms, app stores, price comparison websites) are playing more and more a central role in our social and economic life. They help consumers to find easily online information, and businesses to exploit the benefits of e-commerce. But online platforms also bring new policy and regulatory challenges. In the last years there has been significant market and regulatory disruption caused by digital transformation.

Digital platforms are leading to transnational markets meaning that regulation is increasingly beyond the scope of individual national regulatory authorities. For this reason, the regulation was to be done over supranational organizations and regional organizations (e.g., European Commission). Two-sided and multisided digital platforms (e.g., the 'GAFAMs': Google, Amazon, Facebook, Apple, Microsoft) have

M. I. Ariciu · M. Busu (✉)
Bucharest University of Economic Studies, Bucharest, Romania
e-mail: mihail.busu@fabiz.ase.ro

M. I. Ariciu
e-mail: ariciumaria7@gmail.com

M. Busu (ed.), *Digital Economy and New Value Creation*, Springer Proceedings in Business and Economics, https://doi.org/10.1007/978-3-031-07265-9_9

emerged and grown rapidly. The main reason is that they are offering to consumers innovative services and technology which appear to be free for them, or, in any case, not at a high cost.

Thus, we must take into consideration that the business model of digital platforms relies on customer data to create value that can be monetized on another side of the platform (e.g., to advertisers or content providers). For this reason, digital platforms act as a marketplace, bringing together and reducing transaction costs between distinct groups of customers.

2 The Legal Framework for e-commerce at the European Level

In the last two decades technology and innovation have pushed both the economy and society into new development paths, creating new possibilities, expectations, but also challenges and threats.

The legal framework for e-commerce started over than 20 years ago: The Directive 2000/31/EC of the European Parliament and of the Council of 8 June 2000 on certain legal aspects of information society services, in particular electronic commerce, in the Internal Market (the so-called E-commerce Directive), adopted in the year 2000, provided an efficient legislative scheme for developing the digital market in the European Union (EU). Nowadays, the new market situation imposes new legislative action, and that's the reason why the European Commission (EC) was (and still is) very active in finding new solutions for the current reality of the digital world.

Society has become more and more digitized, and the market structure has changed significantly since the adoption of the E-commerce Directive. For example, platforms create multi-sided structures that connect the demand side with the supply side, meaning that they allow two groups of users to meet. Also, at least some of the platforms were not initially directed towards allowing economic activities by the users (e.g., Facebook—a platform we will present in the next lines), thus supporting the commercialization of online activities.

According to Fenwick and Vermeulen (2019), the type of platforms, by taking into consideration the market sectors, are nowadays: exchange platforms (e.g., Amazon, Alibaba), service platforms (e.g., Airbnb, Booking), content platform (e.g., Netflix, YouTube), software platforms (e.g. Apple iOS, Google, Android), social platforms (e.g. Facebook, Instagram), investment platforms (e.g. Priceland, OpenTable) and smart contract platforms (e.g. Ethereum).

These being said, a new list of legislative acts had to be adopted: the General Data Protection Regulation, which deals with the free movement and processing of personal data; the Geo Blocking Regulation, aiming to remove the barriers created by unjustified Geo Blocking; the Audiovisual Media Service Directive, which strives to protect media consumers from harmful content; the Copyright in the Digital Single

Market Directive, addressing copyright protection in digital and cross-border environments; the Digital Content Directive, which aims to ensure better access to and supply of digital content and digital services; the Fairness and Transparency of Online Platforms Regulation (the 'Platform-to-Business' regulation), which imposes new rules on parties operating online trading platforms and similar services, and taking on the role of an intermediary.

Talking about the last regulation mentioned, new rules are applying to 'online intermediation services' and 'online search engines' that provide their services to business users and corporate websites established in the European Union and offer goods or services to consumers located in the EU. The EC pointed out which type of entities can be covered by the EU Platform-to-Business regulation, no matter if they are established in a Member State or outside the EU: social media for business (e.g. Facebook, Instagram), price comparison tools (e.g. Skyscanner, Booking), e-commerce marketplaces (e.g. Amazon Marketplace, eBay), app stores (e.g. Google Play, Apple App Store, Microsoft Store), and general online search engines (e.g. Google Search). It is also interesting to underline that the EU Platform-to-Business regulation impose to the Member states, through the article 15, the obligation to "ensure adequate and effective enforcement" of mentioned above EU regulation.

As platform scale continues, the EU rules around big tech had to move fast on regulation that would change how tech giants operate in the digital marketplace, to limit the big challenges are faced nowadays especially regarding the antitrust policy: this is the reason of the package of the Digital Markets Act (DMA) and Digital Services Act (DSA). The EU rules around big tech are therefore set to be strengthened in 2022 with those two acts. This package is set to include curbing the uncontrolled spread of unverified content, with the potential for penalties of 10% of an undertaking's annual revenue.

According to the official website of the EC, the DSA and DMA have two main purposes: first, it is "to create a safer digital space in which the fundamental rights of all users of digital services are protected"[1], and the second, "to establish a level playing field to foster innovation, growth, and competitiveness, both in the European Single Market and globally".[2]

The DMA specifically targets the practices of "gatekeepers", the term being used to describe the power that big digital companies have over third parties that use their platforms. The DSA is a complementary legislation which has the goal to protect users online by providing transparency into the working manner of the content algorithms.

But what's a gatekeeper? Which companies will be considered 'gatekeepers', and which are the obligations and restrictions gatekeepers will have to respect?

According to the Commission's DMA proposal, a gatekeeper must operate a 'core platform service', including online intermediation services, online search engines, online social networking services etc. Those core platform services must fulfill some qualitative and quantitative criteria. It is still important to mention that the EC will be able to revise the list of gatekeepers and conduct market investigations.

---

[1] https://digital-strategy.ec.europa.eu/en/policies/digital-services-act-package.

[2] Ibid.

As mentioned above, to better understand the evolution in numbers of the digitalization market, we will take a specific example of one of the biggest tech companies: Facebook. How did this tech company evolve? Can Facebook be caught by the gatekeeper definition in the future EU regulation?

2.1 *The Challenges and the Evolution of One of the Biggest Digital Platforms: Facebook*

First, it is important to underline that Facebook did not exist at the time of introducing the E-commerce directive. So, what Facebook was at the beginning, how it grown, and what it began? After trying to answer to those mentioned questions and referring to the new introduced concept of the gatekeeper, we will try to conclude if Facebook can be caught by this notion and, more important, what does it mean.

Facebook, as a social network, plays an important role in the lives of more than 2.2 billion people worldwide. But nowadays Facebook is not only a platform where one can socialize, the company also sells advertising to marketers. It is interesting to mention that Facebook were initially one-sided platforms, but during the time it has entered on other markets as the market for classified ads (through Facebook Marketplace) and dating (through Facebook Dating) (Condorelli and Padilla 2020).

So, let's see the evolution of Facebook in terms of numbers.

2.2 *Facebook Acquisition Campaign*

Facebook's procurement strategy has led to major changes in competition policy in many jurisdictions. Starting with the acquisition of Instagram in 2012 and culminating in the second largest acquisition that Facebook had, the purchase of WhatsApp for 19 billion euros in 2014, when the mechanisms of the EU and most states failed to determine turnover threshold turnover, even though the transaction was of immense value, Facebook has managed to become one of the most profitable companies in the digital sector that enjoys an indisputable position, at least at this time.

Through its acquisitions over time, Facebook has acquired both essential technical means and people with expertise in mobile platform development, instant messaging, photo sharing, location services, advertising, or data analytics, transforming what was originally standalone applications integrated into the Facebook ecosystem (Glick and Ruetschlin 2019). If most companies were acquired when they did not have a major market exposure, but only the potential to become the next resounding success, Instagram had already become a platform that had developed its own clientele and would capitalize on it, which is likely to remain on the market. Under the original name. If in the case of Beluga (the future Facebook Messenger) it was easy to be renamed and completely absorbed, in the case of Instagram, loyalty and number of users, as well as a significant differentiation from the social network Facebook made Instagram exist today under the same name. Same situation applies to WhatsApp.

In Figure is revealed the number of acquisitions made by Facebook, by year.

From the graph above we could see that the number of acquisitions made by Facebook was volatile in the past 20 years, with minimum values in 2003 and 2017 when there were no acquisitions and a maximum value in 2020 with 18 acquisitions. Figure 1 also reveals that the number of acquisitions were declined after 2010.

In Fig. 2, we could see the distribution of the acquisitions by the primary market of the acquired company.

In the graph above, we could see that the top primary market companies acquired by Facebook were Media Sharing (12), followed by Data & Analytics, Developer Tools, and Platform (8), while the primary markets with the least number of acquired

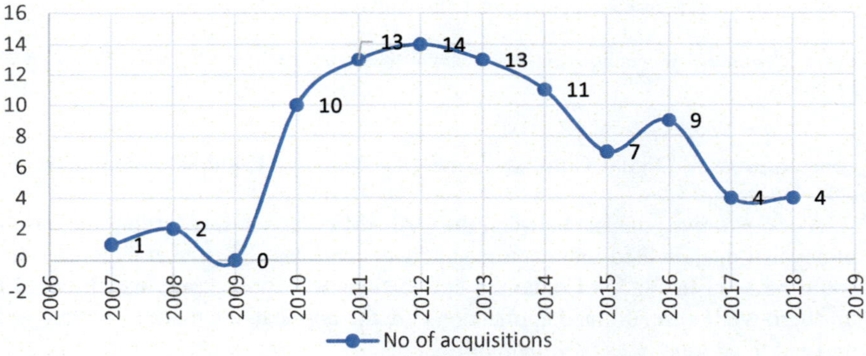

Fig. 1 The number of acquisitions made by Facebook by year. *Source* Thomson Reuters M&A Database

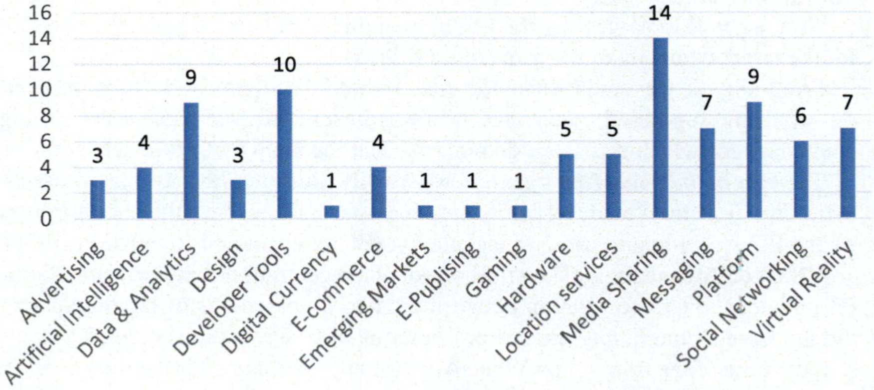

Fig. 2 The primary market of the acquisitions made by Facebook. *Source* Thomson Reuters M&A Database

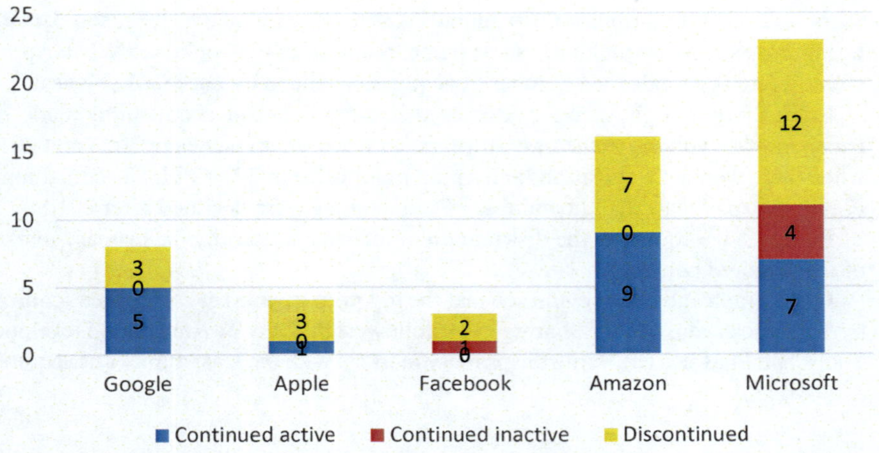

Fig. 3 Continued, discontinued and inactive apps by acquirer. *Source* Affeldt and Kesler (2021a, b)

companies were Digital Currency, Emerging Markets, E-publishing, and Gaming (1).

A comparative analysis between the GAFAM companies, regarding the number of acquisitions, divided into continued active, continued inactive and discontinued, could be seen in Fig. 1. Continued applications are applications which could be updated, while discontinued applications are the applications that can no longer be updated to address bugs or vulnerabilities.

From the above graph we could see that Microsoft had the greatest number of apps acquired (16), followed by Amazon (16) and Google (8). We could also see that half of the total number of the acquired applications (54) were discontinued at the time of acquisitions (27).

Now we will look whether the acquired application is main part of the business for the target company of the acquisition (Fig. 4).

From Fig. 4 we could conclude that Facebook, Apple and Google focused on acquiring app-based companies, while Amazon and Microsoft seem to target companies for which their apps do not represent the main part of the business.

The Facebook/WhatsApp transaction perfectly illustrates how a major transaction in the digital sector is made by companies that exceed the revenue thresholds imposed by the EC According to art. 1 of Regulation (EC) no. Council Regulation (EC) No 139/2004 of 20 January 2004 on the control of concentrations between undertakings (Regulation), the concentration between the two global giants of the digital sector did not have a community dimension (in terms of turnover thresholds). Thus, even if at the time of the transaction WhatsApp had approximately 600 million users, the turnover was small, and the EC could analyze the transaction only under art. 4 para. (5) of the Regulation, respectively following the EC notification by at least three Member States.

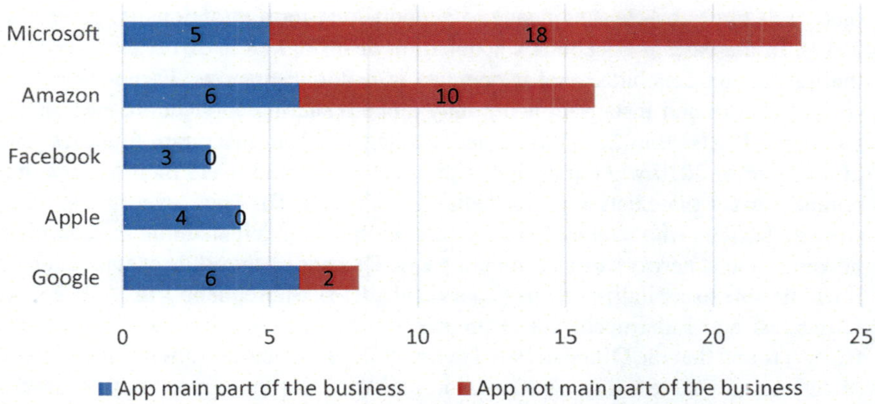

Fig. 4 Acquired apps main part of business by acquirer. *Source* Affeldt and Kesler (2021a, b)

At the time of the transaction, WhatsApp was the largest instant messaging application and the one with the highest growth rate, followed by Facebook Messenger, with approximately 200 million users (Adler 2014). The main distinguishing features of WhatsApp compared to its competitors are the importance given to the protection of personal data and the lack of advertising (Pike 2020). In the year before the acquisition, WhatsApp had a loss of 138 million USD, but it did not matter for Facebook, as it anticipated (correctly) the subsequent capitalization of the user base and, at the same time, removed a potential competitor (e.g., Google) (Morton and Dinielli 2020) from the entry on the social media market. The acquisition of WhatsApp by Facebook in 2014 was, as we mentioned above, the highest in value so far, respectively 19 billion euros, Facebook paying about 42 euros per user. At the time, the price was considered exaggerated, but the increase in network effects caused by the merging of the databases of the two companies allowed Facebook to have an even more important presence on the social networking market.

The EC opened an investigation, but did not block the transaction, as it analyzed it from the perspective of a concentration (only) on the telecommunications services market, where they included WhatsApp. Katz (2021) states that some companies are considering entering new markets in two steps: in the sense that they first build a large and loyal user base using the service provided in the market where they have a significant market share and which is the main activity, following that, in the second stage, to propose—to the same users—a new service in an adjacent market, based on the information collected from the use of the core service (core business). Even if two companies were to operate (exclusively) in related fields, where there are no overlaps, the natural flow of growth would cause them to expand and compete in each other's fields.

But the EC has not identified any documents to support the idea of a two-step entry (a relatively new theory at the time) from WhatsApp into the social media market. On the other hand, in a situation where companies are being bought at an early stage, there will rarely be clear evidence to support long-term intentions (Cremer et al. 2019). In the case of Facebook's acquisition of Instagram in 2012, the analysis of the transaction concluded that it operates in different markets, Facebook being a social network and Instagram being only a photo sharing application, even though Glick and Ruetschlin (2019) identified a statement from Instagram founders given at the level of 2010 when they stated that they wanted to make their application a competitor for the existing social networks (Cain 2010). Thus, even though there were authorities who analyzed the concentration and anticipated that Instagram can become a social network (thus competing with Facebook), they did not appreciate that it will be able to capitalize on this (Caers et al. 2013). Subsequently, in 2019, a report (Argentesi and Filistrucchi 2007) prepared to the Competition Market Authority (CMA) found that the Office of Fair Trading (OFT) had underestimated the potential of Instagram's advertising capitalization and thus lost the opportunity to have an opponent competing with Facebook. However, it is difficult to say with certainty that Instagram would have had the same evolution (the approximate market value in 2018 would have been 100 billion USD, and in 2019 the advertising revenue of Instagram totaled 20 billion USD of the total $ 70 billion of Facebook) if it had not been acquired by Facebook, but it is clear that owning Instagram has allowed Facebook to access more users, increase user engagement, and total revenue unless Facebook bought it. To understand the full picture, we mention that Facebook was fined in 2016 with 110 million euros for misleading the EC, for false claims that there are technical incompatibilities that will not allow the integration of the WhatsApp users database in the already existing Facebook (Carugati 2018). However, the authorization granted in 2014 was not revoked. Now, seeing the evolution of the company, we could say that the EC analysis would probably have had another outcome if the data provided by Facebook had been correct and complete. However, a sound policy must anticipate and counteract the calculations made by a company that may decide to take a risk, providing inaccurate information, internalizing the likelihood and amount of a future fine.

3 An Overview of the Competition on the Digital Market and Cases Related to Facebook

Market competition authorities need to look closely at mergers and acquisitions that may have a major negative effect on consumer welfare and even block them if the subsequent effect on the market is to reduce competition. To determine the competitive equilibrium of the markets, two perspectives must be considered: (i) on the one hand, the interest of the competition authorities in maintaining the competitive level and the interests of the developer, who often wants to make a quick profit and get out of the market once his innovative idea (especially the potential of the idea) has been materialized by selling to a large company. The European Union's policy, based on which the principles of competition law have been developed, seeks to promote the general well-being of consumers and an increase in the efficiency and productivity of markets. These goals can be achieved by establishing a set of rules to be observed by all market players, but also by stimulating a context in which innovation and technological breakthroughs thrive. Authorities do not have unlimited resources and therefore need to focus on the elements that have deviant behavior and a negative impact on consumer welfare. If there are indications of a possible damage to consumers, then the transactions should be analyzed more carefully and approved only if their benign nature is found. A per se ban on the acquisition of innovative companies would also mean encouraging socially unnecessary investments made by the buyer in the desire to replicate the innovative idea (Letina et al. 2020).

In the digital sector, most acquisitions made by large companies have not been investigated because they did not reach the minimum turnover thresholds that would have triggered the notification obligation, because it is specific to this industry for companies to realize their potential only after they arrive to a considerable user base. The acquisition of an innovative company usually takes place before it becomes mature and commercialized. For example, the average age of purchases made by Facebook was two and a half years, and about six and a half years for those purchased by Amazon (Argentesi et al. 2021). The targets are usually small companies at the beginning of the road that have not yet grown exponentially. The ubiquitous digital component in the life of each of us means that at the end of each day we use our phone, laptop, fitness bracelet, navigation applications there is a database about each, a base that grows marginally with each use of digital devices and services. We need to keep in mind that user loyalty and (active) involvement on the platform provides the true value of the platform.

That is why it is important to consider whether the acquisition of a company that has a low annual income, but has a large user base and, implicitly, a huge growth potential should go unnoticed by the competition authorities or (possible) important position on market must be determined according to other criteria. There are opinions that state that there are other factors according to which an economic concentration in the digital sector must be analyzed, in addition to the turnover and the current market share (Bueno et al. 2020).

Innovation is given by the desire of companies to cause imperfect competition in their favor, or the attempt to correct market imperfections will counterbalance the desire of companies to innovate. Innovative companies play an important role in competitive markets, being a constant source of ideas, unprecedented products that can define a new market or ingenious business models, which ultimately lead to a decrease in concentration in certain markets, causing inefficient companies, so that consumers enjoy high efficiency and better products (Ennis et al. 2019). The fact that a company enjoys abnormally high profits creates the desire of potential competitors to enter the market, and the most likely way they will succeed will be because of significant innovation or capital investment. It is these major profits that drive entrepreneurs to start a business.

Thus, competition authorities must provide companies, especially start-ups, with a space in which they can develop freely to compete with existing companies, without fear that they will be the target of behaviors that could drive them away from the market. In a stable (well-defined) market, an innovative company has only one chance: to market a major invention to create a new market to be a leader; otherwise, it would not have the capacity to compete with existing companies and would eventually have to exit the market. Investors-entrepreneurs will think twice about whether they will be able to market their new idea, whether it will limit in any way the possibility of their new company / idea being bought by the big market players. Most of the time, the very illusion of creating an innovative product that would later be needed by a Big Tech company leads entrepreneurs to invest initially.

We mentioned above about the Facebook/WhatsApp merger notified at the European level. In that case, the EC analyzed two main possible theories of harm, one of which was "the possibility that Facebook could use WhatsApp as a potential source of user data for the purpose of improving the targeting of Facebook's advertising activities outside WhatsApp"[3]. The EC also noted that this would require Facebook to match the WhatsApp profile of the users with their Facebook profile, which can be cause major technical obstacles plus the fact that some users would prefer to switch to different consumer communications applications that they perceive as less intrusive. Thus, a surprising aspect is the refusal of the EC to consider the potential effect of the merger on privacy protection issues of the data collection.

It is interesting to notice that it doesn't exist a clear position within the EU on whether privacy-related concerns should be outside the scope of competition law. For example, the Bundeskartellamt noted in its Facebook decision[4] that not all the users were aware that Facebook was able to collect user data from third-party sources such as Instagram or WhatsApp. The Bundeskartellamt noticed as well that Facebook's terms of service didn't respect the GDPR rules and were an exploitative abuse of the dominant position of Facebook in the market for social networks.

Another case related to Facebook is the UK Office of Fair Trading (OFT)' investigation of the Facebook/Instagram merger. Thus, in that case the OFT stated that it doesn't "be the case that the merger may be expected to result in a substantial lessening of competition within a market or markets in the UK"[5]. Therefore, the OFT's investigation has been closed.

Surprisingly (or not), the UK has decided in 2020 to revisit the issue of Facebook/Instagram merger.[6] Also, the EC has opened a formal antitrust investigation in order to conclude, on one hand, if Facebook respected or not the EU competition rules by using advertising data gathered in particular from advertisers in order to compete with them in markets where Facebook is active, and, on another hand, to assess if Facebook ties its online classified ads service "Facebook Marketplace" to its social network, violating the EU antitrust rules. The EC announced that UK's

---

[3] Frederic Jenny, Changing the way we think: competition, platforms and ecosystems, Journal of Antitrust Enforcement, 2021, 9, 1–18, page 7.

[4] Decision of 6 February 2019. For more details, see.
[https://www.bundeskartellamt.de/SharedDocs/Entscheidung/EN/Fallberichte/Missbrauchsaufsicht/2019/B6-22-16.pdf?__blob=publicationFile&v=3].

[5] OFT's Facebook decision, 14 August 2012.

[6] https://assets.publishing.service.gov.uk/media/555de2e5ed915d7ae200003b/facebook.pdf, par. 46.

Competition and Markets Authority (CMA) has also launched its own investigation into Facebook's use of data and that those two authorities will "work closely".[7]

Those cases (and not only) can show us why the EU rules had to be adapted to the actual context of the digitalization world. And now it is the time to answer to the question mentioned in the chapter above: Can Facebook be caught by the gatekeeper definition in the future DMA? Even if the Commission has not named any companies which can be caught by the gatekeeper definition, if we analyze the proposed criteria the answer can be only 'yes'. It was noticed that the 'GAFAM' are always classified as gatekeepers (Mariniello and Martins 2021). Still, it was argued that the EU regulatory framework for gatekeepers as a complement to antitrust law enforcement raises further elaboration concerning its scope, the underlying analysis of platform competition and the remedies proposed.

4 Conclusions

The way in which certain transactions have been carried out in the last decade has led to an imbalance in the market. The novelty of the digital sector and its unprecedented dynamics have created situations in which the legislative framework has been outdated, and the authorities in charge of protecting, maintaining, and stimulating competition have not had the resilience to act appropriately. The significant increase in the complexity of the digital world has accentuated the difference between the classic production paradigm, based on the processing of raw materials that resulted in a salable finished product, and the one in which the product is salable after several stages, namely the sale of advertising. Performs in a different market and is after the processing of data of users of the free platform. Companies that have created the digital market or that have taken over the trend from the beginning have established well-defined market shares.

The advantage of a newcomer to a market must allow the innovator to recoup the costs of research and development, but the fact that there were no rules applicable to the full spectrum of digital platform activity has already allowed disproportionately high profits compared to any other industries or sectors of activity.

Then, within the limits of the competition authorities' resources, investigations could be launched into transactions which have a pronounced ex-post effect not to revoke the authorizations (where the legal framework exists), but to understand what the novelty was exploited by companies and to refine the legislative framework to meet new challenges. To the extent that an authorized or unannounced transaction has led to the creation or consolidation of a dominant position in the market, competition authorities should consider imposing measures to help protect the competitive environment when making new transactions in that company. Of course, insofar as the companies holding a dominant position on a relevant market violate the principles set out in art. 102 of the Treaty on the Functioning of the European Union and

[7] For more details, *see* [https://ec.europa.eu/commission/presscorner/detail/en/IP_21_2848].

resumed in national jurisdictions, regarding the abuse of a dominant position, the authorities will have to intervene effectively. The fact that companies with billions of turnovers could be investigated in newly created and dynamic markets must be seen as an advantage by the competition authorities. Multi-jurisdictional investigations can be carried out through the mechanisms established under the European Competition Network, as the practices assessed are similar in all Member States, and the result may be financial penalties, which may have a significant value in relation to the amount of as well as in the development of unitary practices in the appreciation of the correct behavior in the new markets to increase the security and the predictability.

The competition authorities might conduct a careful and proactive supervision of the activity of the technological giants. Also, the legal framework at the European level must be carefully analyzed accordingly to the digitalization evolution and the implications that it can cause on the related fields as the antitrust law and the data protection policy. It remains to be seen whether the package of DMA and similar instruments will prove capable of bringing about the promised changes in the functioning of digital markets (Colomo 2021). Finally, the Facebook case is a step in the right direction for digital competition policy and could help as an encouragement for initiatives at national and EU level (Scheele 2021).

References

Webography

https://www.techtarget.com/searchcio/news/252512011/EU-closes-in-on-regulating-big-tech-with-Digital-Markets-Act

https://medium.com/mit-initiative-on-the-digital-economy/the-four-biggest-challenges-digital-platforms-need-to-address-a842413ee3c7

https://digitalregulation.org/category/regulatory-governance/

https://www.bundeskartellamt.de/SharedDocs/Entscheidung/EN/Fallberichte/Missbrauchsaufsicht/2019/B6-22-16.pdf?__blob=publicationFile&v=3

https://assets.publishing.service.gov.uk/media/555de2e5ed915d7ae200003b/facebook.pdf

https://ec.europa.eu/commission/presscorner/detail/en/IP_21_2848

European Commission, Online Platforms, Accompanying the document Communication on Online Platforms and the Digital Single Market, Brussels, 25.5.2016 SWD(2016) 172 final, document consulted, between the 21st of February and the 1st of March 2022, on.

https://eur-lex.europa.eu/legal-conent/EN/TXT/PDF/?uri=CELEX:52016SC0172&from=EN

Adler, E.: Tablets and smartphones are used completely differently for shopping and buying. Business Insider (2014)

Affeldt P., Kesler R.: Big tech acquisitions—towards empirical evidence. J. Europ. Compet. Law Pract. **12**(6) (2021a)

Affeldt, P., Kesler, R.: Big tech acquisitions—towards empirical evidence. J. Europ. Compet. Law Pract. **12**(6), 471–478 (2021b)

Anderson, J., Mariniello, M.: Regulating big tech: the Digital Markets Act. Bruegel Blog, 16 February (2021)

Argentesi, E., Buccirossi, P., Calvano, E., Duso, T., Marrazzo, A., Nava, S.: Mergerpolicy in digital markets: an ex post assessment. J. Compet. Law Econ. **17**(1), 95–140 (2021)

Argentesi, E., Filistrucchi, L.: Estimating market power in a two-sided market: the case of newspapers. J. Appl. Economet. **22**(7), 1247–1266 (2007)
Baker, J.: Protecting and fostering online platform competition: the role of antitrust law. J. Compet. Law Econ. **17**(2), 493–501 (2021)
Bueno, C., Zarzur, C., Chakmati, M.: Killer acquisitions: is antitrust prepared to deal with innovative young rivals? Mulheres no Antitruste, 2 (2020)
Caers, R., De Feyter, T., De Couck, M., Stough, T., Vigna, C., Du Bois, C.: Facebook: a literature review. New Media Soc. **15**(6), 982–1002 (2013)
Cain, M.: A Photo-Sharing App With Bigger Aspirations (2010)
Carugati, C.: The 2017 Facebook saga: a competition, consumer, and data protection story. Eur. Competition Reg. L. Rev. 2, 4 (2018)
Colomo, P.: New times for competition policy in Europe: the challenge of digital markets. J. Eur. Compet. Law Pract. **12**(7) (2021)
Condorelli, D., Padilla, J.: Harnessing platform envelopment in the digital world. J. Compet. Law Econ. **16**(2), 143–187 (2020)
Crémer, J., de Montjoye, Y.A., Schweitzer, H.: Competition policy for the digital era. Report for the European Commission (2019)
Ennis, S.F., Gonzaga, P., Pike, C.: Inequality: a hidden cost of market power. Oxf. Rev. Econ. Policy **35**(3), 518–549 (2019)
Evans, D., Tucker, C., Fels, A.: The evolution of antitrust in the digital era: essays on competition policy, Competition Policy International, Vol. I. (2020)
Fenwick, M., Vermeulen, E. (2019). A sustainable platform economy & the future of corporate governance. Lex Research Topics in Corporate Law & Economics Working Paper No. 2019–1. The article can be consulted on. https://papers.ssrn.com/sol3/papers.cfm?abstract_id=3331508
Glick, M., & Ruetschlin, C.: Big tech acquisitions and the potential competition doctrine: The case of Facebook. Institute for New Economic Thinking Working Paper Series, (104) (2019)
Jenny, F.: Changing the way we think: competition, platforms and ecosystems. J. Antitrust Enforc. (2021)
Katz, M.L.: Big Tech mergers: innovation, competition for the market, and the acquisition of emerging competitors. Inf. Econ. Policy **54**, 100883 (2021)
Kokkoris, I., Valletti, T.: Innovation considerations in horizontal merger control. J. Compet. Law Econ. **16**(2), 220–261 (2020)
Kramer, J.: Personal data portability in the platform economy: economic implications and policy recommendations. J. Compet. Law Econ. **17**(2), 263–308 (2020)
Letina, I., Schmutzler, A., Seibel, R.: Start-up acquisitions and innovation strategies (No. 20–03). Discussion Papers (2020)
Mariniello, M., Martins, C.: Which platforms will be caught by the Digital Markets Act? The 'gatekeeper' dilemma. Bruegel Blog, 14 December (2021)
Morton, F.M.S., Dinielli, D.C.: Roadmap for an antitrust case against facebook. Omidyar Network (2020)
Pike, C.: Start-ups, killer acquisitions and merger control (2020)
Sauter, W.: A duty of care to prevent online exploitation of consumers? Digital dominance and special responsibility in EU competition law. J. Antitrust Enforc. **8**, 406–427 (2020)
Scheele, R., Facebook: from data privacy to a concept of abuse by restriction of choice. J. Europ. Compet. Law Pract. 12(1) (2021)

Improving Organizational Performance Through in Terms of Using the Customer Relationship Management system—an Exploratory Study for SMEs in Romania

Mihaela Rodica Ganciu, Andreea Barbu, Iustina-Cristina Costea-Marcu, Dana Corina Deselnicu, and Gheorghe Militaru

Abstract The aim of this research is to understand how the dynamics of customer relationship management system usage have changed small and medium enterprises practices, influencing their organizational performance. Theories that have been used to understand the elements of system practices include the theory of marketing relationships and the theory of resource-based visualization. In this study, customer relationship management resources are classified as customer orientation, technological system resources, knowledge management and organizational capability (established as independent variables). Organizational performance will be te dependent variable. Also, a new factor was introduced, namely the control variable (size of company, seniority and the main field of activity of the company are used in the analysis of this factor). The questions from te survey were stated using the Likert scale (with values from 1 to 7). After using statistical data and case studies through quantitative analysis, the research conducted to the current state of customer relationship management system usage. The significant effect that influences the organizational performance toward using customer relationship management system is represented by organizational capability, which had the most important influence.

Keywords Small and medium enterprises · Organizational performance · Customer relationship management system · Digital technologies · Theory of marketing relationships · Romanian companies

M. R. Ganciu (✉) · A. Barbu · I.-C. Costea-Marcu · D. C. Deselnicu · G. Militaru
Politehnica University of Bucharest, Bucharest, Romania
e-mail: mihaella.ganciu@gmail.com

A. Barbu
e-mail: barbu.andreeab@yahoo.com

I.-C. Costea-Marcu
e-mail: costeamarcuiustina@gmail.com

D. C. Deselnicu
e-mail: d_deselnicu@yahoo.com

G. Militaru
e-mail: gheorghe.militaru@upb.ro

M. Busu (ed.), *Digital Economy and New Value Creation*, Springer Proceedings in Business and Economics, https://doi.org/10.1007/978-3-031-07265-9_10

1 Introduction

As SMEs are considered to be the main driver of global economic development, they have attracted the full attention of researchers. The nature of business in all countries/regions is changing rapidly, and developing countries are modernizing their infrastructure and using the internet for investment. These new interactive media have some issues, such as security and technical issues, lack of trust, lack of face-to-face communication or privacy issues. In addition, as the Internet and electronic communications are rapidly changing traditional marketing methods and technologies, the changes may mean that small and medium-sized enterprises may not fully understand the potential of the digital market.

There is an urgent need to understand the impact of using CRM systems on customer and corporate market performance. To meet the competitive needs of SMEs, companies need to conduct research to investigate the background of successful implementation of CRM systems. In the context of the economic and trade transformation of the twenty-first century, digitalization plays a strategic and decisive role in modernizing the multilateral trading system. It can be said that digitalization is part of the life of small and medium enterprises and can improve the efficiency of business, production, marketing and sales processes. Romania needs an innovative ICT industry and needs to improve digital innovation capabilities in all industries.

The paper aims to answer the questions "*What impact does the technology have on organizational performance? Is there a real benefit for the company in terms of using this customer management system? Which are the critical factors and how they influence SMEs performance through CRM system usage?*".

2 Literature Review and Hypotheses

2.1 *Small and Medium-Sized Enterprises—Role in the Economy Development*

The category of small and medium-sized micro-enterprises (SMEs) includes companies with less than 250 employees, an annual net turnover of up to EUR 50 million and/or total assets of up to EUR 43 million. Especially in the 1970s, the importance of SMEs began to emerge. In the 1980s, they experienced a significant expansion, and their shares in Europe, the United States and Japan increased significantly. However, the gap between regions is still important in Romania and in the EU Member States. In Romania, the small and medium business sector represents approximately 99.7% of the corporate population, and in Europe it is 99.8%. 93% of non-financial companies are micro-enterprises with less than 10 employees (Mascu and Muresan 2019).

The importance of Romanian SMEs is based on the fact that it is an offer for most products and services demanded by the population of a country, these are the main value-generating values of any country, provide jobs for most people, and its performance affects all countries (Nicolescu and Nicolescu 2008). In recent years, even though multinational and transnational businesses in Romania are more present than in the past, small and medium enterprises (SMEs) play the most important role in economic growth and job creation (Marcketti and Kozar 2007). In addition, Romania must prioritize market opportunities and move towards internationalization, given that economic and political stability, combined with high-performance infrastructure, provides a growth-friendly ecosystem. They help reduce the unemployment rate, as it creates jobs for the community and thus increases economic growth (Jabar et al. 2016).

2.2 *CRM System and Organizational Performance*

CRM is a useful system for generating, storing, representing, reproducing and translating information. It is used to manage customer interactions or relationships and to improve the company's understanding of consumer profiles (Gupte 2011). The general process of developing and maintaining a profitable communication with customers is to provide or present a higher value to customers and obtain their satisfaction. For organizations in need of development and growth, Gamson believes that CRM is an indispensable rule. In these organizations, it is very important to determine the key dimensions of CRM.

Organizational performance has two perspectives of financial performance; objective and subjective. A subjective perspective refers to the performance of firms relative to that of competitors (Sin et al. 2005), while an objective perspective is associated with absolute measures of performance (Jaakkola et al. 2009). For this study, a subjective perspective is chosen because information about the organization is considered highly confidential for certain industries, such as the telecommunications industry, where respondents may be reluctant to provide heavy financial data.

Factors considered to have an influence in the implementation of the CRM system by SMEs can be classified into 3 major groups such as: organizational, technical and data quality.

Organizational factors refer to those factors that directly or indirectly influence the structural, operational, human and management parts of the SME business entity and include: Benefits, ICT staffing skills, ICT managerial skills, organizational aspects, internal barriers, customer response/attitude (Wilson et al. 2002). Organizational factors related to operations structures, human resources, customer orientation, rewards and evaluation system, training and administration issues. Becker stated that organizational factors are significantly related to CRM performance (Becker et al. 2009). Below are the implications of the organizational factors found by the authors:

- With a sufficient number of dedicated employees, CRM systems can be learned and accepted accordingly (Nath et al. 2009).
- Through the organization's structure and its operational procedures, CRM implementation could be managed accordingly (Payne and Frow 2006).

– It allows the collaboration of different departments and functions for carrying outCRM activities, such as the level of integration within the organization, the commitment of senior management to the project and the availability of different resources (Croteau and Li 2003).

Technical factors refer to those factors that affect the heavy part of ICT/CRM and include: ICT infrastructure, acquisition costs, implementation and integration, evaluation and selection criteria of the system, sales assistance.

Data quality refers to these factors that directly influence the concept of data quality and how it is conducted in the context of CRM. It includes, the evaluation of data quality tools/process, the classification of customer data types and the classification of customer data source (Wilson et al. 2002).

3 Methodology

Factors affecting the formation of the company's internal CRM practice have been investigated from various theoretical angles and have attracted widespread attention. Theories used to understand elements of CRM practice include resource-based visualization theory (RBV) (Halawi and Aronson 2005), (Keramati and Hamed 2010) and marketing relationship theory (RMT). RBV seeks to take CRM applications from a multidimensional perspective, as RBV seeks to link the performance of high-level companies with various resources that can bring competitive advantages (Coltman 2007).

In this research, CRM resources are divided into CRM technology resources, knowledge management and customers. The RMT meeting takes place to promote adverse reactions from customers and suppliers, thus stimulating market feedback (Rese 2006). This research combines RBV and RMT theories to understand how Romanian SMEs form CRM practices and organizational performance. According to this theory, it derives a set of hypotheses expressed in the research.

Customer orientation involves establishing links between customer needs and customer satisfaction. A good knowledge of customer needs also increases customer loyalty and encourages repeat business activities that are important competitive priorities based on CRM (Sousa 2003). Customer focus develops a strong customer focus, continually delivers superior value to selected key customers, provides a personalized product or service through customer expectations, and overwhelms customer-focused focus (Sin et al. 2005). An important condition that can allow the organization to be truly customer-oriented is to take a cross-cutting approach to how the organization delivers value to its customers (Payne and Frow 2006).

Hypothesis 1: Customer orientation is positively related to organizational performance.

An organization can achieve desirable customer-oriented behaviors. However, CRM cannot be successful even if organizations acquire the most advanced technology and try to generate a customer-centric orientation without fully integrating the project into the organization (Sin et al. 2005). Moreover, Chang and Ku reaffirmed that the structure of the organization plays a key role in the successful implementation of CRM and therefore improves the performance of an organization (Chang and Ku 2009). Ku stressed that the success of CRM does not only require technologies or quality, but also requires an efficient service concept as well as appropriate operational procedures (Ku 2010).

Hypothesis 2: Organizational ability is directly proportional to organizational performance.

With the advancement of information technology (ICT), a modern system can be used to improve the power of organizations to reduce internal costs, better interact with the environment and increase long-term economic profit. It has been found that hotels of different price/quality and size implement different levels of sophistication of the CRM system (Mohammad et al. 2013). To achieve a successful CRM, organizations should go beyond the concept of CRM as a set of technological solutions (Kubi and Doku 2010).

Hypothesis 3: Technical resources are positively related to organizational performance.

In an organization, knowledge management includes learning about the needs and desires of customers, disseminating this knowledge within the organization (Sin et al. 2005). CRM practice uses customer knowledge as a means of improving customer relationships. Knowledge management capacity has four factors such as acquisition, transformation, application and protection. On the other hand, knowledge management has a significant impact on customer loyalty and satisfaction (Wang and Huang 2010). Therefore, much of the research in the field of knowledge management has focused on specific technological solutions, such as discovery, representation or dissemination (Mohamad et al. 2014).

Hypothesis 4: Organizational knowledge management is positively related to organizational performance.

SME size, age, nature of business or company strategy were included as control variables. All of these have been shown to have an impact on adoption decisions and the use of information technology (Harrigan et al. 2010).

Hypothesis 5: Control variable is positively related to organizational performance.

Figure 1 illustrates the conceptual model of the research. Data collection was carried out through a questionnaire with 27 questions. At the end of the meeting to complete the questionnaire, a total of 65 respondents were registered. The significance level of 5% will be used as the recommended threshold for the test study. The sample size will be based on a stratified sample (defined as a framework sample) consisting of small and medium-sized companies.

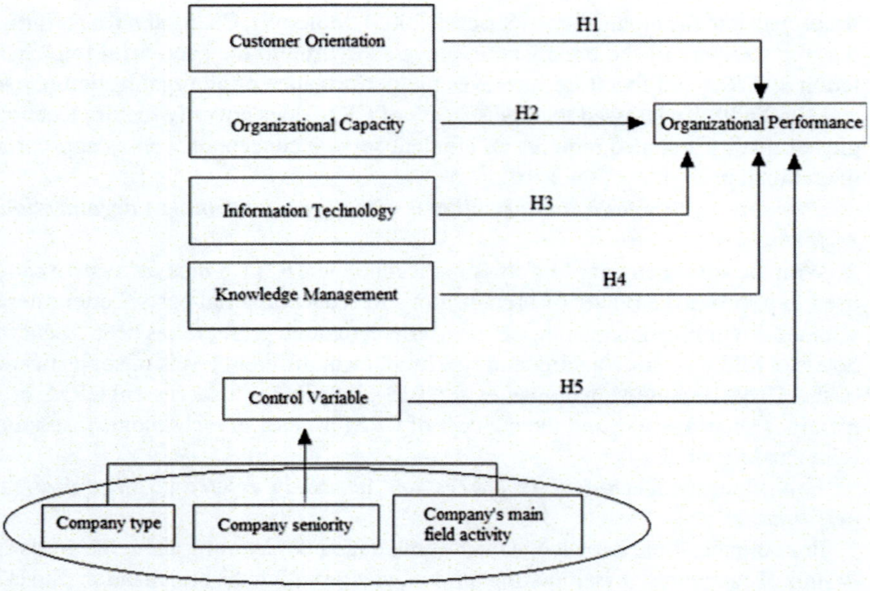

Fig. 1 The conceptual model of research

4 Results and Discussions

This study uses several regressions and correlation analyzes using the statistical program IBM SPSS Statistics. The reason for using the multiple regression tests is to investigate the relationship between independent variables and dependent variables. There are 5 independent variables, namely knowledge management, customer orientation, organizational capability, information technology and control variable. Organizational performance will be the dependent variable.

Table 1 lists some demographic information of the respondents. Descriptive statistics show that the Romanian population is relatively young and well educated. About 95% of respondents are under 40, and 70% of them have at least a university degree. In terms of gender, the distribution of the population between male and female respondents is uneven. Most of the questioned belonged to the female category, ie 41 people (63%), while there were only 24 men (37%).

According to Table 2, Cronbach Alpha coefficient is used to study the internal consistency of the items in the questionnaire or can be used as a method to reduce the items. The distance generated here is not too far (0.923 vs. 0.924), indicating that the average and modification of the original ratio are not much different, so standardization is not much different in Alpha. In this case, the Cronbach Alpha value is 0.923, which is a very good value. Moreover, it shows a fairly high internal consistency related to a certain sample.

Table 1 Demographic profile of respondents

Scales	Atribute	Frequency	Percentage (%)
Gender	Female	41	63.1
	Male	24	36.9
Age	18–23 years	28	43.1
	24–30 years	29	44.6
	31–37 years	5	7.7
	38–45 years	2	3.1
	More than 45 years	1	1.5
Education level	Postgraduate studies	16	24.6
	University studies	46	70.8
	Others	3	4.6

Table 2 Cronbach alpha analysis

Cronbach alpha	Cronbach's alpha based on standardized items	No. of items
0.923	0.924	15

The correlation is usually called bivariate correlation, which is used to specify the simple correlation between two variables (Table 3), rather than the relationship between more than two variables, as is often seen in multiple regression analysis or structural equation modeling. The correlation is usually also called the correlation of the product Pearson or Pearson r. It is believed that Karl S. Pearson has formulas for calculating these correlations. There is the highest positive correlation between organizational performance (OF) and organizational capacity (OC) ($r = 0.713$, $Sig = 0.000$), followed by the link between organizational capacity (OC) and information technology (IT) ($r = 0.695$, $Sig = 0.000$).

As shown in Table 4, the linear regression coefficient $R = 0.813$, which indicates that there is a correlation between the dependent variable and the independent variable. Regarding variability, the value $R2 = 0.662$ or 66.2% explains the fact that 66.2% of the population in the sample is consistent with the correlation between the given variable).

According to the responses of the respondents in Table 5, more than half (88%) of them work in private companies, while 12% of the respondents work in state-owned companies. Regarding the company's social status, 43% of respondents work in medium-sized enterprises.

In Table 6, the answer to the question about the company's years of operation is that most companies have been on the market for more than 15 years (49%), followed by the market between 7 and 15 years (19%), a difference of 33 people from 3 to 7 years (15%) to under 3 years (17%).

Table 3 Pearson correlation matrix

Factors		Organizational capacity	Customer orientation	Information technology	Knowledge management	Organizational performance
Organizational capacity	Pearson Correlation	1	0.510**	0.466**	0.483	0.426**
	Sig		0.000	0.000	0.000	0.000
Customer orientation	Pearson Correlation	0.510**	1	**0.695****	0.595**	0.713
	Sig	0.000		**0.000**	0.000	0.000
Information technology	Pearson Correlation	0.466**	0.695**	1	0.658**	0.614**
	Sig	0.000	0.000		0.000	0.000
Knowledge management	Pearson Correlation	0.483**	0.595**	0.658**	1**	0.576**
	Sig	0.000	0.000	0.000		0.000
Organizational performance	Pearson Correlation	0.426**	**0.713****	0.614**	0.576**	1**
	Sig	0.000	**0.000**	0.000	0.000	

Table 4 Linear regression

Model	R	R square	Adjusted R square	Std. error of the estimate
1	0.813[a]	0.662	0.620	0.74573

Table 5 Company type

The company you work for is:	Frequency	Percentage (%)
Micro enterprise	14	21.5
Small business	23	35.3
Medium enterprise	28	43.0
Private company	57	87.7
Of the state company	8	12.3

Table 6 Company age

How many years has the company been established?	Frequency	Percentage (%)
Between 3 and 7 years	10	15.4
Between 7 and 15 years	12	18.5
More than 15 years	32	49.2
Less than 3 years	11	16.9

Table 7 indicates the position of the employee in the company in which he carries out his activity and resulted mainly in the presence of executive staff (81% of respondents), management positions (General Manager/Department Director) having a smaller number among respondents (19%).

Using the structural equation modeling, the model is matched with the sampled sample (as shown in Fig. 2).

Considering the β coefficient and the standard error, the SPSS AMOS program is used to check/invalidate the hypothetical β coefficient and the standard error. The validity of the acceptance model is presented in Table 8.

Table 7 Employee position in company

Position as employee in the company	Frequency	Percentage (%)
Employee (Execution Staff)	53	81.5
CEO	3	4.6
Department manager	9	13.8

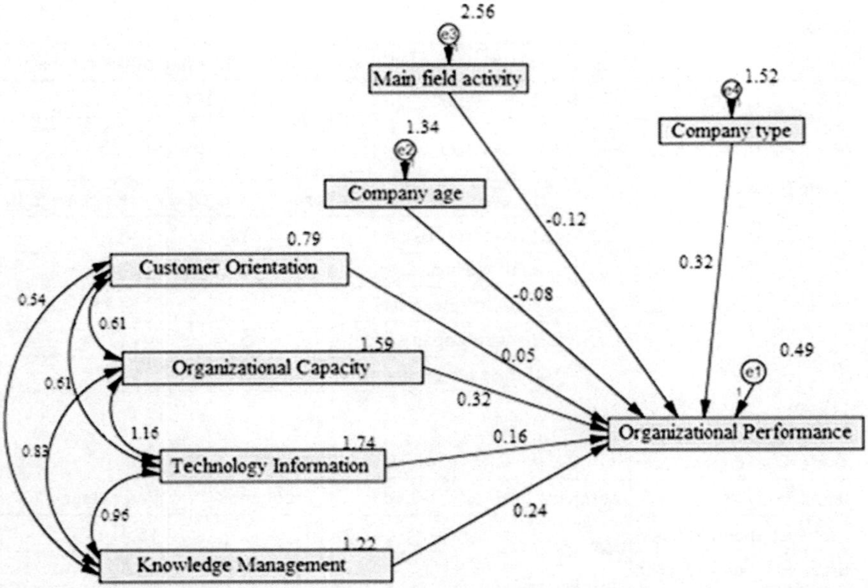

Fig. 2 Research model based on correlation coefficient and standard error

5 Conclusion

A very important aspect is that Romania is at the top of the European ranking, which measures the progress of EU countries towards the digital economy and social development. Romania is a destination for IT services and products. It plays a very important role in regional and global markets. It seems to be waiting between two realities: innovative industries and digital preservation. The results show that, after the adoption and use of the CRM system, all variables, except customer orientation, significantly affect the organizational performance of SMEs. Of these, organizational capacity had the most important impact. Regarding control variables, seniority and field of activity did not have significant effects on the performance of SMEs. However, the type of company has a positive effect on the performance of the company.

It is impossible for organizations to have all the necessary resources to remain competitive without having a close relationship with customers. Therefore, it is essential for organizations to implement CRM resources to develop strong CRM capabilities, subsequently improving the performance of companies. Thus, by managing and maintaining CRM more efficiently, SMEs can satisfy their customers and achieve organizational performance. This field has not been fully studied in Romanian SMEs, despite the importance of a company's performance in customer relations, and research models have not been further investigated in the context of Romanian SMEs.

Regarding regression, the results of multiple and linear regression analyzes show that customer orientation, organizational capability, information technology, and

Table 8 Validation/invalidation of hypotheses

Hypotheses	β coefficient	Standard error	Hypothesis validation/invalidation
H1: Customer orientation is positively related to organizational performance	0.049	0.125	Validate
H2: Organizational capability is positively related to organizational performance	0.319	0.103	Validate
H3: Technological resources are positively related to organizational performance	0.160	0.102	Validate
H4: The knowledge management of the organization is positively related to the organizational performance	0.241	0.112	Validate
H5: Control variable is positively related to organizational performance			
H5.1: Company type is positively related to organizational performance	0.318	0.071	Validate
H5.2: Company age is positively related to organizational performance	-0.122	0.055	INVALIDATE
H5.3: Company's main field activity is positively related to organizational performance	-0.083	0.075	INVALIDATE
The other correlations between the factors			
Information Technology < – > Knowledge Management	0.960	0.218	
Organizational Capability < – > Information Technology	1.156	0.253	
Customer Orientation < – > Organizational Capability	0.609	0.159	
Customer Orientation < – > Knowledge Management	0.537	0.140	
Organizational Capability < – > Knowledge Management	0.829	0.203	
Customer Orientation < – > Information Technology	0.615	0.165	

knowledge management significantly predict organizational performance. In terms of knowledge management, another important organizational factor in IT adoption is the alignment between organizational objectives and IT issues.

This study has some limitations, mainly in terms of sampling and technology examined. Respondents are especially young, and their behavior may be different from the average population. Another limitation is that the Romanian sample contains many full-time employees. However, many students in Romania work full time, which makes the two samples compatible. The consumers surveyed used the system even if they did not fully accept the system. This study is an exploratory study that aims to determine organizational performance based on the rate of adoption of customer relationship management by SMEs in Romania. The research is related to understanding the experience of sample respondents, but is not yet representative of CRM solution users.

References

Becker, J.U., Greve, G., Albers, S.: The impact of technological and organizational implementation of CRM on customer acquisition, maintenance and retention. Int. J. Res. Mark. **26**(3), 207–215 (2009)

Chang, H., Ku, P.: Implementation of relationship quality for CRM performance: acquisition of BPR and organizational learning. Total Qual. Manag. **20**(3), 327–348 (2009)

Coltman, T. R.: Why build a customer relationship management capability? Journa of Strategic Information Systems , 301–320 (2007).

Croteau, A.M., Li, P.: Critical success factors of CRM technological initiatives. Can. J. Adm. Sci. **20**(1), 21–34 (2003)

Gupte, M.: Mapping of information flow in customer relationship management tool. Master thesis, Binghamton University State, New York (2011)

Halawi, L.A., Aronson, J.E.: Resource-base view of knowledge management for competitive advantage. Electr. J. Knowl. Manag. 75–86 (2005)

Harrigan, P., Schroeder, A., Qureshi, I., Fang, Y., Ibbotson, P., Ramsey, E., et al.: Internet technologies, eCRM capabilities and performance, benefits for SMEs: an exploratory study. Int. J. Electron. Commer. **15**(2), 7–46 (2010)

Jaakkola, M., Fr Sen, J., Santala, M., Vassinen, A.: Market orientation and business performance: the mediating effect of core business processes. J. Am. Acad. Busin. **5**(2), 46–61 (2009)

Jabar, F.H., Tajuddin, N., Paino, H.: Internalization of small and medium enterprises. ASEAN Entrepreneurship Conference. Springer Science+Business Media Singapore (2016)

Keramati, A., Hamed, M.: A process-oriented perspective on customer relationship management and organizational performance: an empirical investigation. Ind. Mark. Manag. 1170–1180 (2010)

Ku, E.: The impact of customer relationship management through implementation of information systems. Total Qualit Manag. Busin. Excell. **21**(11), 1085–1102 (2010)

Kubi, B.A., Doku, A.: Towards a sucessful customer relationship management: a conceptual framework. African J. Mark. Manag. **2**(3), 37–43 (2010)

Marcketti, S.B., Kozar, M.J.: Leading with relationship a small firm example. Learn. Organiz. **14**(2), 142–153 (2007)

Mascu, S., Muresan, M.: Enhancing the potential of SMEs by using European funds in Romania. Int. Comp. Manag. **20**(2), 187–197 (2019)

Mohamad, S.H., Othman, N.A., Jabar, J., Majid, I.A., Kamarudin, M.F.: The impact of customer relationship management on small and medium enterprises performance. J. Technol. Manag. Technopr. **2**(2), 11–26 (2014)

Mohammad, A., Rashid, B.B., Tahir, B.S.: Assessing the influence of customer relationship management (CRM) dimensions on organization performance. J. Hospot. Tour. Technol. **4**(3), 228–247 (2013)

Nath, V., Guganani, R., Goswami, S., Gupta, N.: Management practices in selected Indian services industries. J. Mark. Commun. **4**(3), 18–40 (2009)

Nicolescu, O., Nicolescu, C.: Entrepreneurship and management of small and medium enterprises. Economic Publishing House, Bucharest (2008)

Payne, A., Frow, P.: Customer relationship management: from strategy to implementation. J. Mark. Manag. **22**(1), 135–168 (2006)

Rese, M.: The explanatory foundations of relationship marketing theory: a comment. J. Busin. Ind. Mark. **21**(2), 88–91 (2006)

Sin, L.Y., Alan, C.B., Yim, F.H.: CRM: Conceptualization and scale development. Eur. J. Mark. **39**(11–12), 1264–1290 (2005)

Sousa, R.: Linking quality management to manufacturing strategy: an empirical investigation of customer focus practices. J. Oper. Manag. **21**, 1–18 (2003)

Wang, I.C., Huang, C.Y.: The influence of customer relationship management process on management performance. Int. J. Organ. Innov 40–50 (2010)

Wilson, H., Danial, E., Mcdonald, M.: Factors for success in customer relationship management system. J. Mark. Manag. 193–219 (2002)

Creating Value with Big Data in Marketing

Ionuţ Adrian Moşescu, Raluca-Giorgiana Chivu, Ionuţ Claudiu Popa, and Florina Botezatu

Abstract The new technology helps collect and store a considerable amount of information about consumers' preferences and decisions in real-time. Thus, the unprecedented volume, speed and variety of primary data, Big Data, are available from all sources and individual consumers. The term "Big Data" is now widespread and accepted globally. The term Big Data has become increasingly known and used in many industries. Big Data is a new source of product development ideas, customer service, shelf location, distribution, dynamic pricing, etc. Big Data will have an impact on almost every area of marketing. Firms that do not develop the resources and capabilities to use Big Data capabilities effectively will be challenged to develop a sustainable competitive advantage to survive the Big Data revolution. Consumer analysis is at the heart of a Big Data revolution. Marketing specialists are beginning to recognize the importance of Big Data as the new capital and that access to Big Data offers a company new ways to differentiate its products. In this article, exploratory research has been conducted to highlight areas where Big Data allows organizations to fundamentally know about their business and translate it into better decision-making and improved performance in marketing and consumer behavior. Moreover, a global analysis (based on public data) was performed using large volumes of data in marketing and the main reasons they were stored and used.

Keywords Big Data · Marketing · Big Data's growth

I. A. Moşescu · R.-G. Chivu (✉) · I. C. Popa · F. Botezatu
The Bucharest University of Economic Studies, Bucharest, Romania
e-mail: raluca.chivu@mk.ase.ro

I. C. Popa
e-mail: claudiu.popa@mk.ase.ro

M. Busu (ed.), *Digital Economy and New Value Creation*, Springer Proceedings in Business and Economics, https://doi.org/10.1007/978-3-031-07265-9_11

1 The Usefulness of Big Data in Marketing

1.1 Introduction

Big data research is undoubtedly one of the essential areas in the past decade and its impacts on the industries. Although the research of applying data analysis such as data mining, statistics, and information systems has continued for many decades, it was not until the last ten years that the industries and academia started to pay more attention to the development of extensive data studies applications. As it can be observed, in the past decade, the volume of data has immensely increased along with the development of data-generating, extracting, addressing, storage, and output technologies, particularly the connected Internet of Things (IoT) devices along as well as new database technologies, e.g., NoSQL. Consumers and all types of industries worldwide adopt those IoT products such as small phones, tablet pc, and wearing devices. As predicted by the International Data Corporation, it is projected by 2025, and there will be 175 Zettabytes of data increased from 41 zettabytes in 2019 and only two zettabytes in 2010 (Erevelles et al. 2016; Holst 2019; Reinsel et al. 2018). Among them, 79.4 Zettabytes are created by IoT devices, and 87 Zettabytes reside in public cloud environments (Wang and Wang 2020).

The digitalization of organizations activities and businesses has consequently conducted to the increase of the datasets for analyzing purposes, which through big data and analytical tools can generate perspective for providing sustainable value, improving business performance and providing a competitive advantage (Sivarajah et al. 2020).

The most important motive for the exploitation of big data analytics is to generate business knowledge information and understand the relationship between business processes and the business environment to help decision-making in companies. Big data analytics can capacitate better-informed decision-making and improvement of the business processes and understand the customers better. Also, big data analytics using real-time information can supply instant information and create real-time information of the markets and customers that can increase sales. Big data analytics can generate valuable information about customers, and the next challenge for companies is to transform the knowledge into a market advantage (Hallikainen et al. 2020).

1.2 Literature Review—The Usefulness of Big Data in Marketing

The Chartered Institution of Marketing in the United Kingdom describes marketing as “the management mechanism responsible for defining, predicting, and meeting productive consumer requirements.” Since the discovery of accounting spread to Mesopotamia for tracking the growth of crops and herds 7,000 years ago, data has been used to track and manage enterprises (Van Rijmenam and Logue 2021). O’Reilly

Media coined the word "big data" in 2005, and a year later, they coined the term "Internet 2.0," which refers to a broad collection of data (Van Rijmenam and Logue 2021). Big data is the actual result of the modern digital marketing environment for marketing companies (Big data, bigger marketing).

This article would illustrate how big data engineering functions in marketing and how big data could be used in marketing. The benefits of big data on the digital marketing environment would then be assessed.

Big data refers to massive, structured data sets collected by enterprises and governments. "Innovative modes of information collection for increased insight and decision making" (Big data), which involves several kinds of information coming in increasing quantities and at extremely high speeds (Lyons et al. 2018).

Structured data, unstructured data, and semi-structured data are the three categories of data that big data can divide. Structured data is easy to enter, process, query, store, and recover into a fixed format, as Senthilkumar et al. (2018) point out. Unstructured data cannot be fitted or categorized into a net box, and the procedure and processing are complex and time-consuming. Blog posts, text messages, novels, and photographs, for example (Drew 2017). Semi-structured data is structured data that has semantic tags that do not fit into traditional relational databases. Examples include email, XML, and some markup languages (Drew 2017).

The most critical aspect of marketing is the processing of big data. Consumer data has been gathered in several areas, according to Goddard (2019), such as the company explicitly calling for customer data, pulling customer data from social media, email monitoring, and data providers selling data. In the early stages, the bulk of businesses call for consumer details directly (Kunz et al. 2017). Customers must typically fill out paperwork anytime they wish to apply to a service, make their first online purchase, or become a brand member, according to the author. The increased use of social media has become an essential part of people's everyday lives on the internet.

Individuals' online interaction is tracked using social media info, which incorporates posts, views, tweets, and mentions (Segal). As consumers log in to a third-party program using their Facebook account, the user's data is also shared (Kunz et al. 2017). Furthermore, companies are buying consumer data from companies like Acxiom and Oracle, which gather data, analyze it, and sell it through targeted advertising campaigns (Kunz et al. 2017). Big data is also being used in a variety of fields, including marketing. Big data will be used to accelerate product creation and creativity. To recognize the demands of consumers, production evolution necessitates vast amounts of data. Companies will produce a new product closely linked to their consumers' tastes and desires while reducing vulnerability through data mining and analysis (Johnson et al. 2017). As a result, big data has emerged as a promising tool for boosting output efficiency (Saidali et al. 2019).

According to Anastasia, businesses use big data as an intelligence platform to produce innovative products, find ways to market new products, and advance existing product lines. Furthermore, big data can repair significant traces and mistakes (Rubinfeld and Gal 2017). Said et al. suggest that real-time data can be processed instantly, reducing the time spent on analytics, maximizing decision-making performance, and increasing the pace of concept creation to product distribution. According to

Kopanakis (2018), "Amazon fresh and whole foods" are a typical example of using big data to boost product evolution. Amazon whole foods are based on big data processing to consider consumers' primary demand for groceries.

Furthermore, big data can assist in marketing pricing optimization. Saran points out that a product's price is historically dictated by its cost, dynamic pricing, and the customer's and requirements perceived value. On the other hand, big data will include more factors for businesses to consider when determining a price. Companies may, for example, make use of knowledge from completed transactions (Saran et al. 2018). Furthermore, big data can be used to evaluate the right price for a company to increase the profit margin by first looking at the customer's buying history, choices, and actions, then using predictive analysis to determine the price adaptability of the requirement, and ultimately improving the business by looking at the cost price of profit maximization.

Since it can help marketers identify their customers, big data in marketing can help solve advertisers' problems and have marketing information. Big data can be used to efficiently and precisely approach customers with products they want or need (Medal 2017). According to the author, advertising companies gather information about consumers' motivations and track online behaviors to save money and ensure efficiency. One of the most convincing cases of significant companies using big data for personalized ads is Netflix (Kopanakis 2018). Netflix, according to Kopanakis, receives large volumes of data from its over 100 million users, with the data sets relating to recent requests and viewing recordings. To determine what kind of film the subscriber finds most appealing and to make movie recommendations to them.

As a result, big data is being used in a wide range of marketing applications. Around the same time, big data will have several positive effects on marketing. Big data, for example, provides businesses with the ability to predict, improve risk control systems, and boost social media marketing.

To begin with, big data allows businesses to predict and prepare for progress. Marketers will use big data to behave proactively and predict potential patterns (Jobs et al. 2016). Accelerating reporting of a growing volume and variety of data sets and real-time forecasts, the firm would increase its spending and gain a better view of the impact of sales at various stages of spend. Furthermore, by monitoring real-time data and the latest developments in customer behavior, advertisers can develop systematic plans and prepare for more effective actions (Jobs et al. 2016). As a result, the organization can target segmented sub-groups of buyers based on their unique characteristics, allowing them to tailor activities to each client uniquely (Jobs et al. 2016).

Furthermore, big data can be used as a business analytics platform to help banks and the industry handle risk more effectively. People may use big data analytics to recognize possible risks associated with money leading processes in banks, understand industry dynamics, and set different interest rates for each person from various regions (Davenport and Dyché 2013). Similarly, economies are being more intertwined, resulting in a rise in financial risk. According to Sinha et al. (2018), big data should be used to observe the implicit danger involved with all financial transactions using a simulation scenario framework. Big data also helps in social media campaigns. Although social media promotes contact between friends and families, it also acts as the primary data center. According to Erevelles et al. (2016), social media has evolved into a medium for brands to promote and grow their customer base. Likes, comments, shares, mentions, and followers are all customer behavior indicators that provide advertisers and advertisers with actionable insight Erevelles et al. (2016). Facebook's ad campaigns are an excellent example of extensive data-driven social media marketing (Wiesenberg et al. 2017). For the last thirteen years, Facebook has been gathering data and selling the findings to third parties. Marketers use the Facebook website to promote, and they can increase campaign success by indirectly leveraging big data.

Big data, on the other hand, necessitates the use of technology. In many businesses, marketing is the least automated department, and analyzing massive amounts of data is the most significant difficulty in using big data.

To conclude, big data is a significant marketing investment. Through analyzing big data, businesses can gain a strategic edge, reduce operational costs, and retain their client base. Historically, storing vast amounts of data has been difficult, but with the advancement of computational power, this is no longer an issue (Lyons et al. 2018). Nonetheless, the complex technology of using big data has not been widely adopted by the company's majority. As a result, the advancement in technologies in big data is a question that must be considered.

2 Methodology

In this article, exploratory research of the latest studies on the use of large data volumes (Big Data) was carried out to highlight the present fields of application and make predictions regarding the use and utility of large data volumes in the field of marketing. Information is a powerful tool; however, only in terms of the revelations they provide. We cannot do much about the 2.5 quintillion bytes of data we generate every day unless we crunch it with big data analytics. This is where things start to get interesting for companies because we can now forecast consumer behaviour personally with laser-focus precision. We know that as the months, not years, pass, data will only grow more extensive, and technology will become more sophisticated.

3 Results and Discussions

Although the big data industry's scale and share are measured in billions, the overall impact on the global economy is conveniently measured in trillions. Market predictions and industry estimates quickly affirm where big data is headed—led by its sibling, business analytics, and the best business intelligence software applications that it spawns.

- According to Wikibon, worldwide Big Data industry revenues for software and services are predicted to grow from $42 billion in 2018 to $103 billion in 2027, reflecting a CAGR of 10.48%.
- According to Forrester, the global Big Data software market will be worth $31 billion this year, up 14% from last year. The global software industry is expected to generate $628 billion in sales, with $302 billion coming from applications.
- According to a survey by Accenture, 79% of enterprise executives believe that businesses that don't accept Big Data will lose their competitive edge and might even go out of business. Far more, 83%, have undertaken Big Data ventures in order to gain a competitive advantage.
- According to PwC, 59% of executives believe that using AI to boost Big Data in their business will be beneficial.

Advanced analytics, including Big Data, contribute the most to revenue growth in sales and marketing, research and development (R & R&D), supply chain management (SCM), including distribution, workplace management, and operations. The report Analytics Comes of Age, published by McKinsey Analytics in January 2018, provides a detailed analysis of how analytics and Big Data enable entirely new ecosystems and act as a foundational technology for Artificial Intelligence (AI). Analytics and Big Data, according to McKinsey, are making the most significant contributions in the Basic Materials and High Tech sectors. The first of ten charts in this series comes from a McKinsey Analytics report, and it shows how analytics and Big Data are revolutionizing many of Sales and Marketing's foundational business processes (Figs. 1 and 2).

- According to a new McKinsey Analytics survey, nearly half of respondents believe analytics and Big Data have radically changed business practices in their sales and marketing departments. Across sectors, more than 30% claim the same thing about R&D, with respondents in High Tech and Basic Materials & Energy reporting the most functions being changed by analytics and Big Data.
- Revenues from Big Data services and products are expected to rise from $42 billion in 2018 to $103 billion in 2027, representing a Compound Annual Growth Rate (CAGR) of 10.48%. Wikibon forecasts that the global Big Data market will rise at an annual rate of 11.4% between 2017 and 2027, from $35 billion to $103 billion
- According to NewVantage Venture Partners, Big Data adds the most value to businesses by lowering costs (492%) and opening up new opportunities for creativity and disruption (44.3%). Discovering new cost-cutting opportunities by incorporating advanced analytics and Big Data provides the most observable outcomes, making this the most common category in the report. With 27.9% reporting results, 69.4% have started using Big Data to build a data-driven society (Fig. 3).
- The Hadoop and Big Data Market is expected to rise at a 28.5% CAGR from $17.1 billion in 2017 to $99.31 billion in 2022. The most rapid growth is expected in 2021 and 2022, as the demand is expected to increase by $30 billion in a single year. Statista commented on StrategyMRC's findings (Fig. 4).

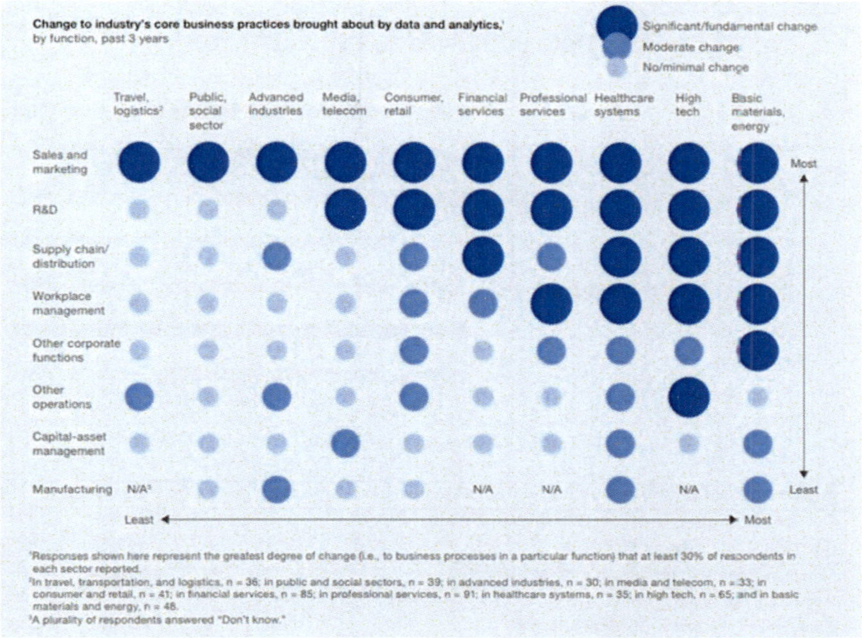

Fig. 1 Changes to industry score business practices about big data. *Source* Analytics (2018)

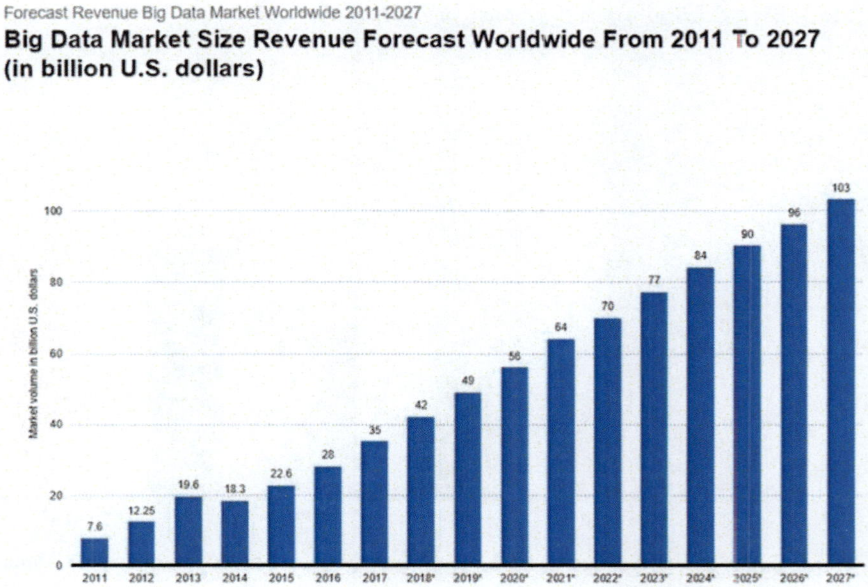

Fig. 2 Big data market size revenue forecast Worldwide From 2011 to 2027 (in billion U.S. Dollars). *Source* Statista.com (2021)

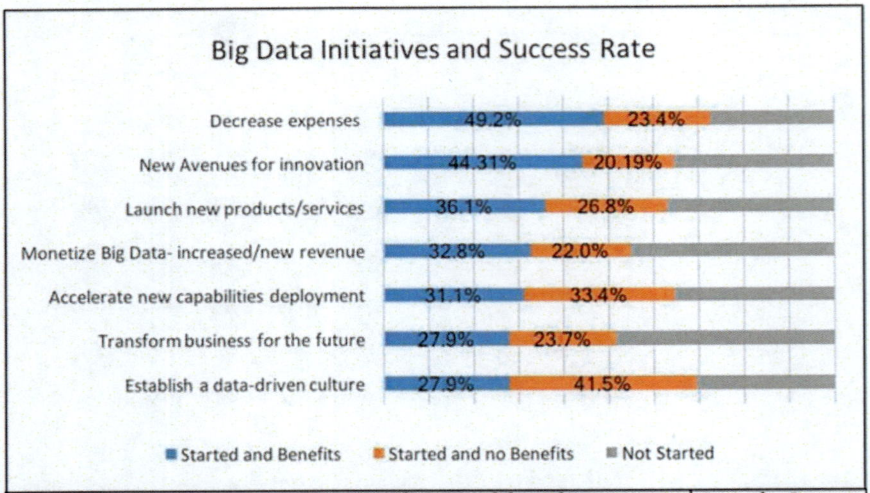

Big Data business initiatives underway; with successful results.	Started	Success
Decrease expenses through operational cost efficiencies	72.6%	49.2%
Establish a data-driven culture	69.4%	27.9%
Create new avenues for innovation and disruption	64.5%	44.3%
Accelerate the speed with which new capabilities and services are deployed	64.5%	31.1%
Launch new product and service offerings	62.9%	36.1%
Monetize Big Data through increased revenues and new revenue sources	54.8%	32.8%
Transform and reposition your business for the future	51.6%	27.9%

Fig. 3 Big data initiatives and success rate. *Source* New Vantage Partners (2019)

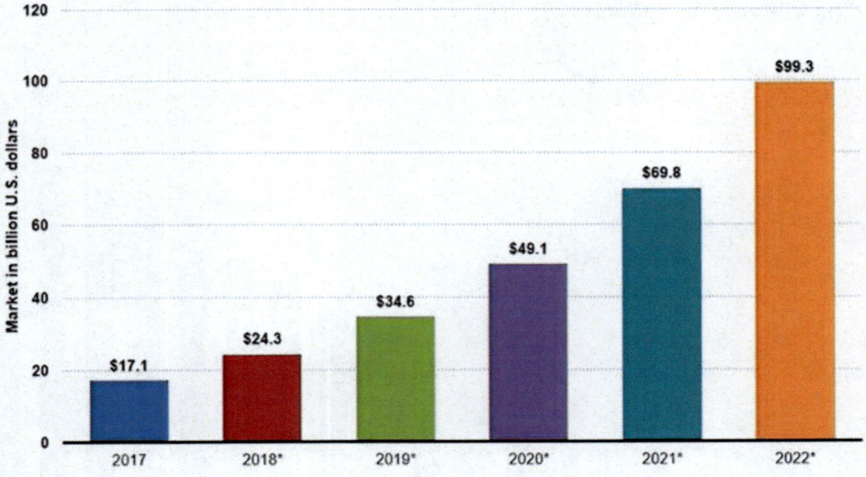

Fig. 4 Size of hadoop and big data market Worldwide From 2017 to 2022 (in billion U.S. Dollars). *Source* strategymrc.com (2017)

- The market for Big Data applications and analytics is expected to expand at a CAGR of 15.49% from $5.3 billion in 2018 to $19.4 billion in 2026. Professional Services in the Big Data industry is expected to rise from $16.5 billion in 2018 to $21.3 billion in 2026 (Fig. 5).
- When the global market for advanced analytics and Big Data-related hardware, services, and applications is compared, the latter category emerges as the vital winner. The software segment is expected to rise at the fastest rate, growing from $14 billion in 2018 to $46 billion in 2027, with a CAGR of 12.6% (Fig. 6).
- Revenue rose by 3% to 4% at a decentralized general-merchandise store that used Big Data to build success community clusters. Big Data drives a transformation in

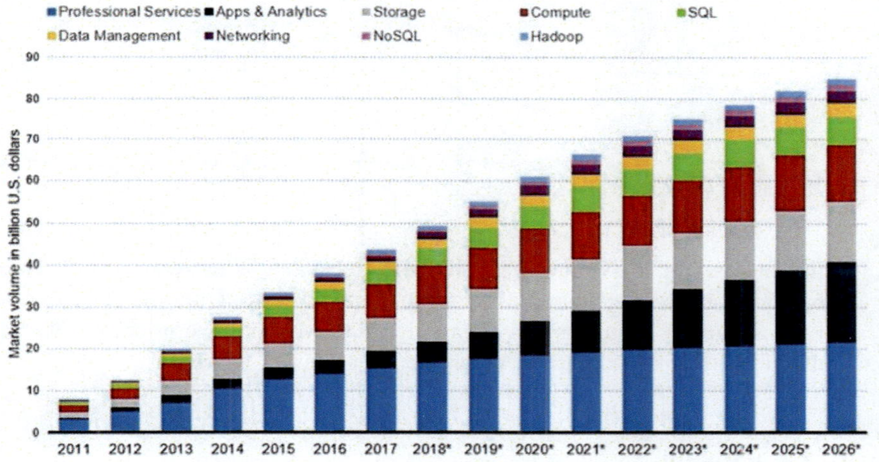

Fig. 5 Big data market forecast Worldwide from 2011 to 2026, by segment. *Source* wikibon.com/ (2017)

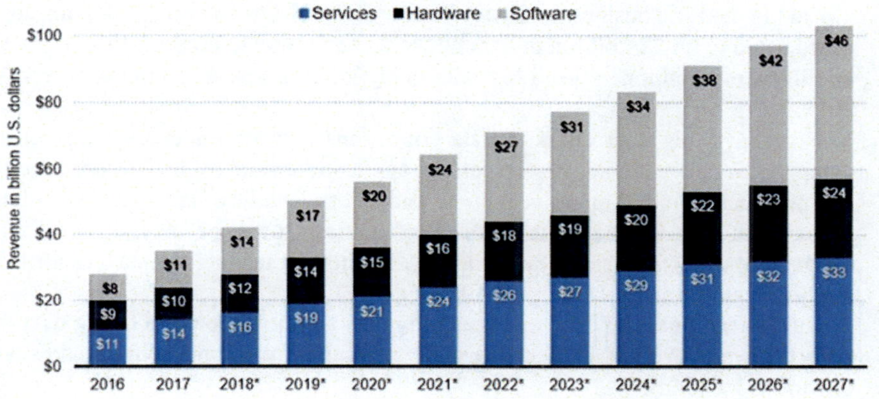

Fig. 6 Big data revenue Worldwide from 2016 to 2027, by significant segment. *Source* wikibon.com/ (2017)

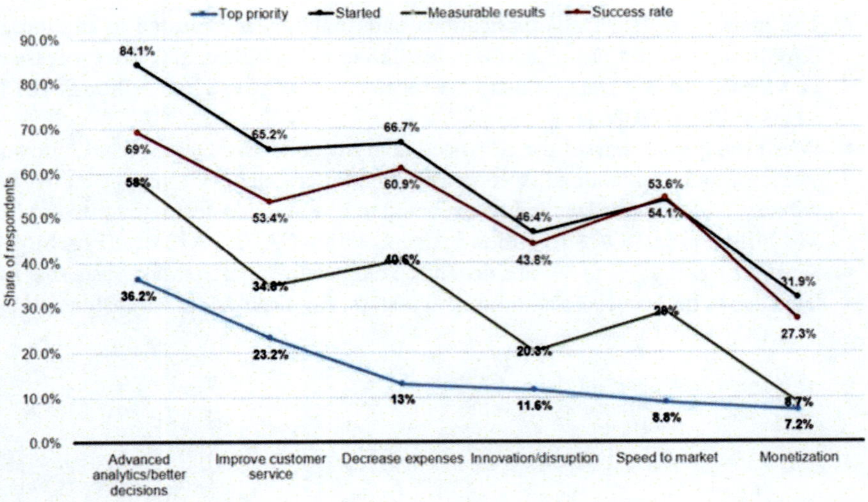

Fig. 7 Big data initiatives and success rates among corporations in the United States and Worldwide, As of 2018. *Source* New Vantage Partners (2019)

the retail sector, allowing for more detailed localization than ever before. Today, big data is being used to improve endcap promotions' ROI, optimize planograms, improve upsell and cross-sell sales efficiency, and optimize prices on products that drive the most foot traffic (Fig. 7).

4 Conclusion

Predictive analytics and data mining are examples of advanced analytics techniques that help derive meaning from data and create new market insights. In 2018, the global big data and business analytics industry were worth 169 billion dollars, and it is projected to hit 274 billion dollars by 2022. According to reports, 45% of market intelligence practitioners used big data analytics as a research tool in November 2018.

Nowadays, big data refers to data collections that are either too large or too complex for conventional data processing applications to manage. It is described as having much volume, many velocities, and a lot of variety. The growing volume and size of data sets are due to the exponential growth of mobile data traffic, cloud computing traffic, and the rapid advancement of technology such as artificial intelligence (AI) and the Internet of Things (IoT).

Customers, financial data, and operating data are the three types of big data that advertisers are involved in. Data is usually compiled from some sources and stored in various locations.

Customer data assists advertisers in gaining a better understanding of their target customers. Names, email addresses, transaction records, and web searches are examples of this sort's apparent data. Indicators of the audience's attitudes gleaned from social media activity, polls, and online forums are just as relevant, if not more so.

Financial data allows you to monitor success and increase productivity. This category covers the company's revenue and marketing figures, prices, and margins. This group will also contain financial details from competitors, such as pricing.

Market operations are linked to operational records. It may have had to do with shipment and logistics, CRM programs, or input from hardware sensors and other sources. This data can be used to increase results and cut costs.

References

Analytics, M.: Analytics comes of age. McKinsey & Company, New York, United States (2018)

Davenport, T.H., Dyché, J.: Big data in big companies. Int. Inst. Anal. **3**, 1–31 (2013)

Drew, R.O.B.B.: Semi-structured data. Int. J. Child Comput. Inter. Elsevier (2018)

Erevelles, S., Fukawa, N., Swayne, L.: Big Data consumer analytics and the transformation of marketing. J. Bus. Res. **69**(2), 897–904 (2016)

Goddard, W.: How do big companies collect customer data. IT Chronicles.[verkkodokumentti][viitattu 3.6. 2020] Saatavilla (2019). https://itchronicles.com/big-data/how-do-bigcompanies-collect-customer-data

Hallikainen, H., Savimäki, E., Laukkanen, T.: Fostering B2B sales with customer big data analytics. Ind. Mark. Manage. **86**, 90–98 (2020)

Holst, A.: Volume of data/information created worldwide from 2010 to 2025 (2019). Availabe at https://www.statista.com/statistics/871513/worldwide-data-created/

https://www.strategymrc.com/ (2017)

Jobs, C.G., Gilfoil, D.M., Aukers, S.M.: How marketing organizations can benefit from big data advertising analytics. Acad. Market. Stud. J. **20**(1), 18 (2016)

Johnson, J.S., Friend, S.B., Lee, H.S.: Big data facilitation, utilization, and monetization: exploring the 3Vs in a new product development process. J. Prod. Innov. Manag. **34**(5), 640–665 (2017)

Kopanakis, J.: Real-world examples of how brands are using big data analytics. Mentionl. Com (2018)

Kunz, W., Aksoy, L., Bart, Y., Heinonen, K., Kabadayi, S., Ordenes, F.V., Theodoulidis, B.: Customer engagement in a big data world. J. Serv. Market. (2017)

Lyons, A.C., Grable, J.E., Joo, S.H.: A cross-country analysis of population ageing and financial security. J. Econ. Ageing **12**, 96–117 (2018)

Medal, A.: How big data analytics is solving big advertiser problems. Entrep. Angel Investor **16** (2017)

New Vantage Partners.: Data and Innovation–How Big Data and AI are Accelerating Business Transformation. NVP, New York (2019)

Reinsel, D., Gantz, J., Rydning, J.: The digitization of the world from edge to core. IDC white paper (November) (2018)

Rubinfeld, D.L., Gal, M.S.: Access barriers to big data. Ariz. l. Rev. **59**, 339 (2017)

Saidali, J., Rahich, H., Tabaa, Y., Medouri, A.: The combination between big data and marketing strategies to gain valuable business insights for better production success. Procedia Manuf. **32**, 1017–1023 (2019)

Saran, R., Robinson, B., Abbott, K.C., Agodoa, L.Y., Bhave, N., Bragg-Gresham, J., Shahinian, V.: US renal data system 2017 annual data report: epidemiology of kidney disease in the United States. Am. J. Kidney Dis. **71**(3), A7 (2018)

Senthilkumar, S.A., Rai, B.K., Meshram, A.A., Gunasekaran, A., Chandrakumarmangalam, S.: Big data in healthcare management: a review of literature. Am. J. Theor. Appl. Business **4**(2), 57–69 (2018)

Sinha, V., Pribanic, B., de Bruin, B., Trincado, M., Grützmacher, H.: Ligand-and Metal-based reactivity of a neutral ruthenium diolefin diazadiene complex: the innocent, the guilty and the suspicious. Chem. Eur. J. **24**(21), 5513–5521 (2018)

Sivarajah, U., Irani, Z., Gupta, S., Mahroof, K.: Role of big data and social media analytics for business to business sustainability: A participatory web context. Ind. Marketing Manag. **86**, 163–179 (2020)

Statista Research Department: Global Big Data market size 2011–2027 | Statista. [online] Statista (2021)

Van Rijmenam, M., Logue, D.: Revising the 'science of the organization': theorizing AI agency and actorhood. Innovation **23**(1), 127–144 (2021)

Wang, W.Y.C., Wang, Y.: Analytics in the era of big data: the digital transformations and value creation in industrial marketing (2020)

Wiesenberg, M., Zerfass, A., Moreno, A.: (2017), Big data and automation in strategic communication. Int. J. Strateg. Commun. **11**(2), 95–114 (2017). https://www.strategymrc.com/

Disclosing Information About Employees According to Global Reporting Initiative Standards

Diana-Nicoleta Nita

Abstract Employees, in terms of human capital, are an important part of a company's engine. Without this part, a company would not be able to function, regardless of its assets. Its presentation is insignificant in most of the company's financial reports. Financial reports represent an important element to get to know in detail a company. The lack of legislative alignment on reporting criteria and requirements has led to the presentation of information by companies in different ways. Global Reporting Initiatives is one of the standards that tracks the impact that a company has on the economic environment, including information about the companies employees. At the international level, researchers have tried to identify the elements that determine the reporting human capital information, but no universal-valid element has yet been identified. In Romania, the research on human capital reporting is in initial stage and in practice it is known that the reporting employees information is limited to several indicators and financial data. The objective of this research is to determine some of the information requested by Global Reporting Initiative about employees and economist perception about this. The research questions are based on Global Reporting Initiative requirements about employees. Using the quantitative method, the questionnaire, was identified the perception and openness of Romanian economists regarding to reporting detailed information about employees according to Global Reporting Initiative Standards. The results show that Romanian economists are not willing to include detailed information about employees. This study case represents the beginning of identifying the perception of Romanian economists about the information proposed by the Global Reporting Initiative. Considering that the research in Romania regarding to employees' disclosure is incipient, the perception of the Romanian economists regarding to change and awareness of the information importance about the employees in the financial reports must be known.

Keywords Human capital · Employees reporting information · Employees reporting information in Romania · Global reporting initiative · Global reporting initiative in Romania

D.-N. Nita (✉)
Academy of Economic Studies, Bucharest, Romania
e-mail: nita.diana.n@gmail.com

M. Busu (ed.), *Digital Economy and New Value Creation*, Springer Proceedings in Business and Economics, https://doi.org/10.1007/978-3-031-07265-9_12

1 Introduction

Human capital, as a component part of intellectual capital, is restricted by the situation provided by current accounting and national and international financial reporting standards. If we have to look at the company behind success, it can be seen that regardless of the assets performance level, the automation degree, debt information, equity, is human factor. However, the limitation of reporting the information about it, determined the presentation of the salary cost or of some indicators. Regardless of the size, industry type, listing status, a company could not function without employees. Employees, as part of human capital, are one of the most important elements of a company. Interesting is that human capital is part of an absolutely vital relation within a company (human-relational-structural capital reciprocity), but scientific research on this aspect is relatively new and it is encounter a lack of individual value knowledge. At the same time, the novelty of the concept is relevant for the position of Romanian financial reporting and traditional accounting. The voluntary presentation of human capital information shows the company's focus. The need to standardize these reports determines a credibility and a high degree of all companies full knowledge.

This research starts from the importance of reporting employees in the financial reports. Regardless of the financial reports users, the employees represent an element that must be analyzed and reported in a period of time. The responsibility of each company to report human capital can become a social responsibility. Employees contribute to the impact that the company has on the economic environment and thereby must be presented in as detailed a way as possible. Human capital is an element of research given the rapid level of processes automation. Human capital is the added value of the process. In this case, its importance must be recognized and presented. The Global Reporting Initiative Standards demonstrate how employee analysis can determine a company's image. Thereby, these standards show how a company can be known through human capital reporting. In Romania, most small and medium-sized companies apply the Order of the Public Finance Ministry no. 1802/2014 in preparing and presentation the financial statements. In the case of those companies, human capital can be known by the cost of personnel, the average and effective number of employees-in the financial statements, and in the explanatory notes-the benefits granted to key employees and indicators attesting information about employees. One reason for this limited reporting are the users of the financial statements. In this case, it is considered that information about human capital is not relevant. However, such an information portfolio can help new investors to know the future potential of the company in which they are going to invest. Over the time, research has been conducted on the element that determines the reporting of human capital in financial reports. The studies were conducted based on the financial reports of companies listed on the stock exchange, because they were publicly available. If at international level, the researches were complex, at national level, in Romania, the researchers are at the beginning. Even in this case, the financial reports of the companies listed on the stock exchange were taken into account. As employees are not only present in such companies, both small and medium-sized companies should be analyzed in order to report human capital. Global Reporting Initiative Standards are standards applicable to small and medium-sized companies as well. This research begins by obtaining the perception of Romanian economists regarding the detailed reporting of employees. Thereby, the hypothesis tested is the acceptance of human capital reporting by small and medium companies, according to the information required by the Global Reporting Initiative Standards.

The analysis starts from the identification of the information provided in Global Reporting Initiative Standards and identification of the Romanian economists' perception regarding to the inclusion of those information in the financial reports. The research is structured in four parts as follows: literature review- the current situation of human capital reporting at national and international level is identified, methodology-research method are presented, results and discussion and conclusions where the main results obtained are presented.

2 Literature Review

2.1 Human Capital. National and International Context

An asset is expected to generate future economic benefits, but intellectual capital is that value that is not recognized in accounting. Castellanos et al. (2004, p. 2) define intellectual capital as a set of knowledge that creates, or is expected to create value in the future for organization. Because it cannot be included in the assets sphere for the reason that it cannot be controlled, owned by the entity, intellectual capital remains at the stage of being; as Klaila et al. (2000, p. 1) mention: "an asset ignored once it has been cataloged". Traditional accounting cannot identify a value of intellectual capital, but Joia (2000, p. 15) says that "treasure is what matters in a society in constant turmoil". The importance of human capital compared to all other intellectual capital components lies in the role it plays in creating value for the company, for what one company is distinct from another (Seetharaman et al. 2004). For companies, human capital is a relevant and heterogeneous strategic asset (Samagaio and Rodrigues 2016). Human capital is defined by six categories of components: talent, knowledge, ability, humanity, desire and managerial leadership Hatamizadeh et al. (2020, p. 4). Also, Carson et al. (2004, p. 9) say that human capital can be defined by two categories of elements: personal attributions and "skills". All of this categories of elements are attributable to employees. People use their own skills to create added value for the organization in two directions: to transfer and convert knowledge and added value increases whenever a transfer or knowledge conversion takes place (Sveiby 2001). Thereby, the intellectual capacity of a company can be known by knowing the employees. This can be achieved by disclosing more information in the financial reports.

To understand a company's intellectual resources, Mouritsen et al. (2001, p. 2) says that users should analyze the "fruits" of the current year, and the company's potential to produce "fruits" in the future. The intellectual capital integration in economic and financial reports is necessary due to the fact that intangible assets are essential for business success given the economic environment growing dynamics (Gogan and Draghici 2013). Nevertheless, firms consider that the information provided by intangible assets is useful for analysts and investors, and therefore decreases the reporting degree of intellectual capital but this situation can be saved by the voluntary intellectual capital disclosing (Schiemann et al. 2015). About disclosing human capital information, Passetti and Cinquini (2014, p. 22) claim that "the low credibility of sustainability reports is a big obstacle that does not permit the valorization of human capital information". In this sense, Ndou et al. (2018, p. 26) believe that "its most used channels for disclosing pertinent quantitative and qualitative intellectual capital information were non-traditional sources, such as its website, online reports and social media pages". Also, Dumay and Guthrie (2017) mention that "a lack of significant innovation in the evolution of ICD indicates that this research stream may have been a victim of its own success" (Cuozzo et al. 2017, p. 3).

Even though intellectual capital reporting has increased over time, there are no signs that investors are interested in this information (Nikolaj Bukh 2003). Suciu et al. (2013, p. 9) said that "reporting, evaluating and disclosing intellectual capital represents reporting, evaluating and disclosing valuable knowledge". In this sense, Nickolaj Buch (2003, p. 7) said that "the incorporation of information about intellectual capital disclosure indicates that companies believe this type of information is important in the capital market's assessment of the company value".

Fontana and Macagnan (2013, p. 3) said that "human capital disclosure has strategic relevance for enterprises because these features, although not recorded in accounting, represent a competitive business edge". Disclosing human capital to the external environment could positively influence the company's reputation (Alvarez Dominguez 2011). But, in this sense, annual reports that include information on human capital may reflect the risk that the entity may use them as "advertising documents" but not in accordance with accounting standards (Abeysekera 2008). Disclosing human capital information can be results-oriented (emphasizing the recognition of employees' contribution to the company and reporting the added value of employees), or process-oriented (emphasizing the entrepreneurial qualities and knowledge of employees related to work) (Abeysekera and Guthrie 2004). In disclosing human capital information, Beattie and Smith (2010, p. 21) said that "HR directors viewed attracting and retaining employees of high caliber to be the most important incentives for disclosing HC information externally". In practice, human capital was presented through information about the number of employees and their cost (Brennan 2001). Probably the reason for human capital non-disclosure is based on variables from the external environment such as: competitors, and from the internal environment such as strategy and Critical accounting theory (Samudhram et al. 2010).

The reason for disclosing information of human capital in the financial reports was analyzed by researchers. Their goal was to identify what influences companies to report information about human capital. Firm size is one of the elements considered to have an impact on Intellectual Capital Disclosure (Branco 2011; Bellora and Guenther 2013; Duff 2018; Ferreira et al. 2012; Fontana and Macagnan. 2013; Garcia-meca et al. 2005; Kateb 2015; Melloni 2015; Taliyang et al. 2011; White 2007), in the sense that large companies are expected to disclose information, either descriptive or qualitative, about intellectual capital. But, from a practical point of view, the objective of this research is to determine if they are willing to provide information in financial reports in small businesses. In Romania, according to the Order of the Public Finance Ministry no. 1802/2014, small enterprises are those enterprises that have less than 50 employees and that have either a net annual turnover not exceeding EUR 8 million or have total assets not exceeding of 4 million euros.

In 2014, a conclusion of the intellectual capital reporting in Romania is that it is at the initial stage (Ienciu et al. 2013). At the same time, the concept of intellectual capital is not as well developed in Romania, compared to more developed countries (Morariu 2014). Accounting in Romania is under the sign of transition that transforms it from a simple tool used to meet the needs of central institutions to a true tool for decision-making (Paladi and Fenies 2016). In Romania, investment in intellectual capital has become urgent, and some stages must be understood not only conceptually but also in terms of mentality (Suciu et al. 2011).

3 Methodology

The objective of the research is to identify the economists and employees perception regarding to inclusion of more detailed information about the company's employees in the explanatory notes of financial statements. The detailed information that should be included are according to Global Reporting Initiative Standards. Thereby, in an indirect way, the level of acceptance of the application of these standards was measured. Only a series of information regarding employees from the Global Reporting Initiative standards were included and measured. To identify whether a new change is considered to be accepted, the quantitative method was used. A questionnaire was applied to respondents, using a Google Forms. This questionnaire was completed in late January and start of February 2021 and it was sent using social media applications. The questions related to the research are following a 3-point scale (offers agree and disagree options along to a neutral option). The answers received were analyzed and statistically interpreted using Data Analysis from Excel. Therewith, it was useful to determine perception regarding to including more details about employees in financial reports in order to be able to analyze a company from the its employees point of view.

The hypothesis, consists in the opening of the Romanian economists regarding to the inclusion in the financial reports of more information about its employees. The information that was taken into account are part of information provided by Global Reporting Initiative Standards.

The research questions are "Which are the Global Reporting Initiative requirements about employees? and "Are Romanian economists agree to report information about employees according to Global Reporting Initiative?".

4 Results and Discussions

Regarding human capital reporting, Global Reporting Initiative has some standards that refers to it. The purpose of these standards is to be used by organizations to report what impacts have on the economy, environment and society. The questions applied support the Global Reporting Standards perspective to analyze a company from employees point a view. 401-Global Reporting Standards (GRI) support that a high rate of employee turnover can indicate levels of uncertainty and dissatisfaction among employees and signal a fundamental change in the structure of an organization's core operations. Also, an uneven pattern of turnover by age or gender can indicate incompatibility or potential inequity in the workplace and turnover results in changes to the human and intellectual capital of the organization which can impact productivity (401-GRI 2016b). Turnover has direct cost implications either in terms of reduced payroll or greater expenses for the employees recruitment (401-GRI 2016b).

The number, age, gender, and region of an organization's new employee hires are important because these values can indicate the company strategy, ability to attract diverse and qualified employees (401-GRI 2016b).

Regarding disclosure of information about parental leave, it is specified in 401-GRI Standard that "many women are discouraged from taking leave and returning to work by employer practices that affect their employment security, remuneration and career path and also many men are not encouraged to take the leave to which they are entitled" (401-GRI 2016b). This type of information supports the company image according to equitable gender choice for maternity and paternity leave (401-GRI Standard 2016b, p. 9).

Employees who participate in training and professional courses support the plan skills acquisition to fulfil strategic targets which influences the organization's human capital and contribute to improving performance (404-2 GRI Standard 2016a, p. 8). Measuring the employee's impact in a company's performance and knowing the value of each employee, the employee performance disclosure helps demonstrate how an organization works to maintain the skills of its employees (404-2 GRI Standard 2016a, p. 9).

Disclosing information about employees' remuneration supports equal opportunity. This type of information supports a company with equal principles applies to recruitment and equal remuneration policies. Also, disclosing this type of information characterizes a company which is retaining qualified employees (405-2 GRI Standard 2016d, p. 7).

A sample of 86 employees from the Romanian economy was used. Of 86 respondents, only 57 work in the financial and accounting departments from public and private institutions. Only this population was considered in order to identify a correct perception.

The respondents from public environments work at the National Agency for Fiscal Administration, Romanian South Region. Answers were obtained from people who are certified by Romanian Body of Expert and Certified Accountants, and also having professional courses. This sample of respondents was chosen to determine a vision on the need to report more information about the employees, in financial reports.

Of 13 questions, 5 were needed to objectively know the respondents. These questions referred to the respondents' age, the experience they have in the field where they work, the positions they occupy in the institution where they work, the institution type and the working field.

For the case study, a sample of respondents aged between 18 and 41 was covered (Fig. 1).

Respondents aged 18–24 are expected to occupy entry-level positions, with operational functions. Respondents aged between 25 and 40, are expected to have more experience and professional training, while respondents over 41 are expected to have occupied managerial positions.

Another necessary question was to know the type of institution where the respondents work. It is important to know the workplace, because employees from the private sector, especially those who provide accounting services, have a wide portfolio of clients and requirements from them. Also, there are those who apply other reporting standards, besides the national reporting standards. At the same time, the perception of those in the public environment is important, because those from National Agency for Fiscal Administration represent a part of the users of the financial reports. The results about the types of institutions can be found in Fig. 2.

Another necessary question was regarding to the positions the respondents occupy within the institutions in which they work. Those who answered that occupy both management and operational positions have two options: either they did not want

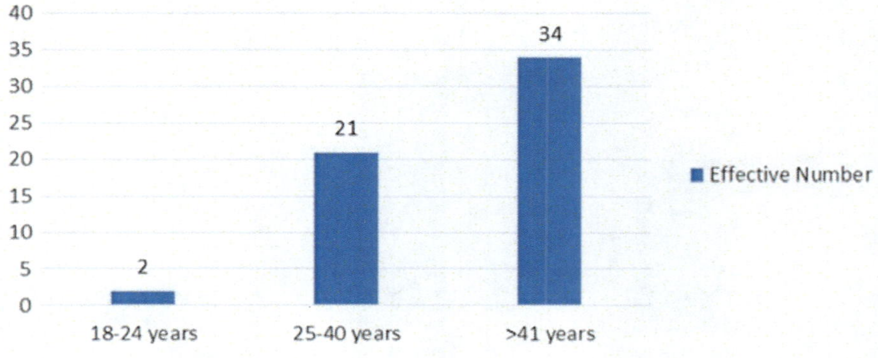

Fig. 1 Age of respondents. *Source* own research

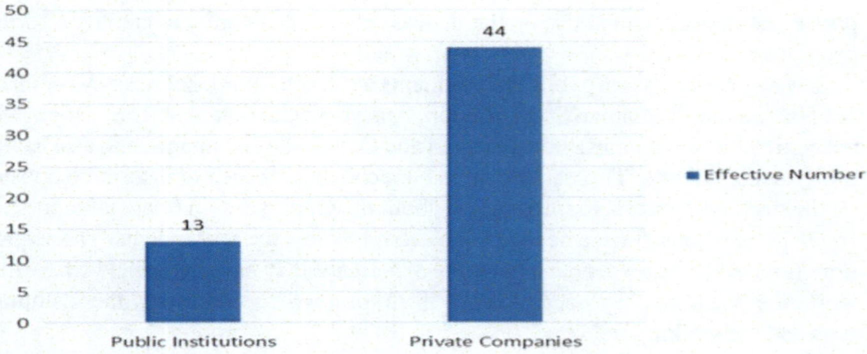

Fig. 2 Type of institutions where respondents work. *Source* own research

their position to be known, or they work in several positions in different companies. Thereby, the results show that most of the respondents included in statistical interpretation occupy operational positions, as seen in Fig. 3.

The other 8 questions used for validating the research hypothesis referred to the inclusion of information about employees in the reports: total number of employees and rate of new employees during the reporting period by age groups, sex, region, type of contract (full-time or part-time) (102-Global Reporting Initiative et al. 2016a); turnover per employee during the reporting period (401-Global Reporting Initiative 2016b); percentage of employees by sex groups, age, educational classification and other criteria relevant to their knowledge (405-Global Reporting Initiative 2016d); benefits granted to all employees, because the quality of benefits for full-time employees is a key factor in retaining employees (401-Global Reporting Initiative 2016b); parental leave (401-Global Reporting Initiative 2016b); the average number of hours and their value per employee regarding the courses and trainings in which

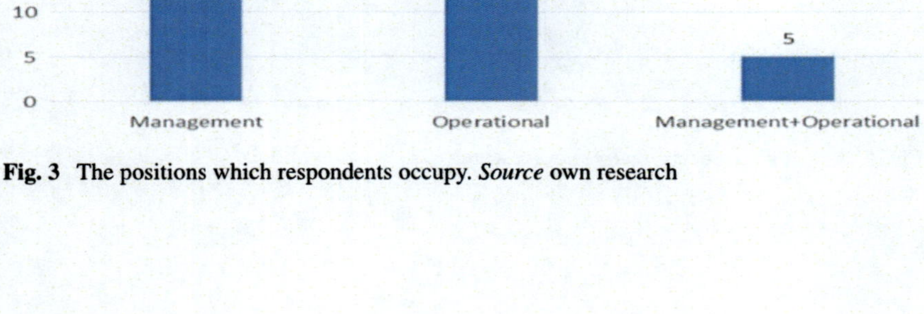

Fig. 3 The positions which respondents occupy. *Source* own research

Table 1 Summary ANOVA

Groups	Count	Sum	Average	Variance
No	8	211	26.375	7.410714286
Neutral	8	93	11.625	20.55357143
Yes	8	152	19	25.14285714

Source Own research using EXCEL-Data Analysis

Table 2 ANOVA

Source of variation	SS	df	MS	F	P-value	F crit
Between groups	870.25	2	435.13	24.58	3.15761E−06	3.466800112
Within groups	371.75	21	17.70			
Total	1242	23				

Source Own research using EXCEL-Data Analysis

the employees participate (404-Global Reporting Initiative 2016c); the remuneration rate for each category of employee (405-Global Reporting Initiative 2016d); the percentage of employees who receive regular performance reviews (404-Global Reporting Initiative 2016c).

An ANOVA-Single Factor was used to analyze the answers (Tables 1 and 2):

H1—Romanian economists agree with the inclusion of information about employees proposed by Global Reporting Initiative Standards in the financial reports of small and medium companies.

A composite score was calculated, totaling the ratings of the 8 questions applied to respondents, which address the openness of the Romanian economists towards disclosing new information according to Global Reporting Initiative Standards. Considering that P-value is 3.16 > 0.05 and F (2, 56) = 24.58, H1 is rejected and H0 is confirmed. There were no significant differences and no statistically significant trend of the difference between the 3 answer variants. The statistical test supports the null hypothesis that Romanian economists are not willing to include more information about their employees according to the requirements of the Global Reporting Initiatives Standards. In this way, the respondents consider that information already presented about the employees is sufficient. The statistical test result is conclusive considering the average age of the respondents, which is predominant, in proportion of 60%, over 41 years. Thereby, this type of population is not psychologically open to new changes, and new responsibilities would lead to "loading the job description" as one of the respondents replied: "What will be imposed in the job description of economists? If we include such information, what other financial bodies would do?" (Answer received from a respondent, over 41 years old, his identity will be kept according to GDPR rules). At the same time, the test result shows that people did not decide. The respondents do not agree with the inclusion of some information about employees provided in Global Reporting Initiative, in small and medium companies in Romania. This result is confirmed also by the fact that many companies in Romania do not have an evaluating model of the employee's impact on the process performed (Nita 2020). In this sense, companies would have nothing to report considering that there is no model of evaluation, of knowing its own employees. However, one good result that was found from the statistic test results is that a large part of the respondents agree with the inclusion of more complex information about employees, according to Global Reporting Initiative Standards. It can be seen that the difference between the answers with yes and no is relatively close. The sample has a rigid perception and is resistant to a possible change, in the sense that they oppose vehemently.

5 Conclusion

If the intellectual capital measurement is not possible at present, its reporting by all companies is a must. Knowing employees and presenting them in financial reports is probably the only way to identify a company's intellectual capacity. It is known that the most complex reports are made by listed companies. Their reports provide detailed information about the company's progress during the reporting period and the most important events. This study case complements the research, determining the perception of Romanian economists regarding to their openness to accepting new requirements from the Global Reporting Initiative Standards. Regardless of size, companies should report additional information about their employees. The contribution they bring to the company is one of major importance, and their knowledge is important to understand the company intellectual potential in the future. One of the elements that this case study presents is the information required by Global Reporting Initiative Standards regarding to employees, and another contribution of this study case is the fact that the perception of Romanian economists about the information disclosure in small and medium companies was identified and measured. One of the important points of this research is that in addition to the large number of respondents who do not agree with the inclusion of employee information in their financial statements, there were respondents who agreed and accepted these requirements from the Global Reporting Initiative Standards. This research is among the first conducted in this regard, in determining the perception of Romanian economists regarding to the requirements of the Global Reporting Initiative on employees. This study case identifies the importance of reporting human capital in the financial reports of all companies. Some major limitations are the relatively small number of respondents and their average age.

References

Abeysekera, I.: Motivations behind human capital disclosure in annual reports. Acc. Forum **32**(1), 16–29 (2008). https://doi.org/10.1016/j.accfor.2006.11.006

Abeysekera, I., Guthrie, J.: Human capital reporting in a developing nation. Br. Acc. Rev. **36**(3), 251–268 (2004). https://doi.org/10.1016/j.bar.2004.03.004

Alvarez Dominguez, A.: The impact of human resource disclosure on corporate image. J. Hum. Resour. Costing Acc. **15**(4), 279–298 (2011). https://doi.org/10.1108/14013381111197225

Beattie, V., Smith, S.J.: Human capital, value creation and disclosure. J. Hum. Resour. Costing Acc. **14**(4), 262–285 (2011). https://doi.org/10.1108/14013381011105957

Bellora, L., Guenther, T.W.: Drivers of innovation capital disclosure in intellectual capital statements: Evidence from Europe. Br. Acc. Rev. **45**(4), 255–270 (2013). https://doi.org/10.1016/j.bar.2013.06.002

Branco, M.C., Delgado, C., Sousa, C., Sá, M.: Intellectual capital disclosure media in Portugal. Corp. Commun. **16**(1), 38–52 (2011). https://doi.org/10.1108/13563281111100962

Brennan, N.: Reporting intellectual capital in annual reports: Evidence from Ireland. Acc. Audit. Acc. J. **14**(4), 423–436 (2001). https://doi.org/10.1108/09513570110403443

Carson, E., Ranzijn, R., Winefield, A., Marsden, H.: Intellectual capital: Mapping employee and work group attributes. J. Intellect. Cap. **5**(3), 443–463 (2004). https://doi.org/10.1108/14691930410550390

Castellanos, A.R., Rodríguez, J.L., Ranguelov, S.Y.: University R&D&T capital: What types of knowledge drive it? J. Intellect. Cap. **5**(3), 478–499 (2004). https://doi.org/10.1108/14691930410550417

Cuozzo, B., Dumay, J., Palmaccio, M., Lombardi, R.: Intellectual capital disclosure: a structured literature review. J. Intellect. Cap. **18**(1), 9–28 (2017). https://doi.org/10.1108/JIC-10-2016-0104

Duff, A.: Intellectual capital disclosure: evidence from UK accounting firms. J. Intellect. Cap. **19**(4), 768–786 (2018). https://doi.org/10.1108/JIC-06-2017-0079

Dumay, J., Guthrie, J.: Involuntary disclosure of intellectual capital: is it relevant? J. Intellect. Cap. **18**(1), 30–45 (2017)

Ferreira, A.L., Branco, M.C., Moreira, J.A.: Factors influencing intellectual capital disclosure by Portuguese companies. Int. J. Acc. Financ. Rep. **2**(2), 278 (2012). https://doi.org/10.5296/ijafr.v2i2.2844

Fontana, F.B., Macagnan, C.B.: Factors explaining the level of voluntary human capital disclosure in the Brazilian capital market. Intangible Capital **9**(1), 305–321 (2013). https://doi.org/10.3926/ic.315

García-meca, E., Parra, I., Larrán, M., Martínez, I.: The explanatory factors of intellectual capital disclosure to financial analysts. Eur. Acc. Rev. **14**(1), 63–94 (2005). https://doi.org/10.1080/0963818042000279713

Global Reporting Initiative: GRI 102: General disclosures 2016a. GRI Standards, GRI (1), 13. www.globalreporting.org (2016a)

Global Reporting Initiative: GRI 401: Employment 2016b. GRI Standards. www.globalreporting.org (2016b)

Global Reporting Initiative: GRI 404: Diversity and Equal Opportunity 2016c. GRI Standards. www.globalreporting.org (2016c)

Global Reporting Initiative: GRI 405: Training and Education 2016d. GRI Standards. www.globalreporting.org (2016d)

Gogan, L.-M., Draghici, A.: A model to evaluate the intellectual capital. Proc. Technol. 9, 867–875 (2013). https://doi.org/10.1016/j.protcy.2013.12.096

Hatamizadeh, N., Ahmadi, M., Vameghi, R., Hosseini, M.A.: Intellectual capital in rehabilitation organizations: Concept clarification. J. Health Res. **34**(3), 195–207 (2020). https://doi.org/10.1108/JHR-04-2019-0077

Ienciu, N.M., Matiş, D.: Intellectual capital disclosure of Romanian listed companies. Res. Acc. Emerg. Econ. **13**, 143–162 (2013). https://doi.org/10.1108/s1479-3563(2013)0000013011

Joia, L.A.: Measuring intangible corporate assets: Linking business strategy with intellectual capital. J. Intellect. Cap. **1**(1), 68–84 (2000). https://doi.org/10.1108/14691930010371636

Kateb, I.: Voluntary human capital disclosure in French annual reports. Int. J. Learn. Intellect. Cap. **12**(4), 323–341 (2015). https://doi.org/10.1504/IJLIC.2015.072196

Klaila, D., Hall, L.: Using intellectual assets as a success strategy. J. Intellect. Cap. **1**(1), 47–53 (2000). https://doi.org/10.1108/14691930010324133

Lardo, A., Dumay, J., Trequattrini, R., Russo, G.: Social media networks as drivers for intellectual capital disclosure: Evidence from professional football clubs. J. Intellect. Cap. **18**(1), 63–80 (2017). https://doi.org/10.1108/JIC-09-2016-0093

Melloni, G.: Intellectual capital disclosure in integrated reporting: An impression management analysis. J. Intellect. Cap. **16**(3), 661–680 (2015). https://doi.org/10.1108/JIC-11-2014-0121

Morariu, C.M.: Intellectual capital performance in the case of Romanian public companies. J. Intellect. Cap. **15**(3), 392–410 (2014)

Mouritsen, J., Lasen, H. T., Bukh, P. N., Johansen, M. R.: Reading an intellectual capital statement: Describing and pre scribing knowledge. J. Intellect. Cap. **2**(4), 359–383 (2001)

Ndou, V., Secundo, G., Dumay, J., Gjevori, E.: Understanding intellectual capital disclosure in online media Big Data: An exploratory case study in a university. Meditari Acc. Res. **26**(3), 499–530 (2018). https://doi.org/10.1108/MEDAR-03-2018-0302

Nikolaj Bukh, P.: The relevance of intellectual capital disclosure: A paradox? Acc. Audit. Acc. J. **16**(1), 49–56 (2003). https://doi.org/10.1108/09513570310464273

Nita, D.N.: The client, in terms of relational capital, an important element to start knowing the human capital value. In: Proceedings of the 11th International Conference of Doctoral Students and Young Researchers, pp. 321–324 (2020)

Paladi, I., Fenies, P.: Performance management in central and eastern European countries: A literature review. Stud. Manag. Financ. Acc. **31**, 215–271 (2016). https://doi.org/10.1108/S1479-351220160000031008

Passetti, E., Cinquini, L.: A comparative analysis of human capital disclosure in annual reports and sustainability reports. In: Value Creation, Reporting, and Signaling for Human Capital and Human Assets: Building the Foundation for a Multi-Disciplinary, Multi-Level Theory, pp. 213–242 (2014). https://doi.org/10.1057/978113747206

Samagaio, A., Rodrigues, R.: Human capital and performance in young audit firms. J. Bus. Res. **69**(11), 5354–5359 (2016). https://doi.org/10.1016/j.jbusres.2016.04.137

Samudhram, A., Sivalingam, G., Shanmugam, B.: Non-disclosure of human capital-based information: theoretical perspectives. J. Hum. Resour. Costing Acc. **14**(2), 106–128 (2010). https://doi.org/10.1108/14013381011062621

Schiemann, F., Richter, K., Günther, T.: The relationship between recognized intangible assets and voluntary intellectual capital disclosure. J. Appl. Acc. Res. **16**(2), 240–264 (2015). https://doi.org/10.1108/JAAR-11-2012-0076

Seetharaman, A., Saravanan, A.S., Low, K.L.T.: Comparative justification on intellectual capital. J. Intellect. Cap. **5**(4), 522–539 (2004). https://doi.org/10.1108/14691930410566997

Suciu, M.-C., Ghitiu-Bratescu, A., Piciorus, L., Imbrisca, C.: Reporting on intellectual capital: value driver in the Romanian knowledge based society. Int. J. Educ. Inf. Technol. **5**(2), 215–224 (2011)

Suciu, M.-C., Piciorus, L., Imbrisca, C.I.: Disclosing intellectual capital in tertiary education: From necessity to reality. In: Proceedings of the 5th European Conference on Intellectual Capital, pp. 419–427 (2013)

Sveiby, K.-E.: A knowledge-based theory of the firm to guide in strategy formulation. J. Intellect. Cap. **2**(4), 344–358 (2001)

Taliyang, S.M., Sultan, U., Abidin, Z., Latif, R.A., Mustafa, N.H., Sultan, U., Abidin, Z.: The Determinants of intellectual capital disclosure among malaysian listed companies. Int. J. Manag. Mark. Res. **4**(3), 25–33 (2011)

The Order of the Romanian Ministry of Public Finance no. 1802/2014. (2014). https://static.anaf.ro/static/10/Anaf/legislatie/OMFP_1802_2014.pdf
White, G., Lee, A., Tower, G.: Drivers of voluntary intellectual capital disclosure in listed biotechnology companies. J. Intellect. Cap. **8**(3), 517–537 (2007). https://doi.org/10.1108/14691930710774894

The Role of Dynamic Capabilities and Digitalization in Achieving Economic Intelligence. The Case of Huawei's Success

Shahrazad Hadad and Iulia Ruxandra Țicău

Abstract The proposed paper draws attention on the topic of Economic Intelligence (EI) to elaborate how dynamic capabilities and digitalization are able to lead to increased competitiveness, effective information management and environmental monitoring. It is essential for companies in emerging economies to constantly absorb external information and apply latest technological advances to keep up with today's competitive environment. In a universe in perpetual change, there is constant need for strong strategies for a firm to evolve. Through the unification of coordinated actions such as research, information storage and circulation and market anticipation, a firm can become economically intelligent. In other words, companies can achieve Economic Intelligence by developing strong dynamic capabilities that imply latest technologies. This paper proposes an analysis of how dynamic capabilities enhanced by digitalization are able to contribute to the development of Economic Intelligence by engaging the case of the international electronics and telecommunications equipment producer Huawei.

Keywords Economic intelligence · Dynamic capabilities · Digitalization · Huawei

1 Introduction

Domestic economy and the business sector have been greatly challenged since the emergence of the concept "economic intelligence". In a market where globalization constantly challenges companies, businesses need coordinated actions to investigate, process and distribute information in order to take strategic economic decisions. The high need to achieve economic advantages have resulted in the development of strong economic intelligence tools, that are able to control the global economic scene. (Spanish Institute for Strategic Studies 2014).

S. Hadad · I. R. Țicău (✉)
Bucharest University of Economic Studies, Piata Romana 6, 010374 Bucharest, Romania
e-mail: ticauiulia16@stud.ase.ro

S. Hadad
e-mail: shahrazad.hadad@fabiz.ase.ro

M. Busu (ed.), *Digital Economy and New Value Creation*, Springer Proceedings in Business and Economics, https://doi.org/10.1007/978-3-031-07265-9_13

Faced with market fluctuations and strong competition, companies have realized that developing unique capabilities may be the key to surviving in the industries (Teece 2007). To absorb information and to adapt to the constantly changing business environment, companies need dynamic capabilities, defined as "the firm's ability to integrate, build, and reconfigure internal and external competencies to address rapidly changing environments" (Teece 2007). The paper however proposes that such capabilities can significantly contribute to the development of a business's Economic Intelligence, as they assume valuable and constant market research and monitorization but also competitive strategy development, through the implementation of new assets and reconfiguration and transformation of companies.

On the level of businesses and organizations, technological development has greatly influenced the way these organize their activity nowadays. We are witnessing a serious shift from the physical way of doing business to the digital one. Everything has now moved online, from collecting customer data, to organizing meetings and signing contracts. Companies need to adapt to new technologies in order to remain efficient and competitive. But these technologies can be incorporated into a company's dynamic capabilities, making these more effective and efficient on the market. This is the reason why paper's approach considers digitalization as a factor that, if embedded in a company's dynamic capabilities, can lead to constructive outputs.

This article, therefore, addresses the contribution of dynamic capabilities upon business's EI, when influenced by latest technology, by observing the strategic activities conducted by an international electronics retailer in China. Based on the resulted findings, we aim to add empirical value to the existing research.

2 Theoretical Background

2.1 *Economic Intelligence (EI)—Definition and Characteristics*

In an economy subject to numerous environmental fluctuations, such as cost pressures, consumer rapid changes in tastes and competition increases, enterprises must constantly analyze their economic environment to anticipate opportunities but also potential threats that may occur (Meddah 2012).

EI can be perceived as a means of providing firms with better competitiveness on the market by focusing on information technologies and communication (Shahraki and Shahraki 2016). The term EI firstly appeared in literature in 1990, when Ruth Stanat emphasizes the need for competitive intelligence and a continuous monitorization of competition on the market (Stant 1990).

EI can be therefore defined as a "set of concepts, methods and tools which unify all the coordinated actions of research, acquisition, treatment, storage and diffusion of information, relevant to the individual or clustered enterprises and organizations in the framework of a strategy" (Menendez et al. 2002).

In fact, EI does not have a single definition but we know that it implies a high degree of competition knowledge characterized by anticipation of future markets, knowledge of other's strategies and superior management of circulation of information within the company (Ursacescu and Cioc 2012).

2.2 *Economic Intelligence (EI) and Dynamic Capabilities (DC)*

Teece firstly introduced the concept of dynamic capabilities in 1997 and later on defined it as a company's "capacities to integrate, build and reconfigure internal and external resources and competences in order to address rapidly changing environments of business." (Teece 2007).

Often being described as unique (Teece et al. 1997), dynamic capabilities can be disaggregated into the following (Fig. 1).

Additionally, definitions of each component, included in dynamic capabilities concept, are outlined in the Table 1.

If we now take a closer look at the characteristics of Economic Intelligence mentioned in the previous subchapter of this paper we may observe that developing DC may be a proper solution to achieving this type of intelligence within

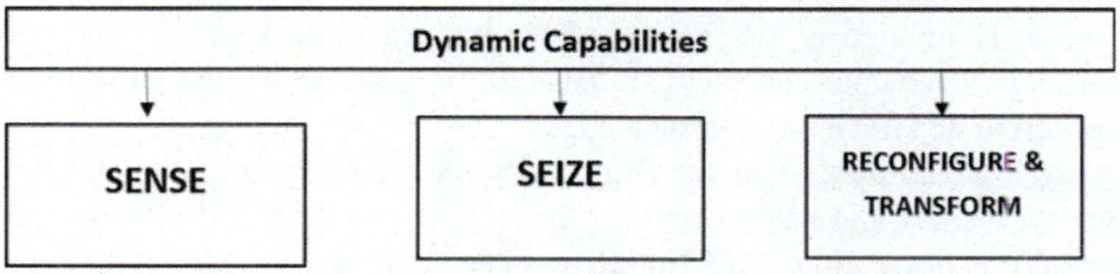

Fig. 1 Dynamic Capabilities components. *Source* Teece et al. (1997)

Table 1 Dynamic capabilities definitions. *Source* Teece (2007)

Dynamic capability	Definition
Sensing and shaping opportunities	"Scanning, creation, learning and interpretive activity" (Teece 2007).
Seizing opportunities	Addressing sensed opportunities through new products, services and processes (Teece 2007)
Reconfiguring and transforming	"Reconfiguration and recombination of assets and organisational structures" (Teece 2007) in line with enterprise growth

the business. DC are able to sense and shapes opportunities and threats by continuous monitorization of the market, to effectively seize the right opportunities and to maintain competitiveness by transforming and reconfiguring the business (Teece 2007). This is a way of forming a sense of intelligence in the firm, which can be defined as a "wisdom arising from the understanding of the basic principles that support knowledge, allowing the creation of scenarios, the modification of strategies and innovation." (Menendez et al. 2002) By developing DC, businesses create a high degree of knowledge within the enterprise, develop and adapt the enterprise to respond to market changes and competitive moves. All these characteristics can significantly contribute to the formation of a sense of EI, a sense which will be further tested in this paper by engaging a relevant example of a successful giant electronics producer.

2.3 Dynamic Capabilities (DC) and Digitalization

In a fast-developing market, the emergence of new digital technologies led to a gradual transition to the digital world, a fact that greatly impacted the main pillars of any business. In today's world, adopting a digital perspective may be a critical choice for successfully growing on the market. The digital age creates new opportunities and future profits, where every player is searching for the use of new technologies to conquer, gain market share and be part of the Intelligent World (Huawei 2021).

This has created new challenges between firms and introduced a need for the majority of companies to digitalize their products, services and processes. Digitalization can be defined as "the process of moving to a digital business" by using digital technologies to change a business model and obtain valuable-producing opportunities and generate new revenue (Gartner 2021).

In 2013, MIT Initiative on the Digital Economy (IDE), which is a visionary team, formed of leaders and researchers exploring how people and business work in time of digital transformation (MIT Initiative on the Digital Economy, n.d.), conducted a survey among 319 large companies located in 31 countries all over the world. The results showed that companies achieving a level of maturity from the point of view of digitalization gain 9% more revenue and become 26% more profitable compared to the rest of competitors in the industry (Huawei 2021).

Organizations feel the pressure to go digital before other competitors do and this forces firms to develop a wide range of capabilities to implement successfully digitalization initiatives (Warner and Wager 2018).

By engaging digitalization in everyday activities, organizations better interact with customers, obtaining new and valuable insights, which contribute to innovating business models (Amit and Zott 2001; Aspara et al. 2013) Digitalization contributes to the creation of a business strategy that is different from traditional forms or changes-it creates an environment that is volatile, complex and ambiguous (Matt et al. 2015; Day and Schoemaker 2016).

Therefore, we consider that companies that seek digitalization are more likely to sense market trends, seize opportunities and threats and transform their way of doing business, while the rest take the risk of remaining behind. By using the lens of dynamic capabilities framework, digitalization stands out by creating increased value and help companies keep up with market growth.

3 Methodology

3.1 Research Methods

The present research was designed to explore the meaning of EI and how it can be enhanced by DC and Digitalization. In order to investigate this matter we have chosen the case of a well-known international electronics and telecommunications equipment producer and seller, that of Huawei Technologies Co. Ltd. Given this research objective, we propose a qualitative analysis, which is aimed at attaining new insights and exploring whether the success of Huawei is connected in any way to developing DC through digitalized structures and Economic Intelligence.

A qualitative research design explores non-numerical data with the purpose of understanding concepts, opinions and experiences. This type of research can be used to explore new insights into a problem but also generate new ideas. (Bhandari 2020) This methods has been considered suitable for our research as it can provide valuable understandings of the connection between DC, Digitalization and EI.

3.2 Data Collection

By engaging two well-known research techniques, those of documentary analysis and secondary data analysis, relevant information was collected and interpreted for our proposed objective.

Documentary Analysis. For realizing in-depth content analysis, the research tool of document analysis was engaged. This type of qualitative research method was realized by „reviewing and evaluating documents" (Bowen 2009) resulting in gaining new meanings, improved understandings and achievement of empirical knowledge (Corbin and Strauss 2008). Document analysis is a systematic evaluation of sources such as advertisements, press releases, agendas, diaries and event programs, under either printed or electronic formats (Bowen 2009). When conducting document analysis, relevant data is selected, appraised and synthesized, so that excerpts, quotations or even entire passages are correctly engaged for researcher's purposes. Documentary analysis has the advantage of being available in the public domain, being accessed with no special permission needed from the author. Additionally, this research methods is 'unobtrusive' and 'non-reactive'. As the researcher does not contribute in any way to the original content, no influence is executed upon the study. Behaviors and events may develop differently if observation is consciously known by the population sample, however, in the case of documentary analysis, this scenario can be completely avoided. (Bowen 2009).

Secondary Data Analysis. Research was also realized through secondary data analysis, a technique based on engaging and reanalyzing data that was previously collected and investigated by other researchers (Punch 2005). It involves academic literature studies, reports, website informal articles. When analyzing secondary data, the researcher is not involved in any way at the moment of data collection (Bryman 2004) and is not responsible for its actual commissioning (Reason and Bradbury 2001). Secondary data can belong to different sources such as statistical data, journals, reference books, technical reports etc. (Chivaka 2018). In our analysis, we have considered academic selected based on their credibility, reliability, sample consistency and relatedness to the topic and research objective. A list of these all these sources can be found in the 'References' section of this paper.

3.3 Operationalization

Operationalization is a process through which a defined concept is split into indicators able to make that concept more measurable. This technique is important in both qualitative and quantitative research, as it makes a concept countable to measure progress or change. (Peters 2021).

As a result of our previously conducted literature review, we have observed that there are various definitions, characteristics and practices of EI. However, from all of these, we have observed that EI is focused on straightening company's competitive advantage by perfecting its strategic plans. All these goals can be resumed as the following phrase "Right information, Right moment, Right place" (Ursacescu and Cioc 2012), which will be further denoted in our analysis as the 3R's of EI. We will additionally consider the three pillars of EI as defined by Shahraki and Shahraki (2016) as guidance in our research paper:

- Monitoring
- Benchmarking
- Knowledge Management

Monitoring consists of "collecting, organizing, analyzing and disseminating collected information." (Shahraki and Shahraki 2016). It is a continuous process that detects changes and signals in a business environment. This phase of EI focuses on managing information collected from the external environment, so evolutions tendencies and competition can be detected. Any organization using this practice can identify opportunities and threats, which can be further used to develop strategies. (Ursacescu and Cioc 2012).

While monitoring targets company's external environment, *benchmarking* focuses on its internal continuity. Being defined as "a level of quality that can be used as a standard when comparing other things" (dictionary.cambridge 2021) benchmarking is developed by identifying performance gaps and transposing observed best practices into the business (Shahraki and Shahraki 2016). Therefore the practice of benchmarking is usually applied to achieve EI as it allows businesses to appropriate best practices of competitors, to receive recommendations of specialized organizations and to adapt these to their existing problems.

Knowledge Management (*KM*) represents a third component for achieving EI. According to Jakobiak (2004), while EI and Competitive Intelligence use information in an 'offensive' manner to create strategies, KM is aimed at optimizing knowledge within the organization. In fact, KM can be defined as "the way in which knowledge is organized and used within a company, or the study of how to effectively organize and use it" (dictionary.cambridge 2021) so that it targets creating value from intellectual assets. Thanks to KM, employees are able to benefit from a specialized knowledge base but also transfer internal information to external resulting in a collective creation and share of valuable knowledge.

4 Empirical Analysis

4.1 History of the Company

Founded in 1987, Huawei is currently a leading information provider and communications technology (ICT) and smart devices, exceeding 197.000 employees in more than 170 countries (Huawei n.d.). Created in Shenzhen, China, by Ren Zhengfei, company's way to success was not easy and smooth, facing strong competition from the beginning. The company entered the market as an importer of telephone switchers. Competition was fierce, especially in 2001 when the internet boomed and many players merged in order to cope with overcapacities, such as Nokia merging with Siemens, Lucent and Alcatel under Nokia brand. By 2013, Huawei became the biggest supplier of telephone network equipment, today internationally competing with Apple and Samsung (Wu et al. 2020) and being ranked fourth largest smartphone brand in the world (Sudarshan 2020). Huawei was chosen as our analysed company as its evolution is remarkable, gaining market share unexpectedly for the crowded market it competes in.

4.2 *Dynamic Capabilities (DC), Digitalization and Economic Intelligence (EI) at Huawei*

In order to remain competitive over the years, Huawei implemented a well-developed sustainable strategy, that covered different areas of interest. It has to be mentioned that DC in the case of Huawei have been mainly selected based on their digitalization relevance as we consider adoption of digitalization an influencing factor in this relation.

Therefore, we will analyze company's moves towards achieving DC, enhanced by digitalization techniques, rooted in the company's strategic changes over the years, to test their contribution in obtaining EI.

Sensing capabilities at Huawei. Through our research, we found that Huawei showed outstanding capabilities to quickly and effectively sense and respond to increasing market needs. In order to achieve its sustainable plan, Huawei showed superior sensing capabilities by rapidly reacting to changes in the environment, and engaging the following:

Migrating R&D to the Cloud. Being currently a "leader in cloud and storage standardization" (martinroll.com, 2018), Huawei has moved its traditional R&D activities to cloud online platforms. Company's IT department made possible to provide service like test cloud, cloud compilation and developer community by decoupling process, data and tools used for product development. Now these services cover the whole R&D process. The advantage of this move towards digitalization significantly diminishes the time to move software and hardware from R&D to the department of production. For instance, creating the operating system of a device, the process of every model was isolated, leading to a great number of hours to finish a large no of models. With the new cloud-based R&D, a compilation platform accessed by everyone performs the process in a few minutes. (Jingwen 2021).

Huawei Institute of Strategic Research. The research institute founded by Huawei helps the company sense future opportunities of cutting-edge technologies. The purpose of the institute is to identify the pathway of technological future, new types of products and business models and innovative opportunities in the industry.

Huawei Innovation Research Program (*HIRP*). Huawei strongly engaged with academic research communities through several research partnerships and by providing constant support for technological related research. As a result, more than 400 universities and research institutes and 900 businesses from 30 countries around the globe received funding from HIRP. In Europe, Huawei spent around US$100 million for such funding (huawei.com n.d.).

AA100Stakeholder Engagement Standards. In order to gain market insights, Huawei categorizes each stakeholder differently, according to a specialized digital system following the AA100 Stakeholder Engagement Standard, using eight groups: business customers, end customers, industry, governments, industry peers, suppliers, NGOs and employees. The company then establishes different relationships and communication channels according to the needs of each group, resulting in a clear understanding of their environmental and social concerns. Next, the company uses a matrix to evaluate the impact of these concerns over the strategy of the firm, by following six principles: influence, responsibility, proximity, dependency, policy, strategic intent and representation (Huawei 2010). This is a way through which Huawei is always aware of the changes in its environment and is able to identify significant concern and prioritize their requirements, in other words sensing external factors. (Wu et al. 2012).

Evaluating sensing capabilities in relation to EI. All these sensing capabilities result in a better understanding of the market's trajectories, increased innovation awareness, increased information knowledge and creation a stronger business ecosystem. Huawei was awarded by the Economist with the 'Corporate Use of Innovation Award' for its ability to develop innovative environments and continuously react to market trends (The Indian Express Ltd. 2010). Sensing capabilities at Huawei, by embedding latest technology, results in a high degree of monitorization of the market but also a better knowledge management levels. Huawei migrated its information to the Cloud and engaged with strategic research program to increase its KM and have better control over the industry. By setting up the Huawei Innovation Research Program (HIRP), company's ability to benchmark is increased, as HIRP is able to explore best practices and developments into the industry and put these as a priority to be transposed into the business. Additionally, by applying the AA100Stakeholder Engagement Standards, the company not only enhances KM, but also monitoring, as it strategically groups stakeholders differently to have better control and evaluate their impact upon the business's future. Taking into consideration the 3R's of EI as defined by Ursacescu and Cioc (2012), those of Right information, Right moment, Right place", we can observe that sensing capabilities contributes to achieving the first category, that of Right information. Sensing consists in continuously detecting valuable trends on the market, valuable for the company's development, therefore selecting the 'right' information for the growth of the business.

Seizing capabilities at Huawei. Considering its notable success in the industry, Huawei seemed to have understood what seizing opportunities and threat mean in a dynamic and complex environment it operates in. By investing in digital assets such as tools and processes, products and services and latest technologies, Huawei showed capabilities to respond to opportunities and threats, resulting in meeting consumer demands and gaining market share.

Unification of global IT resource management. Huawei implemented a cross-cloud organization of resources. Due to this operation, research centers of Huawei all around the globe can commonly create a product as IT departments preliminary developed resource scheduling that cover machines and cloud platforms. Therefore, there is no need to worry about where machines and services will come from (Jingwen 2021).

Effective digital sales. Sales represent a company's department that requires a lot of coordination and discipline in order to work efficiently, otherwise, it can easily end up in chaos and misunderstanding. Before adopting digital sales, Huawei and its sales managers used to have problems in obtaining a clear view of activities in the frontlines. As a result of the introduction of digital sales, the electronics giant coordinates online and offline management and better organizes operations of sales teams. Huawei's sales managers can now observe real-time sales activities of more than 200 branches in around 170 different countries (Jingwen 2021) Therefore, the adoption of digitalization within the sales chain at Huawei significantly increased capabilities and performance and organization of sales teams.

Open Cloud strategy. Huawei realized that IT development is a big opportunity for the future. They tried to open up platforms and standardize computing, storage and network resources. Seizing this opportunities, Huawei developed OpenStack, a mainstream cloud based on open sources OS, that encourages IT vendors to create a shared ecosystem. It is a methods of converging multiple virtualization pools and clouds. Huawei claims that open sources brings several benefits for company, industry and even customer as: software can be shared among industry, there is a sense of open source community created that improves features, and product time-to-market (TTM) is accelerated, fact which can diminish the price of purchase for customers. Therefore committed to be a promoter of the open cloud platforms, Huawei invested in such digital assets and seized the opportunity of the open-source community that is growing more and more everyday under an evolutionary trend. (Huawei.com 2015).

Investments in R&D activities. Around 10% out of its annual revenues and 44% of its base employees have been invested in R&D activities (marketing91.com).

Changes in business model. Huawei started its activity as a B2B company. However, in 2009, the company launched its first smartphone and changed its business model. The main reason behind this decision has its origins in 2003, when Huawei received an order of one hundred thousand custom phones from Hong Kong. However, even though being qualitatively excellent and 3G incorporated, the phones were not selling well, regardless the efforts made by the team. At that point, Huawei sensed the importance of branding in order to reach reputation and seized it by establishing a strong connection with customers. They rebranded from being a positioning focused company, which was about constantly developing its selling points, to a reputation oriented organisation. (Cheung 2018). What is more, after years of practice, Huawei articulated a value-driven model for efficient digitalisation, as far as the electronics giant realized that this is the future. Huawei supports actual business use and scenarios, and the use of these to create value for clients by continuous development. Huawei constantly hosts digitalisation oriented conferences. At the most recent online the "Industrial Digital Transformation Conference" organized on the 24 March 2021, Mr Peng Zhongyang, president of Enterprise BG, Board Member, emphasized three aspects to be followed for every company's to adapt their business models to digital transformation:

1. Business: Customer-centric model;
2. Technology: Seizing 2 key factors: convergence of technologies and Cloud;
3. Ecosystem: Symbiotic digital ecosystem formation- by exploring scenarios, capacity building and a cooperation model (Huawei Enterprise 2021).

By driving its business model focus towards digitalisation, Huawei tries to adapt to latest technologies and change perspectives. This change in the business model represents a way of seizing the increasing importance of shifting to a digitalized way of doing business. Kindström et al. (2013) and Chesbrough and Schwartz (2007) have emphasized that cooperation, co-development and partnering with other stakeholders may be valuable movements towards supporting the business to seize new opportunities. Huawei articulated its value proposition towards sustainable social models. The company works in joint innovation with partners in the finance sector for social purposes. NCBA Bank Kenya is the biggest commercial bank in East Africa, which now benefits by a new digital core system provided by Huawei. This new system implemented financial services for more than 18 mil. users, not only in Kenya but also in neighbouring countries. The initiative boosted region's real economy and promoted social development, by "solving current problems as well as grasp new opportunities" (Huawei Enterprise 2021).

Investments in employee communication channels. Huawei sensed that for its employees to remain committed they need to always stay connected with each other, aspect which in the end can build loyalty and improve teamwork. This opportunity was seized by improving communication within the electronics company with the help of digitalisation. This made it easier to communicate between hierarchical layers by adopting open communication channels such as president mailbox, bulletin board system (BBS) and formal complaint procedure. (Wu et al. 2012) Additionally, Huawei's IT department designed an efficient platform for employees to connect not only between devices, but also to transfer knowledge within teams. In 2017, the platform "WeLink" was launched, which enables users to have meetings, benefit of service applications, share files and in the end share ideas and knowledge with their team, no matter where they are located. The transition to a digitalized way of keeping in touch is now used by around 170.000 employees, an impressive number which shows its efficiency of transferring knowledge. (Jingwen 2021).

Evaluating seizing capabilities in relation to EI. It can be observed that Huawei succeeded in seizing opportunities and obtain success on the market. But doesn't this show us that the company has a high level of Economic Intelligence? The company invested in new digital assets, technologies and processes which significantly contributed to a better knowledge management. By adopting digitalized sales, all the knowledge between sales departments was improved, as information circulates now more easily and efficiently. By adopting an open cloud strategy, information is better organized, contributing to better KM but also monitoring. By changing and expanding its business model, Huawei gained a higher level of EI as new business models are able to reveal innovative ways of organizing knowledge leading to developments in KM. Contributions to KM are made also by investing in communication channels through the platform "WeLink" and through the adoption of open communication channels between hierarchical layers. Considering the definition of EI as stated by Shahraki and Shahraki (2016) that "EI is a means of providing firms with better competitiveness on the market by focusing on information technologies and communication." we can observe that seizing opportunities at Huawei has finally contributed to achieving this objective. Seizing phase seems to find its correspondent in the defining phrase of EI stated by Ursacescu and Cioc (2012) that of "Right information, Right moment, Right place". Seizing stands for the "Right moment" as it engages meeting an opportunity at the moment it arises on the market and in this way gaining success.

Reconfiguring and transforming at Huawei. Reconfiguration and transforming actions have been witnessed in the case of Huawei, as outlined in the following sections.

From centralization to decentralization. From 1993 until 2003, Huawei adopted a centralized way of governing, with the power attributed mainly around the founder and the CEO of that time. However, this pattern was changed in 2003, when the company started to gradually become international and new routines were considered necessary. With the help of a consulting firm, Huawei shared the decision making power between a management team formed out of 8 members, who worked closely with the CEO and were responsible for strategic making decisions. An important aspect is that the chair of the member committee is functioning in rotations, changing every six months. (Wu et al. 2020).

CSR management committee- increases decentralized structure. Huawei created a CSR (Corporate Social Responsibility) management committee that, together with the investment audit committee assures continuous asset orchestration for social responsibility purposes. The CSR management department plays an essential role in decentralizing the power within the company, by acting as a transfer hub for information and knowledge sharing between business units, marketing units and functional units (Wu et al. 2012), as follows (Fig. 2).

Knowledge Management. Frankie Lai, a knowledge manager at Huawei Shenzhen, considered PowerPoint talks as being outdated and changed the traditional way of organising conferences and sharing knowledge. Huawei yearly gathers directors and manager in its headquarters office. In 2016, Frankie ran a Knowledge Market, which redesigned the deployment of conferences as follows (Fig. 3).

Each owner of good practices has a 90 s speech to present key aspects of how value can be added into the company. Next, the knowledge is exchanged in a new venue, where people become 'knowledge hunters' get engaged in conversations and ask critical questions. In the knowledge review phase, people are distributed into smaller meeting rooms and reflect upon what can be used for further action and improvements in the market (Lai 2016). This is a highly efficient way of transferring essential knowledge between employees from different locations and making sure that essential data can be filtered by each individual. We can however notice that, in this case Huawei lowered the importance of digitalisation in its company by focusing on the human factor. The company considers that human to human interaction can create more value when it comes to knowledge transfer. This is an interesting and original change to make, as in today's society, we can easily get lost in a fast-developing digitalized world. Additionally, in order to become international, starting with 2000, Huawei imported know-how from the Western firms but also from the Eastern ones by copying their organisational and structural knowledge. They even developed a program to send students to the US and train in ultimate technology so China could keep up with the outside world. (Wu et al. 2020).

Open innovation-oriented. Digitalization is able to reshape the way industry ecosystems are designed, as industries are now overlapping each other. Therefore, an organization may be unable to compete on its own on the market and it may need to join ecosystems to survive and grow. In a rapidly changing consumer

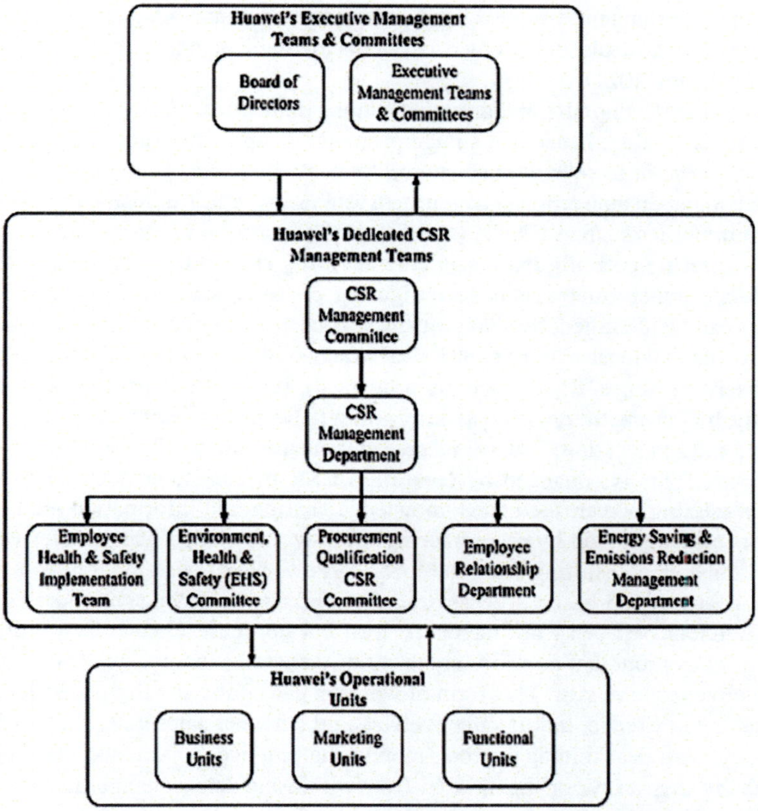

Fig. 2 CSR organisationnel structure at Huawei. *Source* Wu et al. (2012)

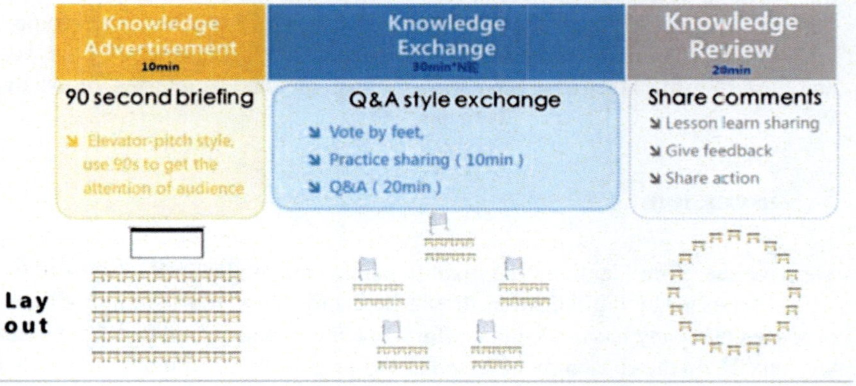

Fig. 3 Huawei knowledge market as implemented by Lai (2016)

demand environment, business need to be more open and adaptable. They need to be part of sustainable and strong ecosystems to establish long-term competitiveness. (Huawei.com, 2021).

To promote digital transformation among industries, Huawei reconfigured its way of innovating. It became a huge promoter of open innovation, being a significant contributor to several open source foundations and communities. They aim at increasing open innovation and open ecosystems in the ICT industry, by contributing to communities such as CNCF, OpenStack, Akraino And Linaro. (Huawei n.d.)

Evaluating reconfiguring and transforming capabilities in relation to EI. In Huawei's transformation and reconfiguring phase significant changes and movements can be identified, by changing the way business is done. It can additionally be noticed that all these actions contributed to achieving a consistent degree of EI, more precisely creating a 'Right place' for growing the business (Ursacesu and Cioc 2012). Firstly, by sharing the decision making power between 8 members, with the chairman functioning in rotation, Huawei increased its management of knowledge within the company. This is enhanced by creating a CSR management department that, by decentralizing power, facilitates knowledge sharing and information management. KM is also improved by reconfiguring the way knowledge was transferred within the company. By shifting from the 'boring' PowerPoint presentations to a human–human interaction based process, transferring knowledge became more consistent and efficient between team members. Last but not least, Huawei is a promoter of Open Innovation, that develops connections, generates innovative ideas and creates disruptive technologies. This form of working can contribute to the development of valuable knowledge, that is effectively shared between enterprises, building strong KM, but also determining a strong monitoring practices for Huawei, as it helps the company keep an eye on the industry from within and detect changes in perspectives. Benchmarking was found to be developed by the way Huawei achieved know-how at the beginning of its business journey. The company carefully dedicated time to transforming observed practices of successful players on the market into own practices, as it engaged a "level of quality that can be used as a standard when comparing other things" (dictionary.cambridge 2021) when in 2000, got inspired by the Western and Eastern firms, borrowing their way of organizing and structuring knowledge.

5 Discussion

We have therefore identified the main dynamic capabilities in the case of Huawei when enhanced by digitalization. It can be clearly observed that, along time, our proposed company has made great efforts to continuously adapt to the environment and survive on the constantly evolving electronics market. From migrating to Cloud resources to changing the way it organizes its stakeholders, Huawei proved to have constantly sensed the external market. If we look at the practices of achieving EI we can observe that these actions helped the company achieve EI, by contributing to improved monitoring and better knowledge management, but also benchmarking, as it allows companies to analyze competitor's performance and compare and determine performance gaps.

Next, Huawei seems to have successfully seized opportunities and threats discovered on its relevant market. The company registered tremendous efforts to invest in digital assets, such as digital technologies, digital processes and digital communication between employees and seized the opportunities and threats through digitalized technologies. This helped the company provide better services and optimize capabilities of the future. Helfat and Peteraf (2015) claim that incremental changes of the business model are necessary for a company to effectively seize opportunities. Huawei validates previous affirmations by constantly developing and innovating its business model, from B2B to B2C, to a value-oriented model, to a social-oriented model. All these investments help the company increase its knowledge management and monitorization abilities and intrinsically enhance its EI.

From Huawei's transformation and reconfiguration phase we mention the transition from a centralized to a decentralized structure, knowledge management technique changes and adoption of open innovation. All these contribute to increasing knowledge management, a change in the way knowledge is organized and transferred within the company. Information is optimized and open for everyone to access. Also, by choosing to achieve know-how from outside the company, by observing other player's good practices, the benchmarking indicator is enhanced.

The summarizing Table 2 shows DC enhanced by Digitalization identified at Huawei and their contribution to the EI practices. For a more clear understanding, the table also includes the 3R's of EI practices as defined by Ursacescu and Cioc (2012) "Right information, Right moment and Right place."

The paper, therefore, introduces a new approach towards EI, seen through the use of DC and digitalization, aimed at enriching the research and obtaining a deeper and more critical analysis of the influence between the three. The diagram illustrated below indicates the new approach of the topic and represents the result of the empirical analysis conducted in the research paper (Fig. 4).

6 Limitations of the Paper

Even though the study is aimed at adding value to current research, by bringing empirical evidence in the EI field, there are a series of limitations that need to be specified.

The first research tool engaged by the paper, that of documentary analysis, may provide insufficient details regarding company's products, as the documents may have been initially created for a different purpose (Bowen 2009). The producer of electronics and telecommunication equipment may not have provided complete characteristics and details of the devices in their advertisements, these may have sometimes been created to advertise the product, and too many details may have been unnecessary. Additionally, according to Yin (1994), an incomplete selection of the right documentation may lead to 'bias selectivity'. Contradicting Bowen's

Table 2 Summary of research results. *Source* Author's own contribution

Dynamic capability		EI contributions
Sense		
• Migrating R&D to the cloud • Huawei Institute of Strategic Research • Huawei Innovation Research Program (HIRP) • AA100Stakeholder Engagement Standards	→	Monitoring Knowledge Management Benchmarking ***Right Information***
Seize		
• Unification of global IT resource management • Effective digital sales • Open Cloud strategy • Changes in business model • Investments in employee communication channels	→	Monitoring Knowledge Management ***Right Moment***
Transform		
• From centralization to decentralization • CSR management committee • Knowledge Management (knowledge transfer, achieving know-how) • Open innovation oriented	→	Monitoring Knowledge Management Benchmarking ***Right Place***

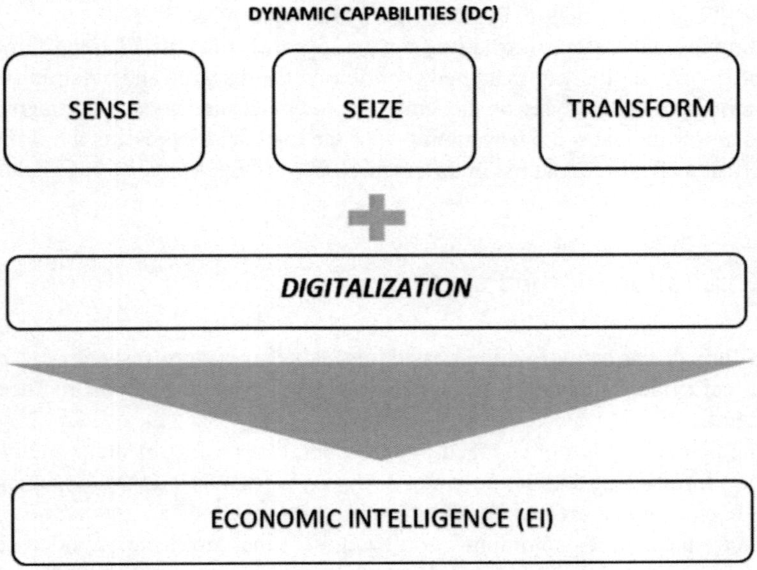

Fig. 4 Visual representation of research results. *Source* Author's own contribution

(2009) view of a positive impact of ‘unobtrusive’ and ‘non-reactive’ characteristic of documentary analysis, Miller and Alvarado (2005) argue that there is a sense of distance created between the researcher and the original analyzed situation, resulting in possible misinterpretations and incorrect perspectives of the setting presented in the document.

Apart from its numerous benefits, secondary data analysis brings a series of critiques as well. Just like in the case of documentary analysis, the researcher is considered to have no control over the data analysed (Bickman and Rog 1998). The amount of data may become overwhelming (Stewart 1984), hard to analyse and not relevant to the research topic (Wolberg 1997). In our case, the amount of secondary data available has been vast and diverse, so that the process of selecting reliable, correct and relevant information from such a large bunch of sources was a tough process with easy to occur errors. Attention to detail and in-depth research was considered essential for a correct understanding of the sources.

However, the combination of the two research tools was aimed at constructing a novel research approach upon the topic, able to capture extensive information through academic articles, informal websites and official advertisements and campaigns. By engaging two well-known research methods in a correct manner, the paper tried overcoming limitations and bring substantial value in the domain of DC and Digitalization and their impact upon the process of achieving EI.

7 Conclusions

Considering all of the above, we can state that there is a degree of EI established once an entity possesses dynamic capabilities and digitalization. By creating competitive advantages, by anticipating future markets and by enhancing knowledge inside and outside the company, Economic Intelligence plays an essential role in any entity’s business journey. In over-saturated markets, with customer tastes constantly changing and unexpected competitive strategies, companies need to be ready to perform the next smart move. However, we noticed that this can be realized by constantly absorbing information from the outside market, by comparing and contrasting, by seizing the right opportunity and acting and transforming, so that competition is surpassed. In other words, companies need dynamic capabilities to survive and compete in a fluctuating environment. Additionally, digitalization is impacting everyday life at an increasing rate. Latest technologies have shifted the majority of businesses to the digital world, changing the way they create, sell and promote products. But doesn’t the digitalization process influence the way companies sense, seize and transform? As a result of our analysis, based on the international electronics and telecommunications equipment producer and seller Huawei, we positively answer our question. Our research adds empirical evidence to current studies by outlining that dynamic capabilities and digitalization contribute to the development of any company’s Economic Intelligence.

References

Amit, R., Zott, C.: Value creation in e-business. Strateg. Manag. J. **22**(6–7), 493–520 (2001)
Aspara, J., Lamberg, J.A., Laukia, A., Tikkaen, H.: 'Corporate business model transformation and inter-organizational cognition: the case of Nokia. Long Range Plan **46**(6), 459–474 (2013)
Bhandari, P.: What is Qualitative Research? | Methods & Examples. Scribbr. https://www.scribbr.com/methodology/qualitative-research/ (2020). Accessed 29 Apr 2021
Bickman, L., Rog, D.J.: Handbook of Applied Social Research Methods. Sage Publishers, London (1998)
Bowen, G.: Document analysis as a qualitative research method. Qual. Res. J. **9**(2), 27–40 (2009)
Bryman, A.: Social Science Research. Oxford Press, Oxford (2004)
Chesbrough, H., Schwartz, K.: Innovating business models with co-development partnerships. Res. Technol. Manag. **50**(1), 55–59 (2007)
Cheung, G.: Building the Huawei brand. https://www.linkedin.com/pulse/building-huawei-brand-glory-cheung/?fbclid=IwAR1dGqNE5wTNvm0yIzI06AbAZUrmoEWNlVwgwe1fhYLKmyLgMJD0N-56pZc (2018). Accessed 15 Apr 2021
Chivaka, R.: Secondary data analysis. https://www.researchgate.net/publication/327060808_secondary_data_analysis (2018). Accessed 7 Aug 2021
Cipriani, J.: 7 things to know about Leica and Huawei's partnership (2018). 7-things-to-know-about-leica-and-huaweis-partnership. Accessed 25 Apr 2021
Corbin, J., Strauss, A.: Basics of Qualitative Research: Techniques and Procedures for Developing Grounded Theory. Sage, Thousand Oaks (2008)
Day, G.S., Schoemaker, P.J.: Adapting to fast-changing markets and technologies. Calif. Manag. Rev. **58**(4), 59–77 (2016)
Dictionary.cambridge: knowledge management. https://dictionary.cambridge.org/dictionary/english/knowledge-management (2021). Accessed 27 Apr 2021
Gartner: Definition of Digitalization—Gartner Information Technology Glossary. https://www.gartner.com/en/information-technology/glossary/digitalization (2021). Accessed 29 Apr 2021
Helfat, C.E., Peteraf, M.A.: Managerial cognitive capabilities and themicrofoundations of dynamic capabilities. Strateg. Manag. J. **36**(6), 831–850 (2015)
Huawei Enterprise: Huawei's Innovative Technology and Scenario-Based Solutions Enable All Industries—Huawei Enterprise. https://e.huawei.com/en/news/ebg/2021/enabling-digital-transformation-in-industries (2021). Accessed 1 May 2021
Huawei Technologies Co., Ltd.: Corporate Social Responsibility Report (2010).
Huawei: Huawei's open cloud strategy. https://www.huawei.com/au/publications/communicate/76/huaweis-open-cloud-strategy (2015). Accessed 25 Apr 2021
Huawei: Core Value: Openness, Collaboration, and Shared Success. https://www.huawei.com/en/about-huawei/corporate-information/openness-collaboration-and-shared-success?fbclid=IwAR0YtfBFhGkGYi5OTJpo6XkKMXYD52O2ODxOtzIqxkxMdD2BgpVEWLjmfN8 (n.d.). Accessed 15 Apr 2021
Huawei: Digitalization: Think, Act, Accomplish. https://www.huawei.com/~/media/CORPORATE/PDF/white%20paper/embrace-digitalization-en-v2.pdf (2021). Accessed 28 Apr 2021
Jakobiak, F.: L'intelligence économique. Les références, Editions d'Organisation (2004)
Jingwen, T.: A Look Inside Huawei's Own Digital Transformation—Huawei Publications. Huawei Enterprise. https://e.huawei.com/en/publications/global/ict_insights/201801221604/digital-huawei/201801241420 (2021). Accessed 1 May 2021
Kindström, D., Kowalkowski, C., Sandberg, E.: Enabling service innovation: a dynamic capabilities approach. J. Bus. Res. **66**(8), 1063–1073 (2013)
Lai, F. (2016). Huawei transforms a global conference with a knowledge market. https://conversational-leadership.net/huawei-conference/ [Accessed 17 Apr 2021]
Matt, C., Hess, T., Benlian, A.: Digital transformation strategies. Business Inf. Syst. Eng. **57**(5), 339–343 (2015)
Meddah, N.: Une démarche d'IE dans mon entreprise. Direccte Lorraine (2012)

Menendez, A., Atanes, E., Alonso, J., et al.: Economic Intelligence. A Guide for Beginners and Practitioners. The CETISME Project (2002). ISBN 84-451-2350-5

Miller, F.A., Alvarado, K.: Incorporating documents into qualitative nursing research. J. Nurs. Sch. **37**(4), 348–353 (2005)

MIT Initiative on the Digital Economy: About Us—MIT Initiative on the Digital Economy. https://ide.mit.edu/about-us/# (n.d.). Accessed 28 Apr 2021

Peters, B.: Qualitative Methods in Monitoring and Evaluation: Concept Formation and Operationalization. American University Online. https://programs.online.american.edu/msme/masters-in-measurement-and-evaluation/resources/concept-formation-and-operationalization (2021). Accessed 1 May 2021

Punch, K.F.: Introduction to Social Research-Quantitative and Qualitative Approaches, 2nd edn. Sage Publications Ltd., London (2005)

Reason, P., Bradbury, H.: Handbook of Action Research. Sage Publications, London (2001)

Shahraki, A.M., Shahraki, E.: Economic intelligence and innovation: case study. Asian Soc. Sci. **12**(12), 220–228 (2016)

Spanish Institute for Strategic Studies. Economic Intelligence in a global world. Strategic Dossier 162 B. Spanish Official Publications Catalogue. Ministry of Defence Publisher (2014)

Stanat, R.: The Intelligent Corporation. American Management Corporation (1990)

Stewart, D.W.: Secondary Research, Information Sources and Methods. Sage Publishers, Beverly Hills California (1984)

Sudarshan: Xiaomi surpassed Huawei to become the World's Third Largest smartphone brand in Feb 2020: Strategy Analytics. https://www.gizmochina.com/2020/03/20/xiaomi-surpassed-huawei-to-become-the-worlds-third-largest-smartphone-brand-in-feb-2020-strategy-analytics/ (2020). Accessed 21 Apr 2021

Teece, D.J., Pisano, G., Shuen, A.: Dynamic capabilities and strategic management. Strateg. Manag. J. **18**(7), 509–533 (1997)

Teece, D.J.: Explicating Dynamic Capabilities' nature and microfoundations of (sustainable) enterprise performance. Strateg. Manag. J. **28**, 1319–1350 (2007)

The Indian Express Ltd.: The Economist's innovation awards recognize Huawei. The Financial Express. Retrieved from: http://www.financialexpress.com/news/the-economists-innovation-awards-recognisehuawei/701193 (2010)

Urascescu, M., Cioc, M.: The economic intelligence practices and their impact on the organization's strategic behavior. Rev. Int. Comp. Manag. **13**(2), 210–223 (2012)

Warner, K.S.R., Wager, M.: Building dynamic capabilities for digital transformation: an ongoing process of strategic renewal. Long Range Plan Int. J. Strateg. Manag. **52**(2019), 326–349 (2018)

Wolberg: Data Analysis Using the Methods of Least Squares: Extracting the most from Experiments. Springer Publisher, New York (1997)

Wu, Q., He, Q., Duan, Y., O'Regan, N.: Implementing dynamic capabilities for corporate strategic change towards sustainability. Strateg. Change J. **21**(5–6), 231–247 (2012)

Wu, X., Murmann, J., Huang, C., Guo, B.: The Transformation of Huawei: From Humble Beginnings to Global Leadership. Cambridge University Press, United Kingdom (2020)

Yin, R.K.: Case Study Research Design and Methods: Applied Social Research and Methods Series, 2nd edn. Sage Publications Inc., Thousand Oaks, CA (1994)

Triadic Models: On the Triad Technology-Efficiency-Culture at the Organization Level

Cezar Scarlat and Corina Anca Stanescu-Agarici

Abstract So different in content, yet triadic models share interesting features as well as deeper investigation potential. The relationships between triad's elements are not only conceptual but operational as well. Three examples are given, as the fourth case is further analyzed in depth. Objective. The paper aims to reveal the research potential of triadic models by examining the triad Technology-Efficiency-Culture at the organization level. The analysis potential of triadic models is emphasized, by systematic investigation of the two-by-two inter-relationships, in a triple S holistic approach. This paper focus is on the last triad case which is both a novel approach and new integrative analysis. Trying to bridge the literature gap, the authors propose a discussion exploring the possible inter-links between the three elements of the triad. Trying to bridge the literature gap, the authors propose a discussion exploring the possible inter-links between the three elements of the triad. The results of this endeavor are twofold encouraging: (i) by underlying the investigation potential of triadic models, in general, and (in particular) the potential of the triadic model technology-efficiency-organizational culture; and (ii) by implications for business managers, in general, and (in this particular case), for company managers while making strategic decisions, considering criteria linked to different business sectors and resources. The study of the triad of technology—efficiency—organizational culture is PhD research in progress that has its inherent limitations. The most serious one is the integrated approach of the triad's elements—which is also the main further research path.

Keywords Triadic models · Triple S holistic approach · Organizational culture

1 Introduction

"Triad" is a word attributed to many different sets of three objects—somehow related—from several areas of knowledge, by both academics and practitioners. Closer to modern business administration, Etzkowitz (2003, 2008) has used the *triple*

C. Scarlat · C. A. Stanescu-Agarici (✉)
Business Engineering & Management, Doctoral School of Entrepreneurship, University "Politehnica" of Bucharest, Bucharest, Romania
e-mail: Corina_agarici@hotmail.com

M. Busu (ed.), *Digital Economy and New Value Creation*, Springer Proceedings in Business and Economics, https://doi.org/10.1007/978-3-031-07265-9_14

helix model to illustrate interlinks identified between university, industry and government. However, more systematic use of *triad*-based models as investigation as well as explanatory tool in business management studies is less common.[1]

1.1 Triadic Models: Triad-Based Models

Triadic models are not simple sets of three elements. According to Scarlat (2017a), *triad-type models* enjoy the following features:

- They should be *systemic*—that means the elements of the triad are parts of the same system;
- They should be *synthetic*—that means the composing elements reveal essential features as well their inter-links;
- They should be *synergic*—that means triad as a whole displays more features and higher value than the sum of features of the three isolated elements.

By this *triple S holistic approach* (Scarlat 2017a) triadic models prove to be useful analytic tools for both academics and practitioners. The triple S holistic approach implies more that individual study of each element of the triad; it includes mostly two-by-two finer analysis as well as study of the triad as a whole—aiming at better understanding of the concept in its entirety.

To note that the three-element set qualifies as a triadic model as long as it satisfactory serves the purpose (to describe the entire system and its behavior) and complies with the triple S approach. In this respect, it is challenging to discuss to what extent the sustainability three-element model (Passet 1979) is or is not a triadic-type model (Scarlat 2017b).

Besides providing three examples of triadic models applicable in the business management studies (stakeholders' triad; project management triad; and technopreneurship triad), this paper aims to launch a discussion on another triad-based conceptual model, to be studied at the organization level: the triad {*Technology; Efficiency; Organizational Culture*}. The study is conducted by systematic investigation of the two-by-two inter-relationships, in a triple S holistic approach (Synthetic, Systemic, and Synergic).

[1] According to *Oxford Advanced Learner's Dictionary*—(https://oxfordlearnersdictionaries.com/definition/english/triad) the noun triad defines a group of three related people or things (as the classic triad of the visual arts: sculpture, architecture and painting). However, *Oxford Dictionary of Business* (Oxford University Press, Second Edition, 1996) does not mention this word.

1.2 *Three Examples of Triadic Models*

Example 1: The Stakeholders' Triad

The stakeholders' triad {business owners; employees; clients} is an illustrative example of triadic model at the company level (profit-based organization).

Besides two-by-two tensions (business owners versus employees over salary level; business owners versus clients over price level; company's employees versus clients over product quality), this triadic model reveals the common interest of business stakeholders. Regardless their narrow interests, the common goal is the success of the firm itself—without which not a single group would be happy. The same basic three-piece structure is used to define the concept of total quality as described by Kelada (1996). Further exploited, the total quality triad explains the concept of total quality management (Kelada 1996, pp. 55–57).

When a fourth element (cultural environment) is added to the model (Scerri and James 2010; James et al. 2015) then the model obviously falls in a different category.

Example 2: The Project Management Triad

A good example of triadic model, satisfying the triple S condition, is well-known by the project management professionals: the triangle-shaped graphical model of project management, emphasizing the key-problem that project managers must solve: compromising the project cost versus its duration—yet reaching the project objective (PMI 2008), in qualitative and quantitative terms. If the project's defining elements as a triad {project cost; project duration; project objective} are clearly part of the system "project" and they synthetically can describe the project, none of them is able to drive the project to its success as the intangible but essential project management.

Example 3: The Technopreneurship Triad and Technowledgepreneurship Triad

Built on Schumpeter's assertion of "entrepreneurship as innovation" (Schumpeter 2013), scholars have deepened and studied the inner relationship between the two related concepts (Acs and Audretsch 2005; Veeraraghavan 2009; Braguinsky et al. 2009; Śledzik 2013). Drucker (1985) also reckons the organic link between innovation and entrepreneurship. Lynn and Lynn (1992) have discussed the concept of "Innopreneurship"—aiming to underline the importance of innovation in entrepreneurial activities.

Technological innovativeness was studied by Antončič et al. (2007) while models needed to analyse innovative technologies (Scarlat 2005) or innovation potential of the firms (Scarlat et al. 2010) were extensively developed even in emerging economies as Romania.

The mutual influence and interdependence between technology and entrepreneurship were studied by many scholars as Cuero Acosta et al. (2014) and Alderete (2014)—so that technology-based entrepreneurship (or technology entrepreneurship) started to be known as Techno-entrepreneurship.

As scholars agree on the intimate interdependence between entrepreneurship, innovation, and technology, corresponding two-by-two studies are carried out: entrepreneurship and innovation; innovation and technology; technology and entrepreneurship.

Following this trend, the virtues of the triple S holistic approach were exploited by Scarlat (2014)—emerging the conceptual model of the *technopreneurship triad*: {entrepreneurship; innovation; technology}.

Nowadays an important part of new businesses are knowledge-based. Accepting that *technowledge* is an appropriate term for technology knowledge, then an even newer triadic model of *technowledgepreneurship triad* was launched (Scarlat 2014): {technowledgepreneurship; technowledge-based innovation; technowledge}.

2 The Triad of: Technology—Efficiency—Organizational Culture

The technology development brings under scrutiny other conceptual triads at the organization level: {Technology; Efficiency; Organizational culture} and—as shown before—{Technowledge; Efficiency; Organizational culture}. The analysis is focused on the first triad only, analyzing its elements in two-by-two sets (with examples from banking and financial services industry).

2.1 Relationship Technology—Efficiency

As technology continues to grow and information technology (IT) becomes a vital survival factor within industries, top managers need to understand and manage IT as a competitive weapon. Researchers as Porter (1980) have contributed to this area by providing several frameworks. None of these frameworks, however, have made real progress towards realizing the potential that may exist for using information technology as competitive or strategic business weapons. Across industries, the digital generation shows a high rate of adoption of online banking services, 27% reporting that they never visit their branch in person, the following 27% only visiting it once a month or less (Rolfe 2015). They are more inclined to reach out to newcomer fintech firms or alternative financial services such as the ones being developed by the GAFAM (Google, Apple, Facebook, Amazon, Microsoft).

This new trend reshapes the previous relationship that banks used to share with their clients. Not only do they have to provide quality online services to this generation of clients that will favour digital products more and more, calling for strong applications with robust security and 24/7 customer service that is typically achieved through the use of chatbots; but this new model is also deconstructing the knowledge that banks have of their clients. The area of the clients preferring to be connected to a human agent is over and the remaining human interaction is weakened. The banking institutions now have different ways to know their customers.

IT can impact an organization by affecting specific competitive forces (e.g. the buyers, suppliers, substitutes, new entrants, or rivals). For example, IT can help a firm deal with its powerful buyers by providing resources to develop buyer information systems, IT can even potentially change the basis for competition by creating an entry barrier, or dismantling an existing barrier, or creating switching costs.

The organization-level variables that are potentially affected by IT include new entrants, entry barriers, buyers and consumers, competitive rivalry or competition, suppliers, search and switching costs, intra-organization efficiency, and inter-organization efficiency.

Specific fields of Artificial Intelligence such as Machine Learning or "deep learning" enable the computer to learn relevant features from the information they are fed by themselves, and to extrapolate meaning from it directly, thus improving the cost-efficiency of this bias-sensitive activity traditionally performed by humans. Two striking examples of the use of this new technology could be the loan or mortgage risk evaluation, or the stock market evolution's prediction. The lending, insurance and financial advising sectors have proved Artificial Intelligence to be extremely promising in the banking industry.

But to be able to manage, aggregate and analyse this massive amount of data in a fast-changing environment, and to offer constantly improving 24/7 services to their customers whose needs and expectations are evolving as the digitalization of our society deepens, banking institutions cannot rely only human workers anymore and the automation of many dimensions of the banking process isn't just a competitive advantage, but has become a necessity to survive in this ever-changing environment.

How and how much the business intelligent systems increase the organizational efficiency is also an area of further research.

2.2 Relationship Efficiency—Organizational Culture

Organizational culture is empirically related to organizational effectiveness and efficiency, while conducive and solid organizational culture motivates employees to achieve organizational success.

Organizational effectiveness and efficiency are not direct related with organizational culture and at the same time the people through organizations are looking forward—especially in the last years, to be guided and accompanied, within regular 1–1 call with their direct managers to achieve their own goals and I the same time the company goals and targets.

A lot of efforts were made to describe and build the organizational culture and its implications and the impact on people, department, and organizational level from many perspectives—especially from performance point of view. Most of these studies have proposed that the culture of organizations is closely linked to many intermediary and outcome variables, causing or having consequences for the performance of organizations.

Organizational culture seems to be indirectly linked to organizational effectiveness and efficiency. The accurate company culture can be described as a predictor of the organizational evolution, effectiveness, and efficiency.

2.3 *Relationship Technology—Organizational Culture*

Leischnig et al. (2016) have shown that in many industries, organizations are redesigning processes and even business models entirely to transform big data and its applications into strategic advantages.

Bughin et al. (2015) revealed a strong positive link between digital performance and a risk-taking culture, that the primary asset in helping organizations attract and retain digital talent is culture, and that a lack of leadership and talent are the greatest challenges companies face in meeting their digital priorities—in the same time, Organizational culture put the spot on the challenges and potentials of digital transformation.

The financial service industry could not provide the level of service it does without the support of advanced information processing and telecommunication technologies. The numbers of checks (over 37 billion annually), credit card drafts (over 3.5 billion annually), and securities trades (over 30 billion shares traded annually) would swamp any manual system that tried to handle them. In a report published in December 2020 by S&P Global Market Intelligence, a survey conducted among financial organizations highlighted that 30% of the digital leaders considered that the increased automation of manual processed that had been set up to answer their Covid-19 remote working policies were to become permanent changes in their institutions (Eagle 2020).

Based on UXDA (2021)—around 400,000 jobs are available in banks to specifically address the digitalization of financial services in positions such as programmers, designers, banking customer experience whose goal is to enhance the online banking experience. Banks are recruiting massively in these specific fields: a total of 1.5 million vacancies related to the banking industry and financial services have been posted on LinkedIn Job Search at the beginning of 2021. But among these 1.5 million, only 16,000 for instance are full-time bank tellers' positions, unrelated to the world of digital banking. One of the important industry where it can be find out the impact of Technology on the competition and on the diversification and gamification is in financial service. The reliance on technology comes from the enablement, as a result of its use, to provide services and process tasks which would not otherwise be provided.

Beyond the results of secondary research and current observations, this research continues to focus on the sector of financial services, aiming at identifying more specific features of this sector by in depth studies—to demonstrate the triple S character of this triadic model, and eventually reveal its uniqueness.

3 Conclusions

Opening a discussion on triad-type models is worthy—as the above examples demonstrate a solid research potential, bringing up new notions (as *technopreneurship, technowledgepreneurship, triple S holistic approach*) or deepening the study of relationships between existing concepts.

The examples of this paper were picked from different areas of business studies (theory of firm, project management, and entrepreneurship) but the triple S approach is applicable for triadic models at large.

The automation of the process massively entered the industry's practices as a value-bringer to many core businesses not only in banking but in many fields such as pharmaceuticals, retail, telecom, real estate and for a large spectrum of activities ranging from employee payroll to report generation including invoice processing, data migration or inventory management for instance. Since then, it has become vital for business operations and a necessity to remain competitive, allowing the companies to delegate mundane rule-based tasks to appropriate software's and freeing the human resources to focus on higher-value work and customer services.

The study of the triad {technology; efficiency; organizational culture}, as research in progress, has its inherent limitations. The most serious one is the integrated approach of the triad's elements—which is also the main further research path.

Trying to bridge the literature gap, the authors propose a discussion exploring the possible inter-links between the three elements of the triad. The results of this endeavor are twofold encouraging: (i) by underlying the investigation potential of the triadic models, in general, and triadic model technology-efficiency-organizational culture, in particular; and (ii) by implications for business managers, in general, and particularly for the top managers—in designing their strategies and making decisions considering criteria from different business sectors and resources.

References

Acs, Z.J., Audretsch, D.B.: Entrepreneurship and Innovation. Discussion Papers on Entrepreneurship, Growth and Public Policy. Max Plank Institute, Jena, Germany (2005)

Alderete, M.V.: ICT incidence on the entrepreneurial activity at country level. Int. J. Entrep. Small Bus. **21**(2), 183–201 (2014)

Antončič, B., Prodan, I., Hisrich, R.D., Scarlat, C.: Technological innovativeness and firm performance in Slovenia and Romania. Post-Commun. Econ. **19**(3), 281–298 (2007)

Braguinsky, S., Klepper, S. Ohyama, A.: Schumpeterian entrepreneurship. DRUID Summer Conference, Copenhagen Business School. SSRN Electronic Journal, 2009/02/19. https://doi.org/10.2139/ssrn.1347063. https://www.researchgate.net/publication/228316652_Schumpeterian_Entrepreneurship/citation/download. Accessed 07 Apr 2021

Bughin, J., Holley, A., Mellbye, A.: Cracking the Digital Code: McKinsey Global Survey Results. McKinsey & Company (2015)

Cuero Acosta, Y.A., Nabi, N.U., Dornberger, U.: Entrepreneurial orientation and its impact on the improvement of technological capability in Colombia. Int. J. Entrep. Small Bus. **21**(2), 231–245 (2014)

Drucker, P.: Innovation and Entrepreneurship. Heinemann (1985)

Eagle, L.: COVID-19 survey Response Illustrates Opportunity Provided by Digital Transformation. S&P Global (2020)

Etzkowitz, H.: Innovation in innovation: the triple helix of university–industry–government relations. Soc. Sci. Inf. **42**(3), 293–338 (2003)

Etzkowitz, H.: The Triple Helix: University–Industry–Government Innovation in Action. Routledge, London (2008)

James, P., Magee, L., Scerri, A., Steger, M.: Urban Sustainability in Theory and Practice: Circles of sustainability. Earthscan, from Routledge. Taylor & Francis Group, London and New York (2015)

Kelada, J.N.: Integrating Reengineering with Total Quality. ASQC Quality Press, Milwaukee, Wisconsin (1996)

Leischnig, A., Woelfl, S., Ivens, B.: When does digital business strategy matter to market performance? Thirty Seventh International Conference on Information Systems, Dublin (2016). https://aisel.aisnet.org/icis2016/ISStrategy/Presentations/13/. Accessed 07 Apr. 2021

Lynn, G.S., Lynn, N.M.: Innopreneurship: Turning Bright Ideas into Breakthrough Business for Your Company. McGraw-Hill School Education Group (1992)

Passet, R.: L'Économique et le vivant. Paris: Payot. Deuxième édition: 1996. Economica (1979)

PMI: A Guide to the Project Management Body of Knowledge (PMBOK® Guide), Fourth Edition. Project Management Institute, Inc., USA (2008)

Porter, M.E.: Competitive Strategy: Techniques for Analyzing Industries and Competitors. Free Press, New York (1980)

Rolfe, A.: Youth attitudes to banking and online banking services (2015). https://www.paymentscardsandmobile.com/youth-attitudes-to-banking-and-online-banking-services/. Accessed 07 Apr. 2021

Scarlat, C.: Innovative model for innovative technologies analysis. Nonconventional Technol. Rev. **2**, 75–82 (2005)

Scarlat, C.: Technopreneurship: An emerging concept. FAIMA Bus. Manag. J. **2**(3), 5–13 (2014)

Scarlat, C.: Triadic models: Triple S holistic approach for inter-relational analysis in business management, entrepreneurship and marketing. Res. J. Soc. Sci. **10**(1), 1–5 (2017a)

Scarlat, C.: Sustainable business development by sustainable technopreneurship. In: Proceedings of the International Conference "Information Society and Sustainable Development". ISSD 2017b, IVth Edition, April 28–29, 2017b, Targu-Jiu, Romania (2017b)

Scarlat, C., Alexe, C., Scarlat, E.I.: Assessing the firm's innovation knowledge potential. Manag. Prod. Eng. Rev. **2**(4), 57–65 (2010)

Scerri, A., James, P.: Accounting for sustainability: combining qualitative and quantitative research in developing 'indicators' of sustainability. Int. J. Soc. Res. Methodol. **13**(1), 41–53 (2010)

Schumpeter, J.A.: Entrepreneurship as innovation. In: Swedberg, R. (Ed.) Entrepreneurship: The Social Science View. Oxford Management Readers, pp. 51–75 (2013)

Śledzik, K.: Trends in Theory and Practice Schumpeter's View on Innovation and Entrepreneurship. Zilina: University of Zilina & Institute of Management by University of Zilina, Slovakia (2013)

UXDA: Banking Industry Cuts Millions of Jobs Forced by AI and Digitization in Banking. UX Design Agency—UXDA. Retrieved 7 March (2021)

Veeraraghavan, V.: Entrepreneurship and Innovation. Asia-Pacific Business Review, January–March 2009, pp. 14–20 (2009)

Multi-Processing Data Analysis for the Residential Load Flexibility in Smart Cities

Simona-Vasilica Oprea, Gabriela (ENE) Dobrița, Anca-Alexandra Ducman, and Cătălin Ceaparu

Abstract Residential consumption is gaining an increasing focus regarding assessing their flexibility and load control. Flexibility potential of smart cities becomes significant as the uncertainties of the power system operation are increasing with the integration of a large volume of Renewable Energy Sources (RES), numerous charging stations of Electric Vehicles (EV), and unexpected climate change. Thus, the load flexibility of residential buildings may contribute to the balance of the power system, flatten the daily load curves, and decrease the energy acquisition costs. The usage of load flexibility also depends on the availability of Demand Response (DR) programs, aggregation of flexible resources, and requirements regarding the implementation of DR programs. In this paper, to extract valuable insights, we propose a multi-processing data analytics framework identifying clusters of residential consumers and the flexibility of the load targeting various DR programs. For this purpose, we will use large and various datasets and multi-processing data facilities. In addition, other varieties of datasets could be correlated: such as ISO affiliation, DR programs, cost of the infrastructure requirements to implement DR, assess and value the flexibility. However, flexibility transactions and DR specific program implementation are complex aspects that also involve new legislative rules and pose policy challenges that will initiate business models and strategies.

Keywords Multi-processing · Flexibility · Controllable · Market value · DR programs · Clustering

1 Introduction

Residential consumption became more promising in terms of offering a flexibility perspective, and they are subject to various DR programs to enhance the stable and

S.-V. Oprea (✉) · G. (ENE) Dobrița · A.-A. Ducman · C. Ceaparu
Bucharest University of Economic Studies, Romana Square 6, 010374 Bucharest, Romania
e-mail: simona.oprea@csie.ase.ro

A.-A. Ducman
e-mail: ducmananca18@stud.ase.ro

M. Busu (ed.), *Digital Economy and New Value Creation*, Springer Proceedings in Business and Economics, https://doi.org/10.1007/978-3-031-07265-9_15

secure operation of the power systems and integrate a larger volume of RES. Most of the DR programs target the flexibility of consumers or prosumers in offering energy to the grid or in load curtailing possibilities when the grid has no reserves to feed a sudden increase in consumption. They mainly rely on controllable appliances and generating sources of prosumers providing programs for shifting, shedding, or shimmying of the consumption at certain hours. Tariffs are playing an important role as they represent a signal to influence the consumers' behavior. In this sense, numerous studies have been performed (Oprea et al. 2019). However, open input data regarding the consumption of residential buildings by categories is not available in many countries, including Romania, limiting the possibility to perform studies and analyses.

OpenEI is an energy-free data source related to renewable energy and efficiency. Its goal is to facilitate novel business models, create innovative frameworks, and inspire new insights into electricity and gas consumption (Office of Energy Efficiency and Renewable Energy (EERE) n.d.). Department of Energy in U.S. developed programs such as Building America program that researches the implementation and adoption of advanced energy technologies in buildings. Such programs promote the energy efficiency of buildings and an engineering approach in the design of new buildings and retrofits. Moreover, they speed up the adoption of high-performance standards for energy systems of buildings, enhancing indoor air quality, comfort, and sustainable development.

The profile-datasets for residential and commercial buildings in the U.S. represent a multi-year statistical consistent reference point by location, designed for energy efficiency scientific research targets for new existing buildings and retrofits (National Renewable Energy Laboratory 2010).

273 residential consumers with hourly consumption data with 8760 records each that total around 2.4 million of records are stored in.csv files with 14 columns as in Table 1. They consist of both electricity and gas consumption data (total and breakdown of main appliances). However, we will study only the electricity flexibility potential since there are different principles, approaches and assumptions regarding gas flexibility potential. Furthermore, the Power-to-Gas (P2G) concept (Quarton and Samsatli 2018) is out of the scope of this paper, as the technology has to gain more maturity.

As mentioned above, the consumption is split into electricity and gas in total and details: heating, cooling, HVAC, interior and exterior lights, interior and other appliances. In addition, gas provides heating for the house and water. However, not every electric appliance could be a reliable controllable resource as a limitation of its capabilities or consumers' comfort constraints. Thus, we selected three of them (electricity based) that can fall into this category: HVAC, water heater and heating.

Table 1 Consumption breakdown for residential sector

No.	Columns in the dataset
1.	Date/Time
2.	Electricity:Facility
3.	Gas:Facility
4.	Heating:Electricity
5.	Heating:Gas
6.	Cooling:Electricity
7.	HVACFan:Fans:Electricity
8.	Electricity:HVAC
9.	Fans:Electricity
10.	General:InteriorLights:Electricity
11.	General:ExteriorLights:Electricity
12.	Appl:InteriorEquipment:Electricity
13.	Misc:InteriorEquipment:Electricity
14.	Water Heater:WaterSystems:Gas

2 Literature Review

Not so long ago, buildings of residential consumers were not considered as controllable resources to provide advanced DR services. Strategies for DR and generation control to improve the flexibility of buildings in smart grid are studied in Chen et al. (2018). The flexibility is extended to a variety of sources such as RES, programmable appliances, buildings thermal efficiency (Brahman et al. 2015; Hurtado et al. 2017) and consumers' behavior. The concept relies on a synergistic approach designed to enhance and assess the flexibility potential of commercial and residential buildings.

However, their potential has been enhanced by ICT progress and sensor technology (Bloess et al. 2018) fostering the implementation of smart city concepts. Additionally, the flexibility necessity from the demand side is increasing with the higher volume of RES that has to be integrated into the power system. In addition, consumers adapt to the advancement of the technology and opportunities of the DR programs (Balvedi et al. 2018; Patteeuw et al. 2016).

Several studies were performed to model and underline the flexibility potential of residential consumers (Finck et al. 2019; Gottwalt et al. 2017), considering thermal storage of the boilers (Dréau and Heiselberg 2016) or heat pumps (Bloess et al. 2018; Fischer et al. 2017).

Flexibility is required for secure operation of the power system and integration of RES. Therefore, in Denmark, newly designed residential buildings must comply with strict regulations regarding low-energy buildings from the thermal storage capacity perspective (Foteinaki et al. 2018). The study assessed the potential curtailed energy from buildings considering the thermal comfort of the inhabitants. The main findings state that the aggregated flexibility of buildings is significant and there is a strong correlation between flexibility potential and factors such as weather and benefits.

The effect of ToU (Time-of-Use) tariffs on DR flexibility of residential consumers is studied by Pallonetto et al. (2016). A case study is performed for a building located in Ireland using an EnergyPlus model simulator (Attia and Carlucci 2015) to evaluate the effectiveness of DR strategies in relation to various ToU tariffs. Implementation of DR strategies led to a significant reduction in generation costs and carbon emissions. Furthermore, the different ToU tariffs effect is analyzed in Oprea et al. (2018), underlining the benefits for consumers and peak reduction depending on the optimization strategy and implemented tariff (Oprea and Bara 2020).

Two methods, LEED (USA) and ITACA (Italy) are compared in Asdrubali et al. (2015) to assess buildings environmental assessment. The case study is performed on buildings located in Italy using five areas such as site, materials, water, indoor environmental quality, and energy. The results show common grounds for the two methods even if they give different importance to the considered areas.

Other studies focus on pilot projects in Italy (Corrado and Ballarini 2016) analyzing the refurbishment trends of residential buildings located in Piedmont area, Lombardy region (Salvalai et al. 2017) and the impact of a DSM strategy in a Mediterranean residential building (Salvalai et al. 2017), or Catania (Italy) considering the energy simulation of buildings impact on the microclimate (Evola et al. 2020).

Such projects and studies are usually carried out in developed countries with strict regulations and available data for testing and validation. Limited studies were performed on building efficiency in Romania (Adam et al. 2019; Muntean et al. 2017). They mainly discuss the thermal efficiency and possible solutions for the existing residential buildings that were built during the communist regime. Some other studies were performed by Romanian researchers, but the input data for buildings was not available. Hence the data was again taken from open sources such as OpenEI.

3 Methodology

The proposed methodology consists in four steps as represented in Fig. 1. In the first step, the source files in .csv format are read in Python and different useful information are extracted from the file names such as: state, location, and consumer ID.

During this pre-processing step, the recording date is validated, and the input string is formatted using a date-time format. In the second step, the pre-processed data is loaded into a MongoDB collection stored as json documents. To reduce the time and allocated resources, in the third step, we used multiprocessing to work with several processes in parallel to load MongoDB collection into multiple cursors and retrieve the records into Dask dataframes. In the last step, we applied k-means clustering method to group consumers into several profiles.

The files were read in Pandas dataframe, columns were renamed, recording_date (former Date/Time) was converted into a date type column using a generic year (such as 2009), new columns such as state and location were created using the .csv

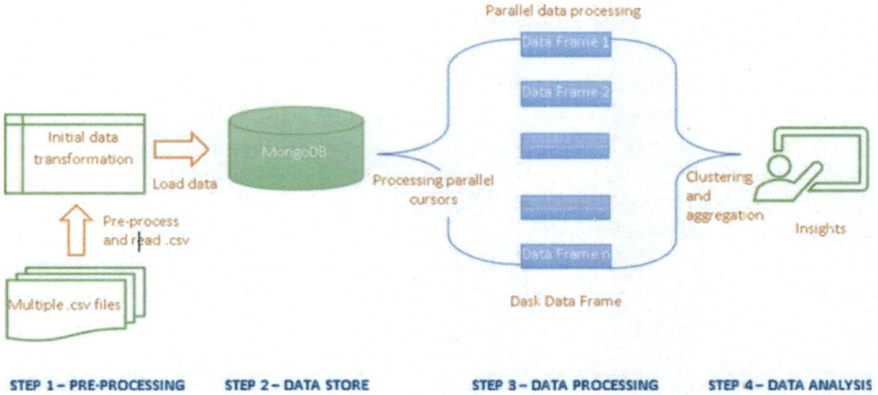

Fig. 1 Data analysis methodology

file name and finally the data was inserted into MongoDB collection. For the data processing step, we used multiprocessing for retrieving the MongoDB collection and for data transformation and preparation. Thus, we start several processes in parallel that extract data chunks from MongoDB using cursors and then load it into Dask dataframe. These dataframes contain multiple Pandas dataframes that are processed in parallel, thus reducing the execution time by half. In this step, we prepared the data for clustering and analytics by transforming the categorical variables into numeric variables, impute missing values and standardize the variables.

Clustering might be used for solving a wide range of problems: from anomaly detection or outlier detection to recommended systems or as a dimensionality reduction method (Al-Wakeel and Wu 2016; Haben et al. 2016). Knowing the behavior of consumers and determining consumption patterns (An et al. 2018; Gianniou et al. 2018), we can improve and streamline the distribution of DR programs for each category. At the same time, determining the groups of consumers can help us to identify anomalies, which could translate into electricity losses or other possible problems that need to be remedied. One algorithm that scales well to a large number of data points is k-means (Motlagh et al. 2019), an iterative algorithm that divides the dataset into a predefined number of subgroups. The partitions are created considering the data similarity and the distance between the mean of the centroid and each data point, basically the objective function is to minimize the Euclidian distance between these two. Steps followed for determining the profile groups are presented in Fig. 2.

The intra-cluster variance is minimized using the following function:

$$F = \sum_{j=1}^{k} \sum_{i=1}^{n} ||x_i^j - c_j||^2 \tag{1}$$

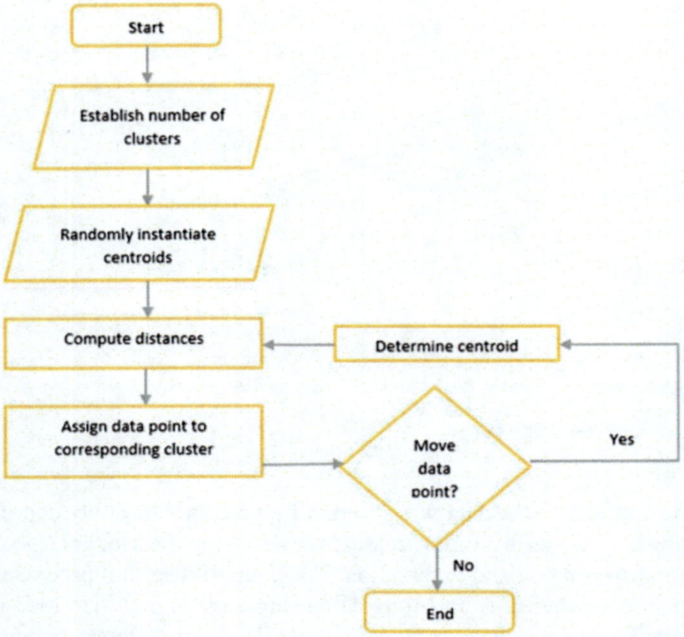

Fig. 2 K-means implementation steps

where: k—Number of clusters; n—Number of observations; x_i^j—Observation i; c_j—Centroid of cluster j.

The well-known drawback of this unsupervised algorithm is that the number of clusters needs to be specified beforehand. The Silhouette Score method is helpful for estimating the best number of clusters. The method returns the measure of how similar a data point is to its belonging cluster. A data point silhouette coefficient is equal to b − a/max(a, b), where a is the mean distance to the other data points in the same cluster, we can refer it as intra-cluster distance, while b is the mean distance to the data points that belong to the nearest cluster, or inter-cluster distance. This coefficient ranges from -1 to 1 and the higher the value, the better the point is placed inside the cluster. If the coefficient is equal to 0, then the point is very close to the cluster boundary. SciKit-Learn's function silhouette_score() is used to compute this coefficient.

4 Results and Discussions

The consumption datasets were uploaded from the .csv files with MongoDB and then read into a dataframe object from pandas' library in Python. The data processing involved the following steps: establishing the same scale of values for the data, keeping only the relevant variables for each type of analysis, normalization and scaling.

The mean hourly total consumption for gas and electricity is shown in Fig. 3. The gas consumption is higher than electricity, house and water heating relying on gas mainly. The gas consumption peak is recorded at 8 a.m., whereas the electricity consumption peak is at 6 p.m.

The breakdown of each category by hour is shown in Fig. 4. The GasHeating component is heavily influencing the gas consumption curve allure, whereas the electricity consumption peak is influenced by HVAC.

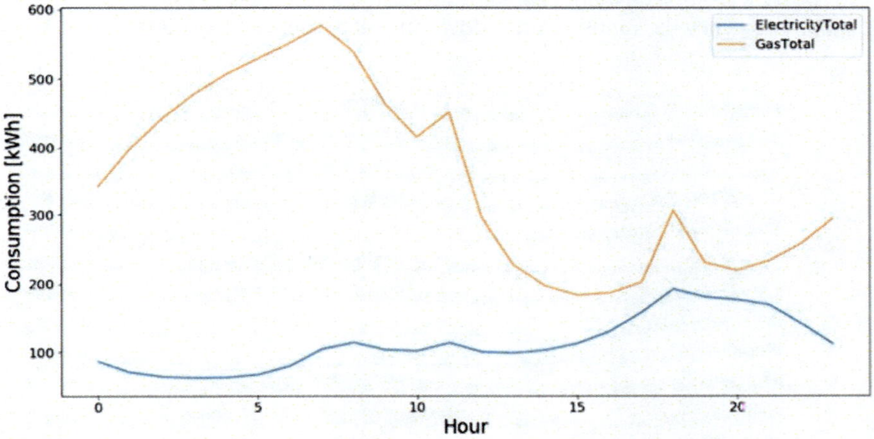

Fig. 3 Hourly total gas and electricity consumption

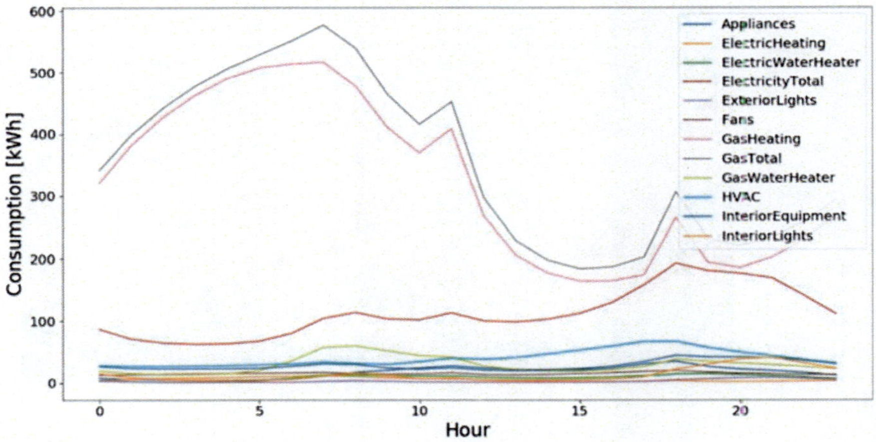

Fig. 4 Hourly total gas and electricity consumption by category

We can determine that the gas peak of consumption is around 8:00 a.m., followed by a sharp decline in GasHeating, which has a large share of the total gas consumption. There is also an increase at a lower rate, around 6:00–7:00 p.m., which coincides with an increase in electricity consumption. As the dataset is a multivariate time series, a line plot was created for each variable to preview the consumption patterns over the entire period for which the data was collected (as in Fig. 5).

As expected, in the summer months, the total consumption of gas was reduced due to a decrease in gas heating consumption. It is interesting to observe that this wasn't the case for the total electricity consumption. While electric heating followed a downward trend, all other electricity appliances were used more often, perhaps because most stayed at home. We also resample the dataframe by month and configure subplots to view the monthly breakdown for each category, as shown in Fig. 6.

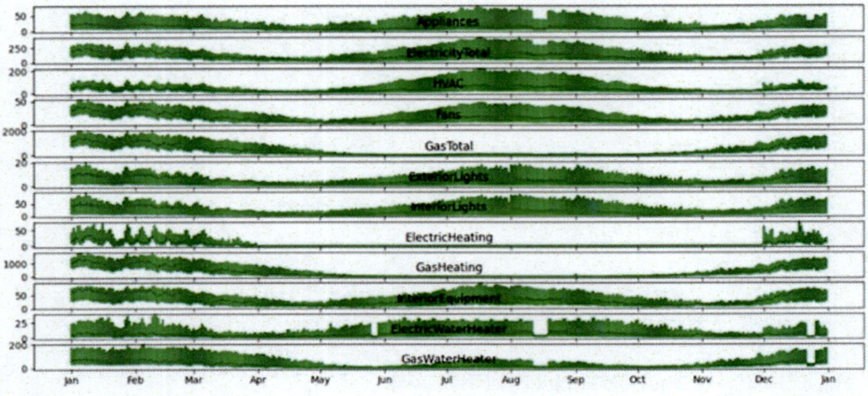

Fig. 5 Feature value representation by month

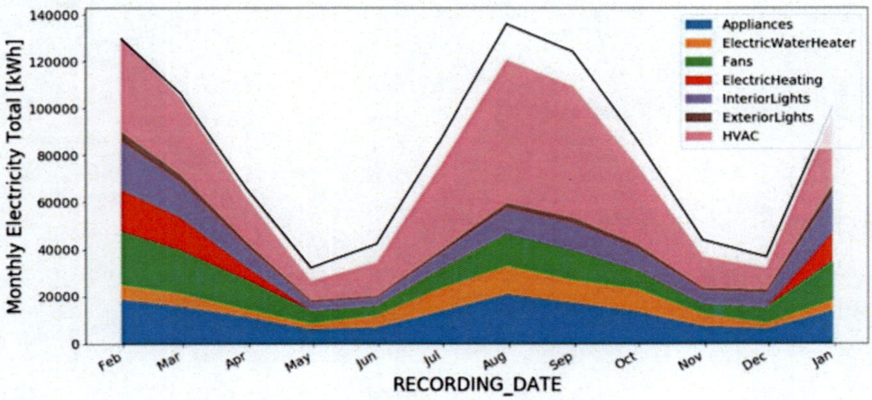

Fig. 6 Electricity total consumption—category breakdown by month

As for gas consumption, the monthly plot is shown in Fig. 7. Gas is used very little in the summer months and heavily in the rest of the year. To reduce the burden of integrating large volumes of RES, programs to convert P2G could be envisioned.

Figure 8 shows the maximum values for each electricity consumer. HVAC, Fans, Interior Equipment and Appliances were the most used, with peaks in January, February, June, July and August. 26% of the total electricity consumption is controllable and could be integrated into DR programs for residential buildings. 63% of the total controllable resources is provided by HVAC, 30% by Water Heater and 7% by Heating.

Residential consumers are allocated to an Independent System Operator (ISO) depending on their state and location. The ISO distribution is represented in Fig. 9.

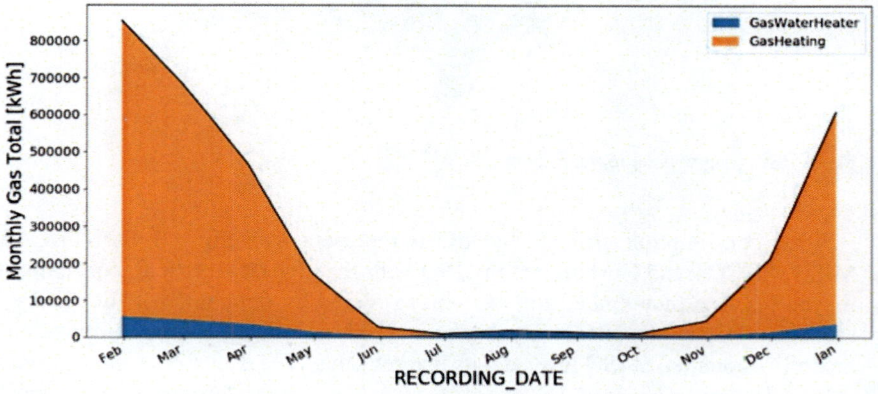

Fig. 7 Gas total consumption—category breakdown by month

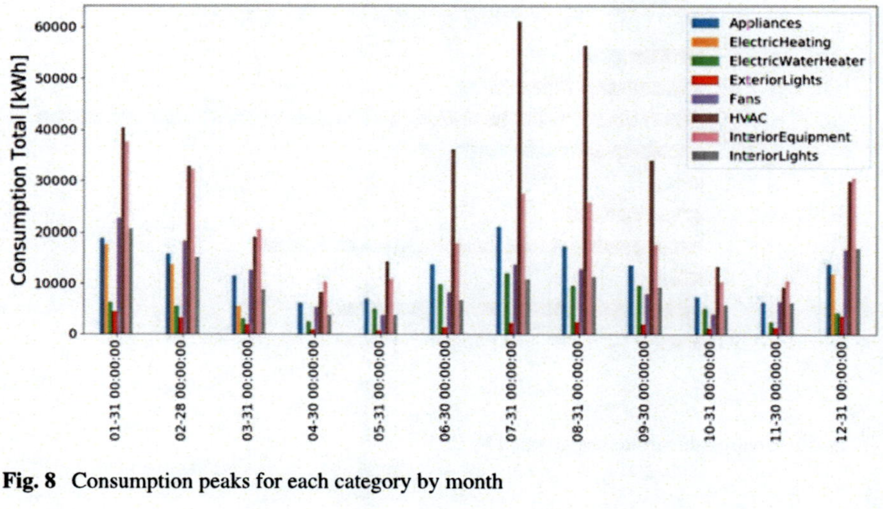

Fig. 8 Consumption peaks for each category by month

Fig. 9 ISO geographical representation

Total consumption grouped by ISO is represented in Fig. 10. SOUTHEAST, MISO, ERCOT and PJM are among ISO with the highest electricity consumption, maybe because they embed the top states ordered by population. A detailed data model converted into a flexibility database could easily facilitate the computation of flexibility potential at ISO level and its market value. The data model should contain at least data regarding load, DR programs, associated costs, tariff rates, geographical representation by location, state, ISO, weather readings to perform forecast, climate

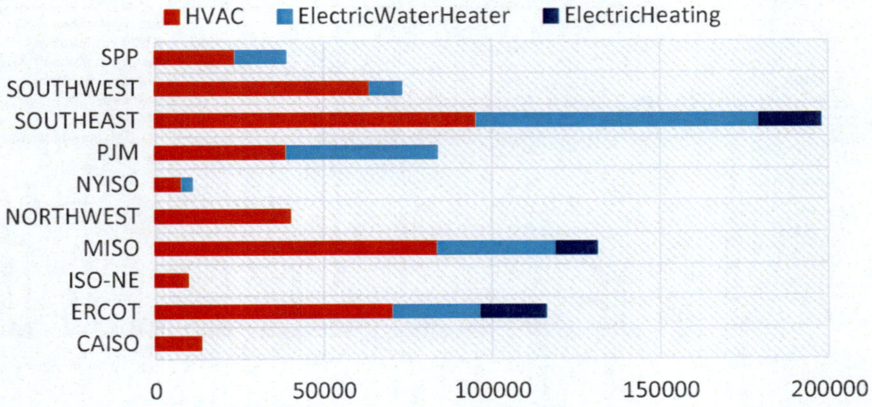

Fig. 10 Controllable resources grouped by ISO

area, etc. The computation could be performed by stored procedures that are called periodically to extract valuable insights from data.

K-means clustering was implemented for determining the electricity consumers' groups, to understand if there are any consumption patterns and responsible groups. Data is normalized using Min–Max Scaler, so each feature is transformed to units from 0 to 1. First, 10 clusters are assumed to determine the highest score for the optimum number of clusters. Hence, the highest score was obtained for two clusters for gas consumption and three for electricity consumption. The obtained scores are the following:

Silhouette Score(n = {2}): {0.7508274719538655}
Silhouette Score(n = {3}): {0.7061200651332357}
Silhouette Score(n = {4}): {0.6884293689805547}
Silhouette Score(n = {5}): {0.6657898081195187}
Silhouette Score(n = {6}): {0.6374764267006435}
Silhouette Score(n = {7}): {0.5962002268939042}
Silhouette Score(n = {8}): {0.5929862418994234}
Silhouette Score(n = {9}): {0.5824524891301696}
Silhouette Score(n = {10}): {0.5890674878457753}

For k = 2, the score was 0.75, which is closest to 1. Figure 11 illustrates the optimum cluster number.

For the electricity consumption, the highest score was obtained with three clusters. Clusters are presented in Fig. 12. K-means algorithm handles numeric data, so the encoding of categorical data is required. Therefore, "State" and "Location" columns were encoded using Label Encoder which belongs to SciKit-Learn library from Python. This will encode labels with values between 0 and the total number of values −1. The fit_transform() method will return the encoded labels after fitting them. To

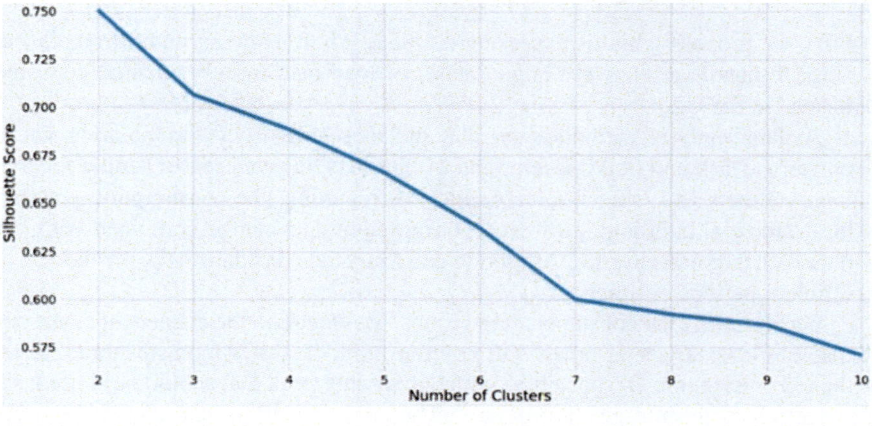

Fig. 11 Silhouette score

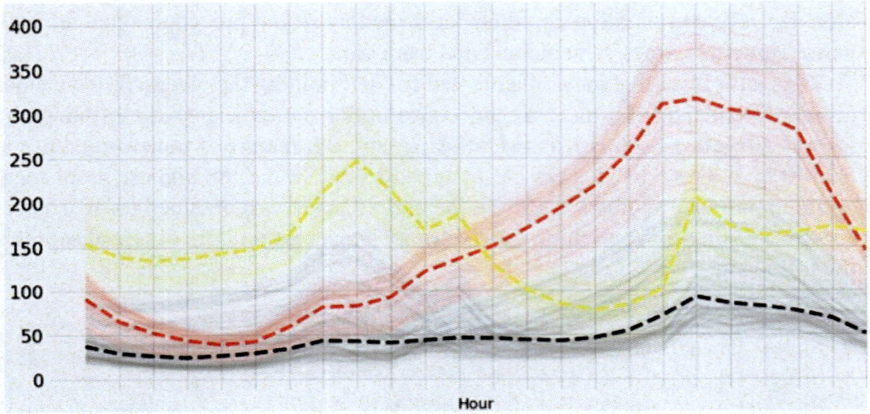

Fig. 12 Clusters for total electricity consumption

obtain the cluster plot, the dataframe index was at first set to RECORDING_DATE column and afterwards, the dataframe was up-sampled to hours.

Figure 12 shows the main groups of consumers: the one from the yellow group seems to have a sudden increase around the morning and evening hours (with a higher morning peak), the black group consumption amplitude is lower, and the red partition shows a steady linear consumption with a smooth growth around evening hours. Clusters do not have similar peculiarities, which indicates a correct number of groups.

5 Conclusion

First, we propose a multi-processing data analysis framework with MongoDB and Dask to handle a relatively large dataset of load data from residential consumers located in the U.S.

A data analysis methodology, that includes several Python libraries such as powerful Dask and Scikit-Learn, and Mongo BD, is proposed to handle relatively large datasets and extract valuable insights regarding the consumption pattern of the residential buildings. However, consumption data can be combined with other data such as weather or ISO affiliation and analyzed considering the breakdown and different periods of time.

By clustering the consumers into groups, we identified the clusters of smart cities that should be targeted from the DR program point of view. Among the three clusters, the red one is more "responsible" with the evening peak that should be shifted, shed or participate to shimmy DR programs. Thus, specific DR programs could be applied differentially to the three clusters according to their consumption daily load profile.

Furthermore, the consumption data recorded by smart meters can be combined with specific DR programs and relevant costs that enable such programs for residential consumers depending on the DR program (such as shift or slide, shed, shimmy, etc.), ToU tariffs and create a robust data model to extract more information from data. Such approach will be considered in a future research study.

Acknowledgements This work was supported by a grant of the Romanian National Authority for Scientific Research and Innovation, CCCDI—UEFISCDI, project title "Multi-layer aggregator solutions to facilitate optimum demand response and grid flexibility", contract number 71/2018, code: COFUND-ERANET-SMARTGRIDPLUS-SMART-MLA-1, within PNCDI III. This paper is also an extension of the scientific results of the project "Intelligent system for trading on wholesale electricity market" (SMARTRADE), co-financed by the European Regional Development Fund (ERDF), through the Competitiveness Operational Programme (COP) 2014–2020, priority axis 1—Research, technological development and innovation (RD&I) to support economic competitiveness and business development, Action 1.1.4-Attracting high-level personnel from abroad in order to enhance the RD capacity, contract ID P_37_418, no. 62/05.09.2016, beneficiary: The Bucharest University of Economic Studies.

References

Adam, M., Muntean, D. M., Popov, M., Grecea, D., Ungureanu, V.: Integrated energy efficient cooling solutions for large prefabricated panels collective dwellings from the 1970s. In: E3S Web of Conferences (2019). https://doi.org/10.1051/e3sconf/20198501004

Al-Wakeel, A., Wu, J.: K-means based cluster analysis of residential smart meter measurements. Energy Proc. (2016). https://doi.org/10.1016/j.egypro.2016.06.066

An, J., Yan, D., Hong, T.: Clustering and statistical analyses of air-conditioning intensity and use patterns in residential buildings. Energy Build. (2018). https://doi.org/10.1016/j.enbuild.2018.06.035

Asdrubali, F., Baldinelli, G., Bianchi, F., Sambuco, S.: A comparison between environmental sustainability rating systems LEED and ITACA for residential buildings. Build. Environ. (2015). https://doi.org/10.1016/j.buildenv.2015.01.001

Attia, S., Carlucci, S.: Impact of different thermal comfort models on zero energy residential buildings in hot climate. Energy Build. (2015). https://doi.org/10.1016/j.enbuild.2015.05.017

Balvedi, B.F., Ghisi, E., Lamberts, R.: A review of occupant behaviour in residential buildings. Energy Build. (2018). https://doi.org/10.1016/j.enbuild.2018.06.049

Bloess, A., Schill, W.P., Zerrahn, A.: Power-to-heat for renewable energy integration: A review of technologies, modeling approaches, and flexibility potentials. Appl. Energy (2018). https://doi.org/10.1016/j.apenergy.2017.12.073

Brahman, F., Honarmand, M., Jadid, S.: Optimal electrical and thermal energy management of a residential energy hub, integrating demand response and energy storage system. Energy Build. (2015). https://doi.org/10.1016/j.enbuild.2014.12.039

Chen, Y., Xu, P., Gu, J., Schmidt, F., Li, W.: Measures to improve energy demand flexibility in buildings for demand response (DR): A review. Energy Build. (2018). https://doi.org/10.1016/j.enbuild.2018.08.003

Corrado, V., Ballarini, I.: Refurbishment trends of the residential building stock: Analysis of a regional pilot case in Italy. Energy Build. (2016). https://doi.org/10.1016/j.enbuild.2016.06.022

Evola, G., Costanzo, V., Magrì, C., Margani, G., Marletta, L., Naboni, E.: A novel comprehensive workflow for modelling outdoor thermal comfort and energy demand in urban canyons: Results and critical issues. Energy Build. (2020). https://doi.org/10.1016/j.enbuild.2020.109946

Finck, C., Li, R., Zeiler, W.: Economic model predictive control for demand flexibility of a residential building. Energy (2019). https://doi.org/10.1016/j.energy.2019.03.171

Fischer, D., Wolf, T., Wapler, J., Hollinger, R., Madani, H.: Model-based flexibility assessment of a residential heat pump pool. Energy (2017). https://doi.org/10.1016/j.energy.2016.10.111

Foteinaki, K., Li, R., Heller, A., Rode, C.: Heating system energy flexibility of low-energy residential buildings. Energy Build. (2018). https://doi.org/10.1016/j.enbuild.2018.09.030

Gianniou, P., Liu, X., Heller, A., Nielsen, P.S., Rode, C.: Clustering-based analysis for residential district heating data. Energy Convers. Manag. (2018). https://doi.org/10.1016/j.enconman.2018.03.015

Gottwalt, S., Gärttner, J., Schmeck, H., Weinhardt, C.: Modeling and valuation of residential demand flexibility for renewable energy integration. IEEE Trans. Smart Grid (2017). https://doi.org/10.1109/TSG.2016.2529424

Haben, S., Singleton, C., Grindrod, P.: Analysis and clustering of residential customers energy behavioral demand using smart meter data. IEEE Trans. Smart Grid (2016). https://doi.org/10.1109/TSG.2015.2409786

Hurtado, L.A., Rhodes, J.D., Nguyen, P.H., Kamphuis, I.G., Webber, M.E.: Quantifying demand flexibility based on structural thermal storage and comfort management of non-residential buildings: A comparison between hot and cold climate zones. Appl. Energy (2017). https://doi.org/10.1016/j.apenergy.2017.03.004

Le Dréau, J., Heiselberg, P.: Energy flexibility of residential buildings using short term heat storage in the thermal mass. Energy (2016). https://doi.org/10.1016/j.energy.2016.05.076

Motlagh, O., Berry, A., O'Neil, L.: Clustering of residential electricity customers using load time series. Appl. Energy (2019). https://doi.org/10.1016/j.apenergy.2018.12.063

Muntean, D. M., Ungureanu, V., Petran, I., Georgescu, M.: Large prefabricated concrete panels collective dwellings from the 1970s: context and improvements. In: IOP Conference Series: Materials Science and Engineering (2017). https://doi.org/10.1088/1757-899X/245/5/052050

National Renewable Energy Laboratory: Building America House Simulation Protocols (2010). Retrieved October 30, 2020, from https://www.nrel.gov/docs/fy11osti/49246.pdf

Office of Energy Efficiency & Renewable Energy (EERE): Commercial and Residential Hourly Load Profiles for all TMY3 Locations in the United States (n.d.). Retrieved October 30, 2020, from https://openei.org/datasets/dataset/commercial-and-residential-hourly-load-profiles-for-all-tmy3-locations-in-the-united-states

Oprea, S.V., Bara, A.: Setting the time-of-use tariff rates with NoSQL and machine learning to a sustainable environment. IEEE Access (2020). https://doi.org/10.1109/ACCESS.2020.2969728

Oprea, S.V., Bâra, A., Ifrim, G.: Flattening the electricity consumption peak and reducing the electricity payment for residential consumers in the context of smart grid by means of shifting optimization algorithm. Comput. Ind. Eng. (2018). https://doi.org/10.1016/j.cie.2018.05.053

Oprea, S.V., Bâra, A., Ifrim, G.A., Coroianu, L.: Day-ahead electricity consumption optimization algorithms for smart homes. Comput. Ind. Eng. **135** (2019). https://doi.org/10.1016/j.cie.2019.06.023

Pallonetto, F., Oxizidis, S., Milano, F., Finn, D.: The effect of time-of-use tariffs on the demand response flexibility of an all-electric smart-grid-ready dwelling. Energy Build. (2016). https://doi.org/10.1016/j.enbuild.2016.06.041

Patteeuw, D., Henze, G.P., Helsen, L.: Comparison of load shifting incentives for low-energy buildings with heat pumps to attain grid flexibility benefits. Appl. Energy (2016). https://doi.org/10.1016/j.apenergy.2016.01.036

Quarton, C. J., Samsatli, S.: Power-to-gas for injection into the gas grid: What can we learn from real-life projects, economic assessments and system modeling? (2018)

Salvalai, G., Sesana, M.M., Iannaccone, G.: Deep renovation of multi-storey multi-owner existing residential buildings: A pilot case study in Italy. Energy Build. (2017). https://doi.org/10.1016/j.enbuild.2017.05.011

Differentiation Strategy and Cost Leadership Strategy: Their Contribution to Achieving Sustainable Financial Performance

Juniarti Juniarti, Clarissa Simanjaya, Marcella Chandra, and Zenia Estella Soesetyo

Abstract This research aims to study the influence of differentiation and cost-leadership strategy on sustainable financial performance with innovation as the moderator variable. This research uses the last 5 years financial report of go-public companies in South-East Asia, with a total of 250 firm years. In the early stage of the research, the companies are grouped into their respective strategy. From that grouping, each strategy is then measured. The researcher uses 5 measurements to test the strategies, 2 for differentiation strategy and 3 for cost-leadership strategy. The finding of this research is that differentiation positively influences sustainable financial performance, with innovation as the moderator variable. However, these findings are not proved in cost leadership strategy. Innovation shows partially influences to the achievement of sustainable financial performance. This research contributes to adding the innovation factor that strengthens the relationship between the successes of a strategy with the sustainable financial performance so that the result can be more consistent.

Keywords Strategy · Differentiation · Cost-leadership · Financial performance · Sustainable financial performance · Innovation

1 Introduction

Both differentiation and cost-leadership strategies have their respective roles in a company's success for sustainability. The success of the strategy chosen by the company is also influenced by innovation. The innovation factor is important in choosing between the two strategies, especially for companies that implement the differentiation strategy. In a differentiation strategy, innovation is a crucial factor for higher performance (Hull and Rothenberg 2008; Miller 1983; Porter 1990). This is due to the fast-paced globalization that causes increasingly fierce competition, which encourages companies to innovate their products, services, and corporate

J. Juniarti (✉) · C. Simanjaya · M. Chandra · Z. E. Soesetyo
Petra Christian University, Surabaya, Indonesia
e-mail: yunie@petra.ac.id

M. Busu (ed.), *Digital Economy and New Value Creation*, Springer Proceedings in Business and Economics, https://doi.org/10.1007/978-3-031-07265-9_16

image with new and different processes that are difficult for competitors to imitate (Zehir et al. 2015). The advantage of the differentiation strategy, which makes products and services difficult to imitate by competitors, is that it increases the chance for the company to be sustainable (Grant 1991). Although innovation is very important in differentiation strategy, this does not mean that cost-leadership strategy does not also require innovation. To benefit from a cost-leadership strategy, companies need to emphasize minimizing costs and undertaking a process of innovation (Frohwein and Hansjurgens 2005). The innovation process is concerned with all of the company's operational activities by improving the quality of offerings and efficient delivery methods than competitors to achieve a competitive advantage (O'Sullivan and Dooley 2009). In addition, the innovation process also allows companies to achieve economies of scale, reduce costs and gain market share (Qin 2007). Thus, innovation is needed in a cost-leadership strategy to achieve cost reduction, larger market share, and better efficiency than competitors (Hilman and Kaliappen 2014).

The company implements strategies to achieve long-term goals, where these strategies influence long-term financial performance. However, the results of previous studies regarding the choice of strategy and financial performance show mixed results. Banker et al. (2014) found that differentiation strategies have an impact on long-term financial performance as measured by future ROA, each using the next 5 years as a function of the company's current performance. This measure refers to the success or persistence of ROA, that is, the extent to which current ROA can be maintained in future periods. Meanwhile, other research results show a relationship between differentiation strategies and financial performance as measured by ROI, ROE, and other financial measures, such as budget variance analysis, working capital ratio, divisional profit, cash flow return on investment, and shareholder value. added (Asdemir et al. 2013; Spencer et al. 2009; Teeratansirikool et al. 2013; Yamin et al. 1999; Yeung et al. 2006). Several other researchers did not find a relationship between differentiation strategies and sustainable financial performance (Altuntas et al. 2014; Oyewobi et al. 2016; Wong et al. 2016). Inconsistency in research results also occurs in research on cost-leadership strategies. Several researchers have proven that cost-leadership strategies are related to short-term financial performance (Hilman and Kaliappen 2014; Kaliappen and Hilman 2013; Nandakumar et al. 2010), however, Li and Li 2008; Banker et al. 2014) did not find a relationship between cost leadership strategies and sustainable financial performance.

The results of previous studies are still not solid, one reason is that these studies have not considered the innovation factor. The changing business environment, changing customer needs and expectations, increasing competition, and rapid technological developments make companies need to innovate to stay ahead of competitors (Bhatt et al. 2010; Chen et al. 2009; Uzkurt et al. 2013; Zaefarian et al. 2017). Research on the effect of innovation on company performance shows positive results (Agarwal et al. 2003; Han et al. 1998). This shows that innovation is an important factor for company growth and sustainability (Han et al. 1998; Kiron et al. 2013). This study adds innovation as a moderating variable for the relationship between differentiation strategies and cost leadership and sustainable financial performance.

Innovation is one of the main sources of achieving competitive advantage (Díaz-Díaz and Saá-Pérez 2014), and is a transformative force that improves the sustainability and economic performance of companies (Cavaleri and Shabana 2018). Innovation is needed by companies in facing the increasing pressure of globalization, changing customer expectations, increasing competition, and rapid technological developments in a changing business environment. Innovation is an important driver of competitiveness and company success in a changing business environment (Chen et al. 2009; Uzkurt et al. 2013; Zaefarian et al. 2017).

2 Literature Review

2.1 Strategy and Innovation

The differentiation strategy is a strategy that involves the addition of a significant aspect from a product and service that are superior, hard to imitate, unique, and of higher quality than the competitors, thus, the resulting additional value can be felt by the customers and creates a competitive advantage (Banker et al. 2014; Green et al. 1993; Porter 1980, 1997; Prajogo 2007). Moreover, the advantage of the differentiation strategy is that customers will be able to customize the product, which depends on the company's rapport with its customers (Banker et al. 2014). An exclusive relationship between a company and its customers becomes a competitive advantage that is hard for competitors to imitate, which enables the company to have a sustainable competitive advantage (Banker et al. 2014; Barney 1986; Ghemawat 1995). Innovations have a central role in the companies that use the differentiation strategy to compete with their competitors (Herzallah et al. 2017). Innovations have to be done continuously to leave no chance for the competitors to imitate the products (Amoako-Gyampah and Acquaah 2008; Asdemir et al. 2013; Banker et al. 2014; Zehir et al. 2015). When a company innovates, its competitors will need time to match the products and services provided by the company because the competitors need to do the research and development phase first. At the time the competitors do their research, the company has already made another innovation (Asdemir et al. 2013). A company's innovation will be reflected in the appearance and technology of high quality and innovative product with a high design or brand image (Banker et al. 2014; Crema et al. 2014; Frambach et al. 2003; Hutchinson et al. 2007; Porter 1980). Because of this, the companies set a high price (Banker et al. 2014; Crema et al. 2014; Frambach et al. 2003; Hutchinson et al. 2007; Porter 1980). The customers of the differentiation strategy are not sensitive to price as long as they receive the value of the product or service. This is the advantage of the differentiation strategy (Black et al. 2000; Green et al. 1993; Prajogo 2007; Porter 1997). On the other hand, the differentiation strategy needs high initial capital, operational costs, and investments to develop the products and services (Banker et al. 2014).

To identify whether a company uses the differentiator strategy or not, previous studies use the survey method (Amoako-Gyampah and Acquaah 2008; Nandakumar et al. 2010; Parnel and Brady 2019; Parnell et al. 2012). This methodology may cause bias because the respondents are influenced by consistency, social desire, and the lack of knowledge, thus this research does not use it (Podsakoff and Organ 1986; Miller and Roth 1994). Other researchers, such as Selling and Stickney (1989), Banker et al. (2011, 2013), Wu et al. (2015), measures the profit margin, which is the sum of adding operating income with R&D expenditure, divided by sales.

In cost-leadership strategy, operational efficiency becomes the foundation of the achievement of this strategy (Banker et al. 2014; De Castro and Chrisman 1995; Porter 1997). This efficiency is conducted by controlling the production costs, increasing capacity, and minimizing other costs (Banker et al. 2014; Prajogo 2007). Thus, this strategy can strive for an average return above its competitors because it uses lower prices (Prajogo 2007). The low cost can be used as a company's main advantage to compete with its competitors (Banker et al. 2014; Teeratansirikool et al. 2013) and will have a positive influence on the company's financial performance (Herzallah et al. 2017).

Companies that apply the cost-leadership strategy also need to innovate to be superior in the competition (Frohwein and Hansjurgens 2005; Seaden and Manseau 2001). Innovation in the cost-leadership strategy can be cone through the application and development of new technology and procedure because it can significantly improve the company's structure and cost efficiency (Francesco, 2014; Seaden and Manseau 2001). Innovation can also safe companies from unhealthy competition from the cost-relationship strategy: the application of a very low price that is irrational and disturbing the company's financial performance (Gunday et al. 2011; Hilman and Kaliappen 2014; Hilmi et al. 2010; Nandakumar et al. 2011; Parnell 2011).

To measure the cost-leadership strategy, Selling and Stickney (1989), Banker et al., (2011, 2013), Wu et al. (2015) use ATO (Asset Turnover) as a measurement tool. However, this research uses three types of ratio used by Asdemir et al., (2013), Balsam et al., (2011), Banker et al., (2014). These three ratios can detect a company's efficiency in its capital utilization (David et al. 2002) because the customers of cost-leadership strategy buy a company's products that are cheaper than its competitors and the company's profits are gained through minimizing costs and assets per unit output (Hambrick 1983b). The first ratio is SALES/CAPEX (Formula 3)—the net sales that are measured with the company's capital expenditure used to buy properties, plants, or equipment. The high value of this ratio shows that the company applies the cost-leadership strategy (Banker et al. 2014). The second ratio is SALES/P&E (Formula 4), which is a ratio between net sales and the net book value of plant and equipment. A higher value shows more efficient use of the company's assets, which shows that the company uses the cost-leadership strategy (Berman et al. 1999; Hambrick 1983a, b; Kotha and Nair 1995; Miller and Dess 1993). The last ratio used by the researcher in this cost-leadership strategy is EMPL/ASSETS (Formula 5), which is the ratio between the number of employees and total assets (Hambrick 1983a, b; Kotha and Nair 1995; Nair and Filer 2002). A higher value of this ratio shows the company uses the cost-leadership strategy (Banker et al. 2014).

2.2 Sustainable Financial Perfromance

Financial performance is the company's financial condition during certain periods that includes modal adequacy, liquidity, solvability, efficiency, leverage, and profitability. Financial performance shows a company's capability in managing and controlling its resources (Fatihudin et al. 2018). To support a company's sustainability, good financial performance in not only the short-term but also the long term is expected. Companies can achieve a sustainable financial performance if they use resources that can encourage the creation of value in the companies' operational in the present and future (Banker et al. 2014). If the companies create future values by using present resources, they can also create a sustainable competitive advantage. With this, companies can have a financial performance that can survive in the long-term (Mohammadi et al. 2019).

In measuring sustainable financial performance, Banker et al., (2014) use future ROA, and Juniarti (2020) uses earning persistence. From the previous studies' measurements, this research will use Earnings Persistence (EP) as a measurement tool to show sustainable earning (Francis et al. 2004; Penman and Zhang 2002). Earning Persistence is a time-series parameter to measure the size of the influence of earnings obtained repeatedly without a loss that the company expects in the future (Aguguom et al. 2019; Pimentel and Aguiar 2016). A high earning persistence shows a strong and sustainable earning, while low earning persistence shows weak, temporary, or unsustainable earning (Ashley and Yang 2004; Francis et al. 2004; Juniarti 2020). This Earning Persistence follows Francis et al., (2004) measurement that uses the autoregressive model of order one (AR1) by observing the earnings in the past 5 years:

$$X_{j,t} = f_{0,j} + f_{1,j} X_{j,t-1} + v_{j,t}$$

Notes:

$X_{j,t}$ A company's net earnings before the extraordinary items in year t is divided by the weighted average of shares outstanding in year t.

$\phi_{1,j}$ Earning persistence as the estimated slope coefficient. Approaching 1 implies very persistent earnings, approaching 0 implies huge temporary earnings.

$v_{j,t}$ J company's specific residue in year t.

2.3 A Subsection Sample

A differentiation strategy involves offering products and services that are different from competitors, that is, offering unique and quality products and services by investing in various activities such as advertising, promotion, customer service, product distribution, and other related activities (Banker et al. 2014; De Castro and Chrisman 1995; Jermias 2008; Liu and Atuahene-Gima 2018; Prajogo 2007).

This makes the company more effective than its competitors, creates higher value for customers, and achieves higher performance and profits (Awwad et al. 2013; Hosseini and Sheikhi 2012; Leonidou et al. 2015; Mathenge 2013; Murray et al. 2011; Tan and Sousa 2015; Yao and Qin 2016). Grant (1991) argues that with this uniqueness, the benefits obtained from a differentiation strategy are more sustainable because the products and services offered by the company are not easily imitated by competitors. Differentiation strategies usually also involve innovation in certain products and adjustments to marketing campaigns that are impossible to imitate quickly (Asdemir et al. 2013). The longer it takes a competitor to respond to a certain comparative advantage, the greater the opportunity for the company to take advantage of a sustainable advantage (Banker et al. 2014). Companies that can withstand the efforts of competitors in imitating their products and services can maintain their competitive advantage, thus enabling the company to have superior performance in the long term (Ghemawat 1995).

On the other hand, the cost-leadership strategy emphasizes operational efficiency through the development of new processes and technologies, economies of scale, and experience (Banker et al. 2014). Kim et al. 2004 argued that companies that implement a cost-leadership strategy are easily trapped into continuously reducing prices because technology tends to be based on a cost structure with low variable costs and high fixed costs. This gives rise to a competitive advantage that is only temporary and makes it impossible to achieve long-term financial performance (Eisenhardt and Martin 2000; Kim et al. 2004), because operational efficiency can be imitated by competitors or cannot be operated due to the emergence of new sources and better (D'Aveni 1994; Hamel 2000). Murray (2011) also argues that imitation by competitors cannot be avoided if the company uses a cost-leadership strategy. Another argument suggests that companies that focus solely on cost-leadership strategies are no longer suitable to meet the different needs and demands of customers in the era of globalization (Baines and Langfield-Smith 2003; Kotha and Vadlamani 1995; Perera et al. 1997).

From the results discussed above, the following hypothesis is formulated in this study.

H1a: The differentiation strategy influences sustainable financial performance.

H1b: The cost-leadership strategy influences sustainable financial performance.

Innovation generally has a positive relationship to performance and is important for companies to maintain their survival (Brown and Eisenhardt 1995; Covin and Miles 1999; Christensen and Bower 1996; Clark and Fujimoto 1991; Han et al. 1998; Hamel and Prahalad 1994; O'Reilly and Tushman 2004; Peters 1990, 1991; Teece et al. 1997; Turulja and Bajgoric 2019; Zahra and Covin 1995). Innovation needs to be done so that companies can explore opportunities and markets as a source of excellence, especially in today's business environment where the business is very complex and changing (Chen et al. 2009; Duane Ireland and Webb 2007; Madhavan and Grover 1998; Miller and Friesen 1983; Montes et al. 2004; Turulja and Bajgoric 2019; Urbancova 2013; Uzkurt et al. 2013; Zaefarian et al. 2017). Therefore, new products and services, new processes, and new ways of organizing work in the company are very important for future success, given that novelty is one of the attractions of the company (Duane Ireland and Webb 2007; Racela and Thoumrungroje 2019). Companies that are still able to survive and innovate in the future can create advantages in terms of competition, are efficient in their operational activities, and can run their business even though they are in an environment with limited resources (Porter 1990; Simpson et al. 2006; Ren et al. 2010). From this innovation activity, companies have the opportunity to get profits, sales revenue, return on investment, and high market share (Saliba de Oliveira et al. 2018; Turulja and Bajgoric 2019). If innovation is carried out continuously, the company can maintain a competitive advantage and have long-term financial performance (Agarwal et al. 2003; Cavalry and Shabana 2018; Duane Ireland and Webb 2007; Han et al. 1998; Hull and Rothenberg 2008; Madhavan and Grover 1998; Miller and Friesen 1983; Montes et al. 2004; Porter 1990; Simpson et al. 2006; Racela and Thoumrungroje 2019; Ren, 2009; Turulja and Bajgoric 2019).

From the results discussed above, the following hypothesis is formulated in this study.

H2: Innovation has a positive influence on sustainable financial performance.

The ability of a company to innovate is also one of the factors that characterize its sustainability and growth potential (O'Reilly et al. 1991). Therefore, innovation is an important thing to implement in strategy, both differentiation and cost-leadership (Cavaleri and Shabana 2018). Companies with a differentiation strategy tend to develop many R and D activities to increase their innovative strength and increase the company's ability to compete with innovations made by competitors (Miller 1987). This company needs innovation based on features that are difficult for competitors to imitate (Amoako-Gyampah and Acquaah 2008). Innovation by companies is important for differentiation strategies and also for achieving higher performance (Hull and Rothenberg 2008; Miller 1983; Porter 1990). In addition to a differentiation strategy, a cost-leadership strategy also requires innovation to be competitive even though this strategy focuses on efficiency and low costs. Companies must emphasize cost minimization and engage with process innovation to gain the advantage of this strategy (Frohwein and Hansjurgens 2005). The aim of implementing innovation in a cost-leadership strategy is to achieve cost reduction, larger market share, and better efficiency than competitors that drive corporate sustainability (Hilman and Kaliappen 2014). Thus, innovation is important to be applied to each strategy, because companies that do not innovate will find it difficult to sustain even though the company has implemented the right strategy (Cavaleri and Shabana 2018; Najafi-Tavani et al. 2018; Prajogo 2016).

From the results discussed above, the following hypothesis is formulated in this study.

H3a: Innovation moderates the relationship between differentiation strategy and sustainable financial performance.

H3b: Innovation moderates the relationship between cost-leadership strategy and sustainable financial performance.

3 Research Method

3.1 Sample

This research uses secondary data from the companies' annual financial report in 2015–2019 from Bloomberg. The population of this research is 5.169 go-public companies in South East Asia from all sectors. The researcher did a reselection to the population to obtain the selected samples when the companies have complete financial report data during the research period. The sample criteria of this research are:

1. Companies listed in the Stock Exchange of each country in South East Asia in 2015–2019.
2. Have their earnings data from 2010–2019. This data is required to measure earnings persistence, which is an indicator of sustainable financial performance.
3. Companies that have R&D data from 2015–2019.

The definition of this research's operational variables can be seen in the Table 1.

3.2 Model Analysis

To evaluate the hypothesis of this research, the analysis model used is as follows:

$$\begin{aligned} EP &= \beta_0 + \beta_1 STRDIF_{i,t-1} + \beta_2 STRCL_{i,t-1} + \beta_3 INV_{i,t-1} + \beta_4 INV * STRDIF_{i,t-1} \\ &+ \beta_5 INV * STRCL_{i,t-1} + \beta_6 SIZE_{i,t-1} + \beta_7 LEV_{i,t-1} \\ &+ \beta_8 CI_{i,t-1} + \beta_9 MS_{i,t-1} + \beta_{10} AGE_{i,t-1} + \varepsilon i,t \end{aligned}$$

Notes:

$EP_{i,t-1}$	*i* company's earning persistence in t period.
$STRDIF_{i,t-1}$	*i* company's differentiation strategy in t − 1 period.
$STRCL_{i,t-1}$	*i* company's cost-leadership strategy in t − 1 period.
$INV_{i,t-1}$	*i* company's innovation in t − 1 period.
$INV*STRDIF_{i,t-1}$	Innovation moderates *i* company's differentiation strategy in t − 1 period.
$INV*STRCL_{i,t-1}$	Innovation moderates *i* company's cost-leadership strategy in t − 1 period.
$SIZE_{i,t-1}$	t − 1. *i* company's size in t − 1 period.
$LEV_{i,t-1}$	*i* company's leverage in t − 1 period.
$CI_{i,t-1}$	*i* company's competition intensity in t − 1 period.
$MS_{i,t-1}$	*i* company's market share in t − 1 period.
$AGE_{i,t-1}$	*i* company's age strategy in t − 1 period.
$\varepsilon_{i,t}$	*i* company's error in t period.

Table 1 Operational variable definition

Variable	Operational definition	Scale
Differentiation strategy	This variable is measured with two measurement tools: SG&A/SALES (Formula 1) and SALES/COGS (Formula 2). The high result of these two formulas shows that the companies are focusing on the differentiation strategy (Banker et al 2014; Kotha and Nair 1995; Nair and Filer 2002)	Ratio
Cost-leadership strategy	This variable is measured with three measurement tools: SALES/CAPEX (Formula 3), SALES/P&E (Formula 4), and EMPL/ASSETS (Formula 5). The high result of these three ratios shows that the companies are using the cost-leadership strategy (Banker et al 2014; Berman et al. 1999; Hambrick 1983a, b; Kotha and Nair 1995; Miller and Dess 1993)	Ratio
Earning persistence	Measured with earning persistence by using the autoregressive model of order one (AR1). If $X_{j,t}$ approaches 1, then the earning is very persistent. If $X_{j,t}$ approaches 0, it implies a huge temporary earning (Francis et al. 2004)	Ratio
Innovation	Measured with the R&D expenditures on total sales ratio (R&D/SALES) (Ameer and Othman 2012; Alshwer and Levitas 2014)	Ratio
Firm size	*Firm size* is measured with *i* company's log total asset in year t (Jermias 2008)	Ratio
Leverage	The use of debt is measured by dividing total debt by the total asset (Dimitrov and Jain 2008; Mercer 2004)	Ratio
Competition intensity	The competition intensity between companies is measured by using the Herfindahl Index (HHI) (Formula 6). The lower index value shows a higher competition intensity (Li et al. 2008; Naldi and Flamini 2018)	Ratio
Market share	Measured by dividing the company's total sales with the industry's total sales (Edeling and Himme 2018). A higher value shows a higher market share (Connolly et al. 1986; Li and Li 2008)	Ratio
Firm age	*Firm age* is measured with the log of a company's age from its foundation until the research year (Hariyanto and Juniarti 2014)	Ratio

Source Authors' own research

3.3 Data Analysis and Technique

The data analysis technique for the panel data regression model is conducted by choosing the best model out of the following: Pooled Least Squared (PLS), Fixed Effect Model, and Random Effect Model. To choose the best panel data model, several tests are conducted: the Chow Test, Hausman Test, and Lagrange Multiplier Test.

Based on the sample selection, the final number of this research's sample is 50 companies, where there is a total of 250 firm-years that are go-public companies in the South East Asia region (Table 2).

The sample includes all sectors and countries in South East Asia. The sample profile covers all sectors and 5 countries in South East Asia: Indonesia, Malaysia, Singapore, Philippines, and Thailand. 5 other South East Asian countries are not listed in the country-based sample profile because the data do not fulfill the sample criteria. As can be seen, the most dominant samples are in the industrial sector, 12 samples. While the least number of sample is in the electronic components and telecommunications and media sectors, with only 1 sample each. For country-based sample profile, Malaysia has the biggest number of samples with 20, and Thailand has the smallest number of the sample with 1. Thailand is not representative in this research sample (Table 2).

The average of EP does not approach 1, which means that the researched companies' earning persistence is not too persistent. STRDIF1 also has a value far below 1, which means the companies are also not reflecting the implementation of the

Table 2 Sample profile

Panel A: based on sectors

No.	Sectors	Sample	(%)
1.	Consumer discretionary, products and services, staples	7	1400
2.	Electronic components	1	200
3.	Health care	3	600
4.	Holding firms	2	400
5.	Industrial	12	2400
6.	Plantation	4	800
7.	Property, real estate, construction	2	400
8.	Services	6	1200
9.	Technology	10	2000
10.	Telecommunications and media	1	200
11.	Utilities	2	400
	Total	50	10,000

Panel B: based on nations

No.	Nation	Sample	(%)
1.	Indonesia	9	1800
2.	Malaysia	20	4000
3.	Singapura	12	2400
4.	Filipina	8	1600
5.	Thailand	1	200
	Total	50	10,000

Source Authors' own research

differentiation strategy when they are measured with formula 1. On the other hand, the average of STRDIF2 has a value above 1, companies with the differentiation strategy that uses the formula 2 have a higher sales average than the cost of goods sold and very much reflect companies with differentiation strategy if measured with formula 2. STRCL 1 has a value far below 1, which means that the companies are not reflecting the use of the cost-leadership strategy if measured with formula 3. STRCL2 has a value far above 1, which means the companies' strategy very much reflects the implementation of the cost-leadership program if measured with formula 4. INV only has a value of 0.04, which means that innovation does not really influence sustainable financial performance.

The average value of SIZE is far above 1, which is 12.68, with a value on the financial report of 35.144.023.277.527. This means that the total asset owned and borrowed by the company is huge. LEV has a value less than 1, 0.18, which means that the companies have a high debt risk. The CI value is only 0.02, far below 1, which means that the companies are not capable enough of facing tight competition. This can hinder the companies in achieving sustainability. MS has a value far below 1, this means that the companies have a small market share and thus do not have a high rate of return. AGE has an average value far above 1, this means that the companies are quite old and have more experience.

The majority of the processed data has a data variation with normal distributions such as EP, STRDIF1, STRDIF2, INV, SIZE, LEV, CI, MS, AGE as can be seen from the low standard deviation, it means that the data do not really represent the distribution inside of a group (Table 3).

Table 3 Descriptive statistic

		Minimum	Maximum	Mean	Std. deviation
EP	250	−2.795	2.614	0.307	0.659
STRDIF1	206	0.000	1.865	0.251	0.290
STRDIF2	200	1.027	3.146	1.532	0.426
STRCL1	250	−420.9	349.562	−19.72	55.94
STRCL2	250	0.120	1202.645	9.989	76.268
INV	250	0.000	0.936	0.049	0.141
SIZE	250	10.861	14.782	12.683	0.947
LEV	208	0.000	0.895	0.180	0.155
CI	250	0.000	0.121	0.020	0.033
MS	245	0.000	0.304	0.036	0.057
AGE	250	0.845	2.267	1.53	0.291
Valid N (listwise)	160				

Source Authors' own research

4 Results and Discussion

To determine the best panel data model, the researcher uses the Chow Test. The Chow Test is used to choose the best model between Pooled Least Square (PLS) and Fixed Effect. If p-value $< \alpha$, which is 0.05, then the Fixed Effect model is chosen. Otherwise, if p-value > 0.05, the chosen model is PLS. Below are the results of the PLS Test using the GRETL application.

This research is conducted by testing each strategy using the described formulas. Table 4 shows the measurement result from SG&A/SALES with 50 companies tested. Based on the Chow Test result above, it is found that p-value > 0.05, thus, the best model for the differentiation strategy with formulas 1 and 2 (SG&A/SALES) and (COGS/SALES) is PLS. The hypothesis test result shows that STRDIF1 has a significant and negative regression coefficient: −3.97610 with a p-value of 0.0042. Table 5 shows the measurement result of formula 2. The measurement result shows that STRDIF2 has a significant and positive regression coefficient: 0.797698 with a p-value of 0.0293. Thus, these two tables prove that H1a is accepted if measured with formula 2. This means that the differentiation strategy influences long term financial performances and can improve the company's future performance (Tables 6 and 7).

The INV variable on STRDIF1 has a significant and negative regression coefficient of −26.4037 with a p-value of 0.0510, which means that innovation negatively influences sustainable financial performance when it is tested with formula 1. Meanwhile, the result on STRDIF2 shows a non-significant and negative regression coefficient of −1280.16 with a p-value of 0.1140, which shows that innovation does not influence sustainable financial performance if measured with formula 2. These results show that innovation itself without connected with the strategy seem meaningless. However, when we moderate the INV with the STRDIF1 and STRDIF2, it proves that innovation strengthen the benefit of differentiation strategy and resulting a higher sustainable financial performance. The evidence that innovation moderate the relationship of differentiation strategy and sustainable financial performance is

Table 4 STRDIF1 PLS test result

	Coefficient	Std. error	t-ratio	p-value	
Const	−5.41438	4.56222	−1.187	0.2425	
STRDIF1	−3.97610	1.30722	−3.042	0.0042	***
INV	−26.4037	13.1109	−2.014	0.0510	*
STRDIF1*INV	136.411	62.1401	2.195	0.0342	**
SIZE	0.598571	0.444177	1.348	0.1856	
LEV	−1.34600	1.40871	−0.9555	0.3452	
CI	−5.97169	2.85929	−2.089	0.0433	**
MS	1.58130	1.66855	0.9477	0.3491	
AGE	−0.238747	0.704850	−0.3387	0.7366	

Source Authors' own research

Table 5 STRDIF1 chow test result

	Coefficient	Std. error	t-ratio	*p*-value	
Const	−12.5081	7.89412	−1.584	0.1236	
STRDIF1	−0.172589	3.65090	−0.04727	0.9626	
INV	326.604	267.083	1.223	0.2309	
STRDIF1*INV	−1280.16	786.347	−1.628	0.1140	
SIZE	0.919340	0.717797	1.281	0.2101	
LEV	0.629352	2.46043	0.2558	0.7999	
CI	25.1878	38.3372	0.6570	0.5162	
MS	−9.73704	15.1695	−0.6419	0.5258	
AGE	0.993111	1.22603	0.8100	0.4243	

*Joint significance of differing group means: F(9, 30) = 1.46912 with *p-value* 0.2045
Source Authors' own research

Table 6 STRDIF2 PLS test result

	Coefficient	Std. error	t-ratio	*p*-value	
const	−3.58285	3.95691	−0.9055	0.3722	
STRDIF2	0.797698	0.310111	2.572	0.0151	**
INV	−7.59195	29.6858	−0.2557	0.7998	
STRDIF2*INV	4.60044	13.1062	0.3510	0.7280	
SIZE	0.0859230	0.389690	0.2205	0.8269	
LEV	3.08429	1.23056	2.506	0.0177	**
CI	2.37017	2.87081	0.8256	0.4153	
MS	2.53418	1.65797	1.528	0.1365	
AGE	0.363957	0.720138	0.5054	0.6169	

Source Authors' own research

Table 7 STRDIF2 chow test result

	Coefficient	Std. error	t-ratio	*p*-value	
const	19.2697	13.3960	1.438	0.1646	
STRDIF2	2.69770	1.15729	2.331	0.0293	**
INV	263.976	406.723	0.6490	0.5230	
STRDIF2*INV	−73.8851	232.312	−0.3180	0.7535	
SIZE	−1.91412	1.19628	−1.600	0.1238	
LEV	4.40121	1.61446	2.726	0.0123	**
CI	−85.1690	59.2065	−1.439	0.1644	
MS	41.2871	25.3232	1.630	0.1173	
AGE	−1.24347	1.77103	−0.7021	0.4900	

*Joint significance of differing group means: F(9, 22) = 1.25677 with *p-value* 0.3133
Source Authors' own research

as shown by the regression coefficient of STRDIF1*INV is significant and positive, with a value of 136.411 and a *p*-value of 0.0342, while the regression coefficient of STRDIF2*INV is positive but non-significant, with a value of 4.60044 and *p*-value of 0.7280. Thus, H3a is accepted. This shows that differentiation strategy needs innovations because it can strengthen the achievement of sustainable financial performance (Hull and Rothenberg 2008; Miller 1983; Porter 1990). This finding is also supported by some previous studies that uncovered that differentiation strategy need to develop many R&D activities to increase their innovative strength and increase the company's ability to compete with innovations made by competitors (Miller 1987). The company needs to innovate based on features that are difficult for competitors to imitate (Amoako-Gyampah and Acquaah 2008). Therefore, when combining with the innovation, differentiation strategy will useful in achieving the higher sustainable financial performance (Tables 4, 5, 6, 7, 8, 9, 10 and 11).

Table 8 STRCL1 PLS test result

	Coefficient	Std. Error	t-ratio	*p*-value	
Const	−1.23132	1.15607	−1.065	0.2890	
STRCL1	−0.000308102	0.00206222	−0.1494	0.8815	
INV	0.611420	0.635918	0.9615	0.3382	
STRCL1*INV	−0.0293655	0.0479078	−0.6130	0.5411	
SIZE	0.0827240	0.0939622	0.8804	0.3804	
LEV	−1.28816	0.530789	−2.427	0.0167	**
CI	3.89686	2.30913	1.688	0.0941	*
MS	0.659576	1.36591	0.4829	0.6301	
AGE	0.356184	0.204495	1.742	0.0841	*

Source Authors' own research

Table 9 STRCL1 chow test result

	Coefficient	Std. error	t-ratio	*p*-value	
const	−3.25731	2.61064	−1.28	0.2148	
STRCL1	0.00271993	0.00356167	0.7637	0.4467	
INV	−36.8278	28.4818	−1.293	0.1987	
STRCL1*INV	−0.687822	1.05588	−0.6514	0.5161	
SIZE	0.230486	0.210682	1.094	0.2763	
LEV	−2.85009	0.828360	−3.441	0.0008	***
CI	8.88490	4.07908	2.178	0.0315	**
MS	0.167979	2.24448	0.07484	0.9405	
AGE	0.624490	0.274262	2.277	0.0247	**

*Joint significance of differing group means: $F(9, 111) = 1.85454$ with *p*-value 0.0663
Source Authors' own research

Table 10 STRCL2 PLS test result

	Coefficient	Std. error	t-ratio	*p*-value	
const	−5.65756	2.76051	−2.049	0.0538	*
STRCL2	0.00458241	0.00848773	0.5399	0.5952	
INV	2.03612	1.03788	1.962	0.0639	*
STRCL2*INV	0.00280943	0.0230811	0.1217	0.9043	
SIZE	0.136524	0.191026	0.7147	0.4831	
LEV	−0.400180	0.658087	−0.6081	0.5500	
CI	7.57690	9.45979	0.8010	0.4326	
MS	3.71669	3.78343	0.9824	0.3377	
AGE	2.79133	0.770914	3.621	0.0017	***

Source Authors' own research

Table 11 STRCL2 test result

	Coefficient	Std. error	t-ratio	*p*-value	
const	4.17939	6.06654	0.6889	0.5051	
STRCL2	0.0134524	0.0089084	1.498	0.1623	
INV	1.33685	0.93763	1.426	0.1816	
STRCL2*INV	−0.00731214	0.0206885	−0.3534	0.7304	
SIZE	−0.137676	0.498398	−0.2762	0.7875	
LEV	−0.764551	0.579221	−1.320	0.2137	
CI	17.9434	24.6456	0.7281	0.4818	
MS	27.3204	10.6480	2.566	0.0262	**
AGE	−2.92422	2.29266	−1.275	0.2284	

*Joint significance of differing group means: $F(9, 11) = 2.01413$ with *p*-value 0.1362
Source Authors' own research

Based on the Chow Test above (Tables 8, 9), it is known that *p*-value > 0.05, thus the best model for the cost-leadership strategy with formula 1 (SALES/CAPEX) is the PLS model. This PLS model also applies to the cost-leadership strategy with formula 2 (SALES/P&E) that also has a *p*-value > 0.05 (Tables 10, 11). From the hypothesis test using the PLS test, STRCL1 has a negative and non-significant regression coefficient of 0.00458241 with a *p*-value of 0.5952, thus **H1b is rejected.** The INV variable on STRCL2 is positive and significant, with a value of 2.03162 with a *p*-value of 0.0639. This proves that innovation influences sustainable financial performance, but it is not absolute, thus **H2 is accepted**. Following the H2 result of this research, innovation can still be a point of consideration for companies to achieve long-term financial performance because by innovating companies can obtain profits, sales income, the return of investment, and high market share (Saliba de Oliveira et al. 2018; Turulja and Bajgoric 2019). STRCL1INV has a negative and non-significant regression coefficient of −0.687822 with a *p*-value of 0.5161. STRCL2INV has a positive and non-significant regression coefficient of 0.00280943 and a *p*-value of 0.9043, which means **H3b is rejected**. The cost-leadership strategy with measurement 3, EMPL/ASSETS, is not included in the descriptive statistics and is not used to determine the best panel data model in the hypothesis test because there is no sufficient data.

In achieving competitive advantage, companies need to apply an appropriate strategy (Asdemir et al., 2019). The most commonly used strategies are the ones introduced by Porter (1980), differentiation strategy, and cost-leadership strategy. Differentiation strategy prioritizes innovation, as can be seen from the uniqueness and quality of a product to obtain a long-term superior performance (De Castro and Chrisman 1995; Frambach et al. 2003; Ghemawat 1995; Hutchinson et al. 2007; Jermias 2008; Porter 1980; Prajogo 2007). The more unique a product compared to the competitors, the more superior the company from its competitors. In the cost-leadership strategy, innovation is needed to reduce costs to get ahead of the competitors (Frohwein and Hansjurgens, 2005). By innovating, the reduction of costs, a higher market share, and efficiency can be achieved. Thus, the companies that implement this strategy can be sustainable (Hilman and Kaliappen 2014).

The control variable of this research shows diverse influences, SIZE and MS do not influence sustainable financial performance on both differentiation strategy and cost-leadership strategy in all measurements. The AGE variable has a relationship with a long-term financial performance involving how the long-standing companies are more trusted for the investor, have more experience, and a more skilled workforce, thus it can be assumed that long-standing companies will receive higher earnings than the newly built ones (Hariyanto and JUniarti 2014; Mehmood et al., 2019). The leverages with negative influence have higher risks and cannot sustain their long-term performance (McGuire et al. 1988; Waddock and Graves 1997). The CI variable shows a negative and significant regression coefficient value on STRDIF1 of −5.97169 and a *p*-value of 0.0433. The direction of the regression coefficient that is positive but non-significant is shown on STRDIF2 and STRCL2, of which each has the value of 2.37017 with *p*-value 0.4153 and 7.57690 with a *p*-value of 0.4326 respectively. Meanwhile, on STRCL1, the CI variable shows a regression coefficient direction that is positive and significant, with a value of 3.89686 and a *p*-value of 0.0941. This signifies that competition intensity influences sustainable financial performance only on STRCL1.

5 Conclusion

In this research, the differentiation strategy is found to significantly influence sustainable financial performance on STRDIF2 (H1a is accepted), and cost leadership does not significantly influence sustainable financial performance (H1b is rejected). Innovation is found to have a not so solid influence on sustainable financial performance because it is only proven in the STRCL2 test (H2 is accepted). It means that innovation can still influence a company's future success. Aside from that, innovation also moderates the relationship between differentiation and sustainable financial performance (H3a is accepted) but does not moderate the relationship between cost-leadership and sustainable financial performance (H3b is rejected). This research result contradicts the research of Altuntas et al., (2014) and Wong et al., (2016) who find that differentiation strategy does not have a significant influence, and Li and Li (2008) do not find any influence. This research is also different from the research of Banker et al., (2014) who find that the differentiation strategy and cost-leadership strategy has a positive influence on performance, but is in line with the context that differentiation strategy is more sustainable than the cost-leadership strategy.

The implication of this research on managers is that the managers need to utilize scarce resources to make it hard for competitors to imitate and to be able to survive in the long-term. Managers can utilize common resources and create a good financial performance, but it will only last in a short term. Furthermore, it is also important for the managers to keep innovating to be able to compete in a dynamic environment and the ever-growing pace of technology development. However, this differentiation strategy is quite risky because it needs huge costs to obtain scarce resources and innovate. Because of that, managers need to be careful in implementing the best and the most suitable strategy for the company.

This research is limited to the samples of go-public companies that are listed on South East Asia's stock exchange. The amount of sample is low because there is a limited number of companies that have 5 consecutive years of R&D data and 10 consecutive years of EPS data. Aside from that, innovation is not an absolute on either differentiation or cost-leadership strategy. The strategy measurement in this research also shows diverse and inconsistent results. Thus, this opportunity can be taken by the next researcher to find the most suitable measurements to calculate the strategies. The research can also use other measurements that can give a more absolute result. The next studies are expected to broaden the research samples to improve the validity of sustainable financial performance measurement. Different moderator variables can also be used to identify the relationship between the strategy and sustainable financial performance.

References

Agarwal, S., Erramilli, M. K., Dev, C. S.: Market orientation and performance in service firms: role of innovation. J.Serv. Market. **17**(1), 68–82 (2003). https://doi.org/10.1108/08876040310461282

Aguguom, A. T., Dada, S. O., Nwaobia, A. N.: Earnings persistence and firm performance: implications of analysts' accurate forecast ability from the emerging market of nigeria. Int. J. Acc. Res. **07** (2019). https://doi.org/10.35248/2472-114X.19.7.197

Alshwer, A.A., Levitas, E.: How innovation can affect ownership structure: the case of transient and dedicated institutional investors. Financ. Strateg. **31**, 291–319 (2014). https://doi.org/10.1108/S0742-332220140000031008

Altuntas, G., Semerciöz, F., Mert, A., Pehlivan, Ç.: Industry forces, competitive and functional strategies and organizational performance: evidence from restaurants in Istanbul, Turkey. Proc. Soc. Behav. Sci. **150**, 300–309 (2014). https://doi.org/10.1016/j.sbspro.2014.09.066

Ameer, R., Othman, R.: Sustainability practices and corporate financial performance: a study based on the top global corporations. J. Bus. Ethics **108**, 61–79 (2012). https://doi.org/10.1007/s10551-011-1063-y

Amoako-Gyampah, K., Acquaah, M.: Manufacturing strategy, competitive strategy and firm performance: an empirical study in a developing economy environment. Int. J. Prod. Econ. **111**, 575–592 (2008). https://doi.org/10.1016/j.ijpe.2007.02.030

Asdemir, O., Fernando, G.D., Tripathy, A.: Market perception of firm strategy. Manag. Financ. **39**(2), 90–115 (2013). https://doi.org/10.1108/03074351311293972

Ashley, A. S., Yang, S. S.: Executive compensation and earnings persistence. J. Bus. Ethics **50**(4), 369–382 (2004). Retrieved from https://www.jstor.org/stable/25075203

Awwad, A. S., Al Khattab, A. A., Anchor, J.: Competitive priorities and competitive advantage in Jordanian manufacturing. J. Serv. Sci. Manag. **6**(1), 69–79. https://doi.org/10.4236/jssm.2013.61008

Baines, A., Langfield-Smith, K.: Antecedents to management accounting change: a structural equation approach. Acc. Organ. Soc. **28**(7–8), 675–698 (2003). https://doi.org/10.1016/S0361-3682(02)00102-2

Balsam, S., Fernando, G. D., Tripathy, A.: The impact of firm strategy on performance measures used in executive compensation. J. Bus. Res. 64(2), 187–193 (2011). https://doi.org/10.1016/j.jbusres.2010.01.006

Banker, R. D., Flasher, R., Zhang, D.: Strategic positioning and asymmetric cost behavior. In: AAA 2014 Management Accounting Section (MAS) (2013). https://doi.org/10.2139/ssrn.2312852

Banker, R. D., Hu, N., Pavlou, P. A., Luftman, J.: CIO reporting structure, strategic positioning, and firm performance. MIS Q. **35**(2), 487–504 (2011). https://doi.org/10.2307/23044053

Banker, R.D., Mashruwala, R., Tripathy, A.: Does a differentiation strategy lead to more sustainable financial performance than a cost leadership strategy? Manag. Decis. **52**(5), 872–896 (2014). https://doi.org/10.1108/MD-05-2013-0282

Barney, J. B.: Organizational culture: can it be a source of sustained competitive advantage? Acad. Manag. Rev. **11**(3), 656–665 (1986). https://doi.org/10.2307/258317

Berman, S. L., Wicks, A. C., Kotha, S., Jones, T. M.: Does stakeholder orientation matter? The relationship between stakeholder management models and firm financial performance. Acad. Manag. J. **42**(5), 488–506 (1999). https://doi.org/10.2307/256972

Bhatt, G., Emdad, A., Roberts, N., Grover, V.: Building and leveraging information in dynamic environments: the role of IT infrastructure flexibility as enabler of organizational responsiveness and competitive advantage. Inf. Manag. **47**(7–8), 341–349 (2010). https://doi.org/10.1016/j.im.2010.08.001

Black, E. L., Carnes, T. A., Richardson, V. J.: The market valuation of corporate reputation. Corp. Reputation Rev. **3**(1), 31–42 (2000). https://doi.org/10.1057/palgrave.crr.1540097

Brown, S. L., Eisenhardt, K. M.: Product development: past research, present findings, and future directions. Acad. Manag. Rev. **20**(2), 343–378 (1995). https://doi.org/10.2307/258850

Cavaleri, S., Shabana, K.: Rethinking sustainability strategies. J. Strateg. Manag. **11**(1), 2–17 (2018). https://doi.org/10.1108/JSMA-08-2016-0050

Chen, J.-S., Tsou, H.T., Huang, A.Y.-H.: Determinants of firm performance: Service delivery innovation: antecedents and impact on firm performance. J. Serv. Res. **12**(1), 36–55 (2009)

Christensen, C.M., Bower, J.L.: Customer power, strategic investment, and the failure of leading firms. Strateg. Manag. J. **17**(3), 197–218 (1996)

Clark, K.B., Fujimoto, T.: Product Development Performance: Strategy, Organization, and Management in the World Auto Industry. Harvard Business School Press, Boston (1991)

Connolly, R. A., Hirsch, B. T., Hirschey, M.: Union rent seeking, intangible capital, and market value of the firm. Rev. Econ. Stat. **68**(4), 567–577 (1986). https://doi.org/10.2307/1924515

Covin, J. G., Miles, M. P.: Corporate entrepreneurship and the pursuit of competitive advantage. Entrepreneurship Theory Prac. **23**(3), 47–63 (1999). Retrieved from https://doi.org/10.1177%2F104225879902300304

Crema, M., Verbano, C., Venturini, K.: Linking strategy with open innovation and performance in SMEs. Meas. Bus. Excel. **18**(2), 14–27 (2014). https://doi.org/10.1108/MBE-07-2013-0042
D'Aveni, R.A.: Hypercompetition : Managing the Dynamics of Strategic Maneuvering. The Free Press, New York (1994)
David, J.S., Hwang, Y., Pei, K.-W., Reneau, J.H.: The performance effects of congruence between product competitive strategies and purchasing management design. Manag. Sci. **48**(7), 866–886 (2002)
De Castro, J.O., Chrisman, J.C.: Order of market entry, competitive strategy, and financial performance. J. Bus. Res. **33**, 165–177 (1995)
Díaz-Díaz, N. L., de Saá Pérez, P.: The interaction between external and internal knowledge sources: An open innovation view. J. Knowl. Manag. **18**(2), 430–446 (2014). https://doi.org/10.1108/JKM-07-2013-0257
Dimitrov, V., Jain, P. C.: The value-relevance of changes in financial leverage beyond growth in assets and GAAP earnings. J. Acc. Auditing Financ. **23**(2), 191–222 (2008). Retrieved from https://doi.org/10.1177/0148558X0802300204
Duane Ireland, R., Webb, J.W.: Strategic entrepreneurship: Creating competitive advantage through streams of innovation. Bus. Horiz. **50**(1), 49–59 (2007). https://doi.org/10.1016/j.bushor.2006.06.002
Edeling, A., Himme, A.: When does market share matter? New empirical generalizations from a meta-analysis of the market share-performance relationship. J. Market. (2018)
Eisenhardt, K. M., Martin, J. A.: Dynamic capabilities: what are they? Strateg. Manag. J. **21**(10–11), 1105–1121 (2000). Retrieved from https://doi.org/10.1002/1097-0266(200010/11)21:10/11%3C1105::AID-SMJ133%3E3.0.CO;2-E
Fatihudin, D., Jusni, Mochklas, M.: How measuring financial performance. Int. J. Civil Eng. Technol. **9**(6), 553–557 (2018). Retrieved from https://www.researchgate.net/publication/326141100
Frambach, R. T., Prabhu, J., Verhallen, T. M.: The influence of business strategy on new product activity: the role of market orientation. Int. J. Res. Market. **20**(4), 377–397 (2003). https://doi.org/10.1016/j.ijresmar.2003.03.003
Francis, J., LaFond, R., Olsson, P. M., Schipper, K.: Costs of equity and earnings attributes. Acc. Rev. **79**(4), 967–1010 (2004). https://doi.org/10.2308/accr.2004.79.4.967
Frohwein, T., Hansjürgens, B.: Chemicals regulation and the Porter hypothesis: a critical review of the new European chemical regulation. J. Bus. Chem. **2**(1), 19–36 (2005)
Ghemawat, P.: Competitive advantage and internal organization: Nucor revisited. J. Econ. Manag. Strat. **3**(4), 685–717 (1995). https://doi.org/10.1111/j.1430-9134.1995.00685.x
Grant, R.M.: The resource-based theory of competitive advantage: implications for strategy formulation chemical regulation. Calif. Manage. Rev. **22**, 114–135 (1991). https://doi.org/10.2307/41166664
Green, R. F., Lisboa, J., Yasin, M. M.: PORTER's (1980) generic strategies in portugal. Eur. Bus. Rev. **93**(2) (1993). https://doi.org/10.1108/EUM0000000001911
Gunday, G., Ulusoy, G., Kilic, K., Alpkan, L.: Effects of innovation types on firm performance. Int. J. Prod. Econ. **133**(2), 662–674 (2011). https://doi.org/10.1016/j.ijpe.2011.05.014
Hambrick, D. C.: Some test of the effectiveness and functional attributes of Miles and Snow's strategic types. Acad. Manag. J. **26**(1), 5–26 (1983a). https://doi.org/10.2307/256132
Hambrick, D.C.: High profit strategies in mature capital goods industries: a contingency approach. Acad. Manag. J. **26**(4), 687–707 (1983b). https://doi.org/10.5465/255916
Hamel, G.: Leading the Revolution. Harvard Business School Press, Boston (2000)
Hamel, G., Prahalad, C.K.: Competing for the Future. Harvard Business School Press, Boston (1994)
Han, J. K., Kim, N., Srivastava, R. K.: Market orientation and organizational performance: is innovation a missing link? J. Market. **62**(4), 30–45 (1998). https://doi.org/10.2307/1252285
Hariyanto, L., Juniarti: Pengaruh family control, firm risk, firm size, dan firm age terhadap profitabilitas dan nilai perusahaan pada sektor keuangan. Bus. Acc. Rev. **2**(1), 141–150 (2014)

Herzallah, A., Gutierrez-Gutierrez, L. J., Rosas, J. F.: Quality ambidexterity, competitive strategies, and financial performance: an empirical study in industrial firms. Int. J. Oper. Prod. Manag. **37**(10), 1496–1519 (2017). https://doi.org/10.1108/IJOPM-01-2016-0053

Hilman, H., Kaliappen, N.: Do cost leadership strategy and process innovation influence the performance of malaysia hotel industry? Asian Soc. Sci. **10**(10) (2014). https://doi.org/10.5539/ass.v10n10p134

Hilmi, M., Ramayah, T., Mustapha, Y., Pawanchik, S.: Product and process innovations: evidence from Malaysian SMEs. Eur. J. Soc. Sci. **16**(4), 556–565 (2010)

Hosseini, S.M., Sheikhi, N.: An empirical examination of competitive capability's contribution toward firm performance: moderating role of perceived environmental uncertainty. Int. Bus. Res. **5**(5), 116–131 (2012). https://doi.org/10.5539/ibr.v5n5p116

Hull, C.E., Rothenberg, S.: Firm performance: the interactions of corporate social performance with innovation and industry differentiation. Strateg. Manag. J. **29**, 781–789 (2008). https://doi.org/10.1002/smj.675

Hutchinson, K., Alexander, N., Quinn, B., Doherty, A.M.: Internationalization motives and facilitating factors: qualitative evidence from smaller specialist retailers. J. Int. Mark. **15**(3), 96–122 (2007). https://doi.org/10.1509/jimk.15.3.96

Jermias, J.: The relative influence of competitive intensity and business strategy on the relationship between financial leverage and performance. Br. Acc. Rev. **40**(1), 71–86 (2008). https://doi.org/10.1016/j.bar.2007.11.001

Juniarti.: Does mandatory CSR provide long-term benefits to shareholders? Soc. Responsib. J. (2020). https://doi.org/10.1108/SRJ-03-2019-0114

Kaliappen, N., Hilman, H.: Enhancing organizational performance through strategic alignment of cost leadership strategy and competitor orientation. Middle-East J. Sci. Res. **18**(10), 1411–1416 (2013). https://doi.org/10.5829/idosi.mejsr.2013.18.10.12410

Kim, E., Nam, D.-I., Stimpert, J.L.: The applicability of Porter's generic strategies in the digital age: assumptions, conjectures, and suggestions. J. Manag. **30**(5), 569–589 (2004). https://doi.org/10.1016/j.jm.2003.12.001

Kiron, D., Kruschwitz, N., Reeves, M., Goh, E.: The benefits of sustainability-driven innovation. MIT Sloan Manag. Rev. **54**(2), 68–73 (2013). Retrieved from https://sloanreview.mit.edu/wp-content/uploads/2012/12/f0bfde1b37.pdf

Kotha, S., Nair, A.: Strategy and environment as determinants of performance: evidence from the Japanese machine tool industry. Strateg. Manag. J. **16**(7), 497–518 (1995). https://doi.org/10.1002/smj.4250160702

Kotha, S., Vadlamani, B.L.: Assessing generic strategies: an empirical investigation of two competing typologies in discrete manufacturing industries. Strateg. Manag. J. **16**, 75–83 (1995). https://doi.org/10.1002/smj.4250160108

Leonidou, L.C., Fotiadis, T.A., Christodoulides, P., Spyropoulou, S., Katsikeas, C.S.: Environmentally friendly export business strategy: its determinants and effects on competitive advantage and performance. Int. Bus. Rev. **24**(5), 798–811 (2015). https://doi.org/10.1016/j.ibusrev.2015.02.001

Li, C. B., Li, J. J.: Achieving superior financial performance in China: differentiation, cost leadership, or both? J. Int. Market. **16**(3), 1–22 (2008). Retrieved from https://www.jstor.org/stable/27755568

Li, J.J., Poppo, L., Zhou, K.Z.: Do managerial ties in china always produce value? competition, uncertainty, and domestic vs. foreign firms. Strateg. Manag. J. **29**, 383–400 (2008). https://doi.org/10.1002/smj.665

Liu, W., Atuahene-Gima, K.: Enhancing product innovation performance in a dysfunctional competitive environment: the roles of competitive strategies and market-based assets. Ind. Market. Manag. (2018). https://doi.org/10.1016/j.indmarman.2018.01.006

Madhavan, R., Grover, R.: From embedded knowledge to embodied knowledge: new product development as knowledge management. J. Mark. **62**(4), 1–12 (1998). https://doi.org/10.2307/1252283

Mathenge, J.: The effect of innovation on competitive advantage of telecommunication companies in Kenya. Research Project Submitted in Partial Fulfilment of the Requirements for the Degree of Master of Business Administration, University of Nairobi, Kenya (2013)

McGuire, J.B., Sundgren, A., Schneeweis, T.: Corporate social responsibility and firm financial performance. Acad. Manag. J. **31**(4), 854–872 (1988). https://doi.org/10.2307/256342

Mercer, M.: How do investors assess the credibility of management disclosures? Acc. Horiz. **18**(3), 185–196 (2004). https://doi.org/10.2308/acch.2004.18.3.185

Miller, A., Dess, G.G.: Assessing Porter's (1980) model in terms of its generalizability, accuracy and simplicity. J. Manag. Stud. **30**(4), 553–585 (1993)

Miller, D.: The correlates of entrepreneurship in three types of firms. Manag. Sci. **9**, 770–791 (1983). https://doi.org/10.1287/mnsc.29.7.770

Miller, D.: The structural and environmental correlates of business strategy. Strateg. Manag. J. **8**(1), 55–76 (1987). Retrieved from https://www.jstor.org/stable/2486146

Miller, D., Friesen, P.H.: Strategy-making and environment: the third link. Strateg. Manag. J. **4**(3), 221–235 (1983). https://doi.org/10.1002/smj.4250040304

Miller, J. G., Roth, A. V.: A taxonomy of manufacturing strategies. Manag. Sci. **40**(3), 285–304 (1994). Retrieved from https://www.jstor.org/stable/2632800

Mohammadi, P., Fathi, S., Kazemi, A.: Differentiation and financial performance: a meta-analysis. Compet. Rev. **29**(5), 573–591 (2019). https://doi.org/10.1108/CR-10-2018-0067

Montes, F.J., Moreno, A.R., Fernández, L.M.: Assessing the organizational climate and contractual relationship for perceptions of support for innovation. Int. J. Manpow. **25**(2), 167–180 (2004). https://doi.org/10.1108/01437720410535972

Murray, J. Y., Gao, G. Y., Kotabe, M.: Market orientation and performance of export ventures: the process through marketing capabilities and competitive advantages. J. Acad. Market. Sci. **39**(2), 252–269 (2011). https://doi.org/10.1007/s11747-010-0195-4

Nair, A., Filer, L.: Cointegration of firm strategies within groups: a long-run analysis of firm behavior in the Japanese steel industry. Strateg. Manag. J. **24**(2), 145–159 (2002). https://doi.org/10.1002/smj.286

Najafi-Tavani, S., Najafi-Tavani, Z., Naudé, P., Oghazi, P., Zeynaloo, E.: How collaborative innovation networks affect new product performance: product innovation capability, process innovation capability, and absorptive capacity. Ind. Market. Manag. (2018). https://doi.org/10.1016/j.indmarman.2018.02.009

Naldi, M., Flamini, M.: Dynamics of the Hirschman–Herfindahl index under new market entries. Econ. Pap. J. Appl. Econ. Pol. **37**(3), 344–362 (2018). https://doi.org/10.1111/1759-3441.12222

Nandakumar, M., Ghobadian, A., O'Regan, N.: Generic strategies and performance—evidence from manufacturing firms. Int. J. Product. Perform. Manag. **60**(3), 222–251 (2011). https://doi.org/10.1108/17410401111111970

Nandakumar, M., Ghobadian, A., O'Regan, N.: Business-level strategy and performance: the moderating effects of environment and structure. Manag. Decis. **48**(6), 907–939 (2010). https://doi.org/10.1108/00251741011053460

O'Reilly, C. A., Chatman, J., Caldwell, D. F.: People and organizational culture: a profile comparison approach to assessing person-organization fit. Acad. Manag. J. **34**(3), 487–516 (1991). https://doi.org/10.2307/256404

O'Reilly, C.A., Tushman, M.L.: The Ambidextrous Organization. Harvard Bus. Rev. **82**(4), 74–81 (2004)

O'Sullivan, D., Dooley, L.: Applying Innovation. SAGE Publications, Inc. (2009). https://doi.org/10.4135/9781452274898

Oyewobi, L. O., Windapo, A. O., Rotimi, J. O., Jimoh, R. A.: Relationship between competitive strategy and construction organisation performance: the moderating role of organisational characteristics. Manag. Decis. **54**(9), 2340–2366 (2016). https://doi.org/10.1108/MD-01-2016-0040

Parnell, J.A.: Strategic capabilities, competitive strategy, and performance among retailers in Argentina, Peru and the United States. Manag. Decis. **49**(1), 139–155 (2011). https://doi.org/10.1108/00251741111094482

Parnell, J. A., Lester, D. L., Long, Z., Köseoglu, M. A.: How environmental uncertainty affects the link between business strategy and performance in SMEs: evidence from China, Turkey, and the USA. Manag. Decis. **50**(4), 546–568 (2012). https://doi.org/10.1108/00251741211220129

Parnell, J., Brady, M.: Capabilities, strategies and firm performance in the United Kingdom. J.Strateg. Manag. **12**(1), 153–172 (2019). https://doi.org/10.1108/JSMA-10-2018-0107

Penman, S., Zhang, X.-J.: Modeling Sustainable Earnings and P/E Ratios with Financial Statement Analysis. Working Paper, Columbia University, Graduate School of Business (2002)

Perera, S., Harrison, G., Poole, M.: Customer-focused manufacturing strategy and the use of operations-based non-financial performance measures: a research note. Acc. Organ. Soc. **22**(6), 557–572 (1997). https://doi.org/10.1016/S0361-3682(96)00048-7

Peters, T.: Get innovative or get dead (Part I). Calif. Manag. Rev. **33**(1), 9–26 (1990)

Peters, T.: Get innovative or get dead (Part II). Calif. Manag. Rev. **33**(2), 9–23 (1991)

Pimentel, R. C., Aguiar, A. B.: The role of eearnings persistence in valuation accuracy and the time horizon. Revista de Administração de Empresas **56**(1) (2016). https://doi.org/10.1590/S0034-759020160107

Podsakoff, P. M., Organ, D. W.: Self-reports in organizational research: problems and prospects. J. Manag. **12**(4), 531–544 (1986). https://doi.org/10.1177/014920638601200408

Porter, M.E.: Competitive Strategy: Techniques for Analyzing Industries and Competitors. The Free Press, New York (1980)

Porter, M.E.: The Competitive Advantage of Nations: With a New Introduction. The Free Press, New York (1990)

Porter, M.E.: Competitive strategy. Meas. Bus. Excell. **1**(2), 12–17 (1997). https://doi.org/10.1108/eb025476

Prajogo, D.I.: The relationship between competitive strategies and product quality. Ind. Manag. Data Syst. **107**(1), 69–83 (2007). https://doi.org/10.1108/02635570710719061

Prajogo, D.I.: The strategic fit between innovation strategies and business environment in delivering business performance. Int. J. Prod. Econ. **171**(2), 241–249 (2016). https://doi.org/10.1016/j.ijpe.2015.07.037

Qin, Z.: Process innovation, cost leadership and market power analysis based on electronic information industry. In: Research Project of Zhejiang Philosophy and Social Science (2007)

Racela, O. C., Thoumrungroje, A.: When do customer orientation and innovation capabilities matter? An investigation of contextual impacts. Asia Pac. J. Market. Logist. **32**(2), 445–472 (2019). https://doi.org/10.1108/APJML-03-2019-0143

Ren, L., Xie, G., Krabbendam, K.: Sustainable competitive advantage and marketing innovation within firms: a pragmatic approach for Chinese firms. Manag. Res. Rev. **33**(1), 79–89 (2010). https://doi.org/10.1108/01409171011011580

Saliba de Oliveira, J. A., Basso, L. F., Kimura, H., Sobreiro, V. A.: Innovation and financial performance of companies doing business in Brazil. Int. J. Innov. Stud. **2**(4), 153–164 (2018). https://doi.org/10.1016/j.ijis.2019.03.001

Seaden, G., Manseau, A.: Public policy and construction innovation. Build. Res. Inf. **29**(3), 182–196 (2001). https://doi.org/10.1080/09613210010027701

Selling, T. I., Stickney, C. P.: The effects of business environment and strategy on a firm's rate of return on assets. Financ. Anal. J. **45**(1), 43–52+68 (1989). Retrieved from www.jstor.org/stable/4479186

Simpson, P.M., Siguaw, J.A., Enz, C.A.: Innovation orientation outcomes: the good and the bad performance: evidence from Australia. J. Bus. Res. **59**, 1133–1141 (2006). https://doi.org/10.1016/j.jbusres.2006.08.001

Spencer, X.S., Joiner, T.A., Salmon, S.: Differentiation strategy, performance measurement systems and organizational performance: evidence from Australia. Int. J. Bus. **14**(1), 83–103 (2009)

Tan, Q., Sousa, C.M.: Leveraging marketing capabilities into competitive advantage and export performance. Int. Mark. Rev. **32**(1), 78–102 (2015). https://doi.org/10.1108/IMR-12-2013-0279

Teece, D. J., Pisano, G., Shuen, A.: Dynamic capabilities and strategic management. Strateg. Manag. J. **18**(7), 509–533 (1997). Retrieved from https://www.jstor.org/stable/3088148

Teeratansirikool, L., Siengthai, S., Badir, Y., Charoenngam, C.: Competitive strategies and firm performance: the mediating role of performance measurement. Int. J. Product. Perform. Manag. **62**(2), 168–184 (2013). https://doi.org/10.1108/17410401311295722

Turulja, L., Bajgoric, N.: Innovation, firms' performance and environmental turbulence: is there a moderator or mediator? Eur. J. Innov. Manag. **22**(1), 213–232 (2019). https://doi.org/10.1108/EJIM-03-2018-0064

Urbancova, H.: Competitive advantage achievement through innovation and knowledge. J. Compet. **5**(1), 82–96 (2013). https://doi.org/10.7441/joc.2013.01.06

Uzkurt, C., Kumar, R., Kimzan, H.S., Eminoglu, G.: Role of innovation in the relationship between organizational culture and firm performance: a study of the banking sector in Turkey. Eur. J. Innov. Manag. **16**(1), 92–117 (2013). https://doi.org/10.1108/14601061311292878

Waddock, S. A., Graves, S. B.: The corporate social performance-financial performance link. Strateg. Manag. J. **18**(4), 303–319 (1997). Retrieved from http://www.jstor.org/stable/3088143

Wong, W.P., Soh, K.L., Chong, C.L.: Differentiated service consumption and low cost production: striking a balance for a sustainable competitive advantage in Malaysia. Int. J. Prod. Econ. **181**, 450–459 (2016). https://doi.org/10.1016/j.ijpe.2015.09.029

Wu, P., Gao, L., Gu, T.: Business strategy, market competition and earnings management: evidence from China. Chin. Manag. Stud. **9**(3), 401–424 (2015). https://doi.org/10.1108/CMS-12-2014-0225

Yamin, S., Gunasekaran, A., Mavondo, F.T.: Relationship between generic strategies, competitive advantage and organizational performance: an empirical analysis. Technovation **19**(8), 507–518 (1999). https://doi.org/10.1016/S0166-4972(99)00024-3

Yao, Q., Qin, H.: Marketing capability, competitive advantage, and business performance. Int. J. Technol. Policy Manag. **16**(3), 195–213 (2016). https://doi.org/10.1504/IJTPM.2016.079242

Yeung, J.H., Selen, W., Sum, C.-C., Huo, B.: Linking financial performance to strategic orientation and operational priorities: an empirical study of third-party logistics providers. Int. J. Phys. Distrib. Logist. Manag. **36**(3), 210–230 (2006). https://doi.org/10.1108/09600030610661804

Zaefarian, G., Forkmann, S., Mitręga, M., Henneberg, S.C.: A capability perspective on relationship ending and its impact on product innovation success and firm performance. Long Range Plan. **50**(2), 184–199 (2017). https://doi.org/10.1016/j.lrp.2015.12.023

Zahra, S. A., Covin, J. G.: Contextual influences on the corporate entrepreneurship-performance relationship: a longitudinal analysis. J. Bus. Ventur. **10**(1), 43–58 (1995). https://doi.org/10.1016/0883-9026(94)00004-E

Zehir, C., Can, E., Karaboga, T.: Linking entrepreneurial orientation to firm performance: the role of differentiation strategy and innovation performance. Proc. Soc. Behav. Sci. **210** (2015). https://doi.org/10.1016/j.sbspro.2015.11.381

Evolutions of the Forest Fund Against the Background of Climate Change and the Effects on the Carbon Stock and the Environment

Adela Sorinela Safta and Lavinia Popescu

Abstract The investment will provide a holistic aspect of generating the cost of productive costs for agricultural systems in the field of new climate change. The development of phased climate change effects in the functioning of the developing regions has a clear and important record of information on new environmental conditionality in the new Community Agricultural Policy. As a matter of course, the role is protected by the protection of the population - forest areas from the area of the company which squats the protection of the area with agricultural crops and the climatic factors of climate change. The evolution of the industrial system of agricultural systems has been one of the main points of research and observes the general level of agricultural systems, thus excluding the role of forestry supporters. The study presented and proposed in the continuation of the role of silviculture, by finding the effects and benefits of regeneration and the protection of forestry perpetrators in the context of climate change. A good mentor for good practice in the field of agricultural research will have effective climate change in agriculture.

Keywords Climate change · Agricultural systems · Forest fund · Environment

1 Introduction

Existing natural environments, climate, and agricultural practices worldwide as well as there is an acute need to maintain these areas soils, plants forming a circle in which pesticides have their role. The degradation of the environment, manifested in the last decades by radical modifications of the geosystems on large spaces, with the installation of some chronic ecological imbalances, requires the taking of urgent ecological reconstruction measures. The relevance of this study is based on the importance of the

A. S. Safta (✉) · L. Popescu
Bucharest University of Economic Study, 40987 Bucharest, Romania
e-mail: saftaadela19@stud.ase.ro

L. Popescu
e-mail: popesculavinia14@stud.ase.ro

M. Busu (ed.), *Digital Economy and New Value Creation*, Springer Proceedings in Business and Economics, https://doi.org/10.1007/978-3-031-07265-9_17

elements of good practice in the management of forest areas and forest curtains for the protection of the natural habitat at the foot of the development areas in Romania.

It is true that climate change also entails new innovative production models precisely to find new sources that are sustainable, meant to reduce costs for sustainability (European Commission 2021). This new approach is of particular interest to farmers alike and to researchers, the interference between needs and standards shows that the existence of an ongoing debate on how best to adapt agricultural practices to climate change is still relevant today.

Against the background of these debates, the whole world is facing higher levels of vulnerability (Popescu et al. 2020).

Our aim is to provide a diagram of the risks of a sustainable agricultural economy in the context of the demands of climate change, but also of its global consequences. From this perspective, we stress in part why it is important to prioritize vulnerabilities in the agricultural sector, the relationship of costs and subsidies to prioritize the conservation of forest areas, meadows through the use of agricultural management in mountainous and hilly areas.

2 Literature Review

Existing natural environments, climate, and agricultural practices worldwide as well as there is an acute need to maintain these areas soils, plants forming a circle in which pesticides have their role. The relevance of this study is based on the importance of the elements of good practice in the management of forest areas and forest curtains for the protection of the natural habitat at the foot of the development areas in Romania.

The inventory of methods used in agricultural systems does not necessarily reflect a standard of zonal development, but as a form of conservation instinct, nature reserves the right to prove its usefulness, the forests leaving room for meadows. This new approach is of particular interest to farmers alike and to researchers, the interference between needs and standards shows that the existence of an ongoing debate on how best to adapt agricultural practices to climate change is still relevant today.

Against the background of these debates, the whole world is facing higher levels of vulnerability (Aldag Tecimen 2017).

Our aim is to provide a diagram of the risks of a sustainable agricultural economy in the context of the demands of climate change, but also of its global consequences. From this perspective, we stress in part why it is important to prioritize vulnerabilities in the agricultural sector, the relationship of costs and subsidies to prioritize the conservation of forest areas, meadows through the use of agricultural management in mountainous and hilly areas.

The main objective of the presentation is the external factors that encumber the agricultural system in the case of forest areas. Forest bottom, buffer zones and soil. Romania is a country with a high biological diversity, the natural and seminatural ecosystems represent approximately 47% of the country's surface. The forests occupy a total area of 3.0043.946 hectares, representing 96.7% of the total area of the state-owned forest fund. From conserving and increasing biological diversity by reducing negative impacts and rebuilding ecosystems and damaged habitats to banning the use of nonselective pesticides and rebuilding ecosystems degraded by overexploitation, each regional development area is addressed separately. Restoring bushes and creating protective curtains, as well as preventing the reduction of biological diversity are some of the sensitive issues that make the management plan to be customized according to the local geographical area.

It has become vital to investigate alternative agricultural methods that reduce greenhouse gas emissions, biodiversity loss, deforestation, and soil erosion (Andrei et al. 2015).

3 The Distribution of the National Forest Fund

About two-fifths of EU land is cultivated, which demonstrates the direct impact that agriculture has on the natural environment and thus on forest systems. But how each system contributes to the spatial construction of the environment is highlighted in the impact of climate change on agriculture and, equally, agriculture on climate change. Field protection curtains—forest areas in plain areas (steppe and forest-steppe) that have the role of protecting areas with agricultural and forestry crops against harmful climatic factors (strong winds, etc.).

In 2018, public property represented 64.3% of the total area of the National forest Fund, and private property represented 35.7%, being managed mostly by private forestry structures (95.6%). The distribution of the National Forest Fund in development regions and counties is uneven, depending on the physical-geographical conditions and the economic-social development of the area (Aznar-Sánchez et al. 2019).

At the level of development regions, in 2018, 26.0% of the total regenerated area was achieved in the North-East region, 18.0% in the North-West region, 16.6% in the Center region, 11.9% in West region, 9.5% in the South-East region, 9.0% in the South-Muntenia region, 8.6% in the South-West Oltenia region, and 0.4% in the Bucharest-Ilfov region. Forest protection curtain—formations with forest vegetation located at a certain distance from each other or with an objective to protect it against the effects of harmful factors and/or for the climatic, economic, and aesthetic sanitary improvement of the land.

The degradation of the environment, manifested in the last decades by radical modifications of the geosystems in large spaces, with the installation of some chronic ecological imbalances, requires the taking of urgent ecological reconstruction measures (Huang et al. 2019).

An important environmental factor that can be directly influenced and that effectively contributes to preventing and combating environmental degradation is forest vegetation.

The forest vegetation is a real biological barrier both against pollutants of any kind and against harmful climatic factors (Ramírez et al. 2019).

By creating forest protection curtains, the climatic, economic, and aesthetic sanitary improvement of the land is achieved. The installation of antierosion forest curtains is necessary in all areas where the slope of the land is higher than 5% and where the phenomena of surface erosion, deep erosion (ravines and gullies), on torrential alluvium deposits as well as on lands degraded by displacement phenomena. When setting up forest curtain networks for field protection, the aim is to remove as little as possible from the agricultural circuit (maximum 5%). For this purpose, the land configuration will be used (secondary roads, border roads, muddy roads, embankments of some dams, canal banks, unproductive lands for agriculture (Ruttan 2019).

The surface of the forest fund in private property has an increasing tendency to the detriment of the surface in public property, due to the continuation of the process of restitution of the forests.

An important environmental factor that can be directly influenced and that effectively contributes to preventing and combating environmental degradation is forest vegetation. The forest vegetation is a real biological barrier both against pollutants of any kind and against harmful climatic factors (OECD 2011).

The National forest Fund on December 31, 2019, registered an increase of 9165 ha. This increase is mainly due to the redevelopment of forested pastures and the introduction into the forest fund of degraded lands and nonforested lands, established under law1 (Sauerbeck 1993). The approach to risk in our opinion that affects biodiversity and the resilience of natural capital and on the other hand, the sustainability of agricultural ecosystems can be applied separately depending on the ecosystem yield process being cyclical, and hence the distinct features of agricultural management approaches. Thus, the benefit of agri-environmental measures and subsidies is reflected in the balanced analysis of production methods.

4 Methodology

We also considered the analysis of stocks by climate, soil type, degree of agricultural activities per ha, and soil works in the methodology for determining carbon stocks, which took into account the study of subsidies and the impact of agricultural activities on agricultural holdings compared to production as shown above. In the case of the IPCC approach (Westhoek et al. 2014), we utilized the soil carbon stock calculation rule for the reference use of CSR land, as well as the actual use of CSA land based on plant activity on the ground (cultivated per ha) and above the ground. (completed project)

$$\mathrm{CSi} = \mathrm{SOC} + \mathrm{CVEG} \times \mathrm{A}$$

where

CSi = amount of carbon per unit area associated with land use.

SOC = amount of organic carbon in the soil) measured as mass of carbon per hectare.

CVEG = soil organic carbon stock (measured as mass of carbon per hectare).

A = coefficient of the area in question (measured as number of hectares per unit area

$$SOC = SOCST \times FLU \times FMG \times FI$$

where:

SOCST = the standard amount of organic carbon in the soil in the 0–30 cm layer at the surface of the land (measured as mass of carbon per hectare);

FLU = land use factor reflecting the difference between the amount of organic carbon in the soil associated with the type of land use, compared to the standard amount of organic carbon in the soil;

FMG = the management factor that reflects the difference between the amount of organic carbon in the soil associated with the management practices in principle, compared to the standard amount of organic carbon in the soil;

FI = input factor reflecting the difference between the amount of organic carbon in the soil associated with different levels of carbon inputs to the soil compared to the standard amount of organic carbon in the soil.

Factors is in Table 7.1 Annex V to Directive 2009/28/EC of the European Parliament and of the Council of 23 April 2009 (europa.eu), L140/16, OJ EU, 5.06.2009

$$SOC1a = 63 \times 0,8 \times 1 \times 0,95 = 47,88$$

$$SOC2a = 63 \times 0,8 \times 1 \times 1 = 50,4$$

$$SOC3a = 63 \times 0,8 \times 1 \times 1,37 = 69,04$$

$$SOC4a = 63 \times 0,8 \times 1 \times 1,04 = 52,42$$

then

$$CVEG = CBM + CDOM$$

with CDOM the value 0 may be used, except in wooded areas with a crown of more than 30%—excluding forest plantations

$$\text{CBM} = \text{CAGB} + \text{CBGB};$$

$$\text{CAGB} = \text{BAGB} \times \text{CFB};$$

$$\text{CBGB} = \text{BBGB} \times \text{CFB}$$

The figure for BAGB = weight of live biomass above ground (measured as mass of dry matter per hectare) for agricultural land, perennial crops, and forest plantations; BBGB reflects the average weight of alive biomass in the soil during the production cycle. The value of 0.47 is utilized for CFB, resulting in

$$\text{CS1} = 47,88 + 0,47(\text{BAGB} + \text{BBGB}) \times \text{A}$$

There are some practical implications in the second part of the investigation, which served as the foundation for this research. To begin with, agriculture has been triggered by the continued attractiveness of subsidy money and has a tendency to rely on the financing provided as an incentive without at least gradually adding an index to calculate the rise. In addition to the compelling studies from the Recommendations for determining the C stock of soil, the approach utilized for 1:1 scale evaluations was a major problem. To estimate the increase in application of environmental requirements, we had to introduce a correlation indicator. At the humidity index, we introduced the atmospheric pressure, which has the effect of favoring soil moisture in climatic vulnerabilities on the storage and storage of CSAU at ground level. As a result, we discovered a stagnation of 5–8 months per year in which the storage C at ground level does not stagnate, following the Spor index to be an average compared to 4–6 months preferably: 4 months when it is presumed to be a long day, so the dynamics-humidity variable is low during the summer. As a consequence, the formula is as follows

$$Sphor = P^1 + CSAU + \sum_{k^1}^{n} \frac{E^1}{t}$$

n = number of periods for which data was collected in a given t,

k_1 = extension coefficient to a number of measurements.

C_{SAU} = effective density C at the value SAU (per ha) in Kg/m^3 C_{SAU}.

P = soil nutrient power measured as a coefficient of weight C in mass, qualified as an index measured progressively at A surface area at depth, a_1 = 0–10 cm^3, a_2 = 0–30 cm^3, a_3 = 0–40 cm^3.

E^1 = item independent of the atmospheric pressure variable that determines the humidity.

Table 1 Forms of agricultural land management (own documentation)

Land area cut per hectare	Degree of land occupation (%)
Surface 1 ha the plot 0, 3 ha*	70
Surface 1 ha the plot 0, 03 ha*	63
Surface between 10–30 ha	75
Surface 30 ha 2 different cultures	75
Surface > 30%, 3 diff. cultures	71
Proportion SAU Total area ≥ 5%	≥5

Sphor = Increased absorption of C in the soil at variable atmospheric pressure.

The nutritional power of the soil must be measured at the given area by conversion to the volume of a sphere, r being the relative correspondence to a, being calculated at the slope coordinate 0–5°, 5–10°, 10–15° and over 15°.

Thus the tolerance r at 1 ha is 0.72/0.90/1.08/1.14. These determinations take into account Topometric Tables bying. Niculescu. The elements of probability have in view the atmospheric pressure exerted at the level of the atmosphere that favors the humidity coefficient in air other than the humidity at ground level, so not as a linearly determined coefficient. Result (Table 1).

$$P = A \times \sqrt{\frac{4}{3}\pi r^3}$$

5 Results and Discussions

The capacity to minimize greenhouse gas emissions has increased as forest systems have evolved. Agriculture, according to Tilman et al. (2001), is at the root of the problem, as it is the world's largest driver of global environmental change. However, agriculture and forestry can be assessed by simulating the growth of the two systems, and forest areas actually protect agricultural production soils (European Commission 2021). Finally, net SCS techniques can produce synergies and trade-offs between diverse sources of GHGs and other environmental consequences, such as biodiversity, air quality, and water quality. Net SCS techniques, more than most other agricultural mitigation strategies, have a beneficial impact on the agricultural sector's overall environmental performance. Policies that encourage the adoption of these practices must consider these interconnections in order to maximize the environmental gains and minimize the environmental trade-offs.

Improving the efficiency of agricultural-environmental policies by gradually increasing the basic requirements of agricultural practices and, where appropriate, imposing conditionality on farmers' ability to manage current risks and adapt to evolving risks, particularly in the context of climate change. Natural regeneration accounted for 17,972 hectares (66.5%), which is 676 hectares higher than in 2017, while artificial regeneration accounted for 9071 hectares (33.5%), which is 1665 hectares fewer than the previous year. The forest fund contains, by its legal character, the method in which the forest fund is represented depending on the form of ownership, as illustrated in Table 2. In 2018, public property contributed for 64.3% of the entire area of the National Forest Fund, with the National Forests Authority—Romsilva managing the majority of it, while private property accounted for 35.7%, with the structures private forestry managing the majority of it. The National Forest Fund is distributed unevenly among development areas and counties, based on the area's physical-geographical circumstances and economic-social development.

The importance of achieving proper forest fund management entails, first and foremost, the importance of forest protection curtains, which are formations of forest vegetation located at a certain distance from each other or with the goal of protecting it from harmful factors and/or improving the lands' climatic, economic, and aesthetic sanitary conditions. We emphasized technological modalities for relaying forest curtain management via their involvement in climatic zones that more or less exacerbate the vulnerabilities of bordering agricultural ecosystems in the technique.

In the midst of these disputes, the entire planet is becoming increasingly vulnerable (Singh et al. 2020).

One strategy to exploit the potential of geographical ecosystems to enhance the area's biosphere, including agricultural land, is to advance agricultural production practices to protect natural areas, conserve ecosystem functions, and destruction of forest vegetation outside the forest fund.

As a result, nature-based solutions can help society become more resilient to the consequences of climate change by enhancing the synergy between ecosystem management and climate mitigation (Szymczyk 2010).

For instance, maximizing the restoration of forest protection zones protects the agricultural region from the risk of excess rainwater. As a result, there are several options for increasing carbon sequestration while also safeguarding the biosphere.

Fruit and vine crops, which are commonly grown on such ground, require special attention since soil erosion and, implicitly, the threats of nutrient loss via runoff are more frequent and strong. Buffer strips are grassy areas near to protection zones where chemical and organic fertilizer applications are not permitted. The buffer strips' width is measured from the physical block's outside edge to the protective

Table 2 Forest fund

Whitewood	beech	oak tree	miscellaneous	mountain	hill	plain
30,97%	38,94%	13,74%	16,34%	59,70%	33,80%	6,50%

Source https://insse.ro/cms/sites/default/files/com_press/com_pdf/silvicultura_r2018.pdf

zone's interior edge. The average slope of the physical block close to the watercourse is referred to as the land slope (Tilman et al. 2001). The protection zone width is determined by the watercourse width, the kind and destination of the water resource, or the hydrotechnical arrangement as follows:

Illegal logging, grazing, fires, frequent defoliation, natural catastrophes, technical faults in installation, maintenance and administration, and so on are the primary reasons of forest protection curtain deterioration. In comparison to the previous year, the forest area expanded by 9127 hectares in 2019. As shown in Fig. 1, the allocation of the national forest fund by development areas and counties is unequal, based on the area's physical-geographical features and economic-social development.

Natural regeneration accounted for 17,972 hectares (66.5%), which is 676 hectares higher than in 2017, while artificial regeneration accounted for 9071 hectares (33.5%), which is 1665 hectares fewer than the previous year.

Forest curtains that no longer correspond to the role for which they were created, being subject to the phenomenon of degradation, are covered with regeneration or improvement works (Tilman et al. 2001).

Natural regeneration accounted for 17,972 hectares (66.5%), which is 676 hectares higher than in 2017, while artificial regeneration accounted for 9071 hectares (33.5%), which is 1665 hectares fewer than the previous year.

In 2018, the forest fund was concentrated in a significant proportion in the Central (19.3% of the total forest fund) and North-East (18.2%) development regions, followed by the Western development regions (16.1%), North-West (15.2%), South-West—Oltenia (12.4%), South—Muntenia (10.4%), South-East (8.4%), and Bucharest—Ilfov (0.4%). Table 1 shows that in 2018, the North-East region received 26.0% of the total regenerated area, 18.0% in the North-West region, 16.6% in the Center region, 11.9% in the West region, 9.5% in the South-East region, 9.0% in

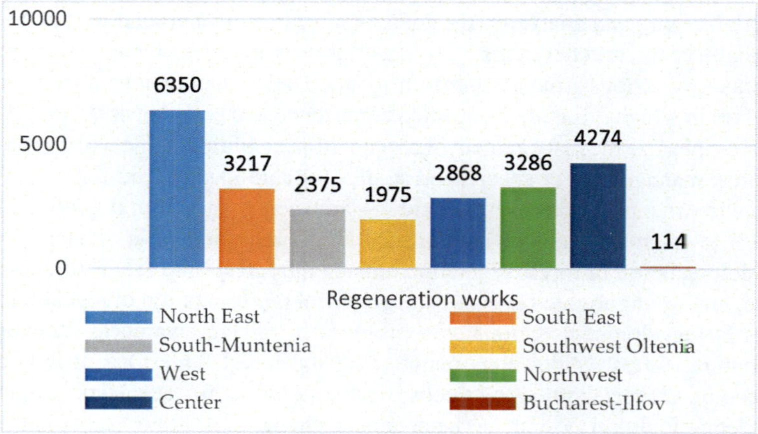

Fig. 1 Forest regeneration works by development regions in 2019. *Source* insse data silvicultura

the South-Muntenia region, 8.6% in the South-West Oltenia region, and 0.4% in the Bucharest-Ilfov region of the total forest fund at the country level.

When drafting relief forms, the maximum altitude is taken into consideration, when it is changeable, the allocations designated for the forest services, for the owners of forest funds, are calculated according to the calculation methodology. As the agricultural sector improves, there's a good chance that GHG emissions from agriculture will rise again—especially if livestock numbers rise and/or crop output intensifies again.

As a result, some causes, such as the impacts of climate change, may be handled using alternative means provided by nature, and the position of protective curtains is part of the standard of sustainable water management in the major lines of study. The examination of land use changes, especially in terms of sustainable development in new types of agricultural management, such as organic farming, are also markers of improved climatic effects in the region.

Technical forestry standards for the establishment, maintenance, and management of forest vegetation of forest protection curtains play a significant part in the natural balancing conditions generated by excellent practices and connected to existing vegetation.

The Romanian National Forest Fund covers 6,559 thousand hectares, or 27.5% of the country's land surface. In comparison to the European average of 32%, the distribution of forests by relief types is typical.

The changes in the agricultural industry's worldwide progress have forced the inclusion of new topics on the work agenda, such as climate change and the strain on agriculture as a result of changing climatic impacts.

In this setting, agricultural policy and rural development strategic goals require reorientation and resizing, and environmental resources are limited, with a considerable recovery period in the usage phase (Tuomisto et al. 2016; Westhoek et al. 2014).

Establishing and analyzing the sources of injuries is a crucial part of measuring the health of forest ecosystems, as well as a priority in terms of diagnosing symptoms produced by various groups of harmful agents. The soil carbon inventory, particularly the form in which it is stored, capacity, persistence and bulk density, and soil textural class, emphasize the management of regeneration activities in line with the provisions of forest management or specialized studies for carbon sequestration.

Environmental impact assessment by advancing agricultural production techniques to maintain areas of natural habitat, conservation of ecosystem functions, and deforestation of forest vegetation outside the forest fund are of strategic importance, and we highlighted in Fig. 2 the graphical representation of regenerated forest areas by development regions as a progress in raising awareness of maintaining the natural forest habitat, thus maintaining the forest fund. One of the causes of the ecological reconstruction of ecosystems, including forest systems as part of the broader agricultural system, has been climate change. The importance of agricultural ecosystems with the highest carbon stocks and the ability to store carbon, particularly through practices appropriate to local conditions, such as cultivating permanent

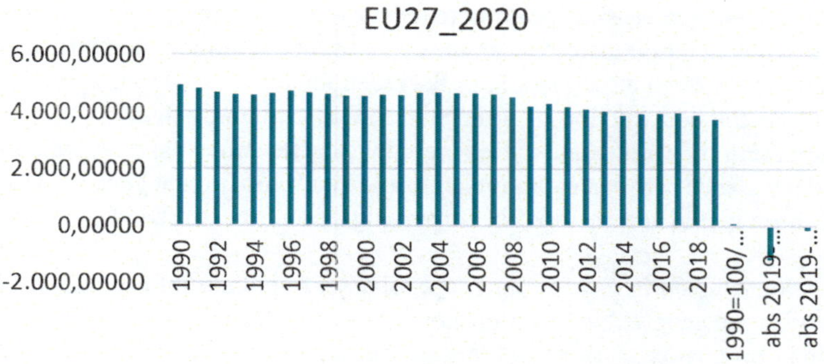

Fig. 2 Emission Land. use–foresttotal E_All_Data Net emissions/removals (CO_2) gigagrams EU27. *Source* Owner data from Eurostat

pastures, maintaining traditional agricultural practices, and avoiding the use of chemical fertilizers and pesticides on valuable pastures, should not be overlooked, in our opinion (Tables 3 and 4).

When drafting relief forms, the maximum altitude is taken into consideration, when it is changeable, the allocations designated for the forest services, for the owners of forest funds, are calculated according to the calculation methodology. As the agricultural sector improves, there's a good chance that GHG emissions from agriculture will rise again—especially if livestock numbers rise and/or crop output intensifies again.

Table 3 National Forest Fund for development regions, at the end of 2019

Forms of ownership	2014	2015	2016	2017	2018
Forest fund—total	6545	6555	6559	6565	6583
Public property	4253	4249	4245	4233	4232
of the state	3217	3203	3194	3189	3186
of the administrative-territorial units	1036	1046	1051	1044	1046
Private property	2292	2306	2314	2332	2351
of natural and legal persons	2203	2215	2225	2239	2248
of the administrative-territorial units	89	91	89	93	103

*Source*https://insse.ro/cms/sites/default/files/com_press/com_pdf/silvicultura_r2019.pdf

Table 4 Trends in carbon stock in living forest biomass 1990–2010

Carbon stock in living forest biomass (year)	1900	2000	2005	2010
UM: million metric tons)				
Total	600	599	601	618

As a factor in increasing the effort to reduce greenhouse gases, the implementation of forest regeneration plans adds to the concern for the introduction into the productive circuit, through afforestation, of degraded land, unsuitable for agricultural use, as well as the construction of protective forests curtains, achieving the objectives of increasing the forest fund by regenerating forests and expanding their area. The interdependence between carbon sequestration and climate change has led to the idea of the need to find new methods of conserving agriculture, as stated by Popescu et al. (2020).

The conclusion is that the interdependence of agricultural ecosystems depends largely on the geographic amphitheater in which they are located, and our attention to forest protection areas shows that the biosphere balance can be maintained by frequent application of advanced techniques. Protection against pests with protection products so that the soil has to increase the properties even in the conditions of climate change nature can support through itself obstacles and risks, the role of subsidies being even to support this mechanism of balancing ecosystems.

As he claims (Popescu et al. 2020), currently, the sources of calculation of climate change must be rethought, the Sphor index being a step to re-establish a simple analysis in another spectrum. We cannot infinitely calculate the effects of climate change only as fixed sources when they are counted worldwide as a whole (Westhoek et al. 2014). Addressing the risks in our opinion that affect biodiversity and the resilience of natural capital and on the other hand, the sustainability of agricultural ecosystems can be applied separately depending on the ecosystem yield conservation process, and hence the distinct features of agricultural management approaches. Climate change has been one of the causes of the ecological reconstruction of ecosystems, including forest systems as part of the entire agricultural system. In our opinion, the role of agricultural ecosystems that have the highest carbon stocks and can store especially through practices appropriate to local conditions, by cultivating permanent pastures, maintaining traditional agricultural practices by avoiding the application of chemical fertilizers and pesticides on valuable pastures should not be neglected. High nutrients, maximizing the restoration of forest protection areas protects the agricultural area, so there are multiple ways to increase carbon sequestration, protecting the biosphere.

The results of this study have several directions that could be a real benefit of agri-environmental measures and subsidies that are reflected in the balanced analysis of sustainable production methods following three pillars: economic, environmental, and social. In terms of the agricultural context, the economic pillar of sustainability is often the most prevalent. Hence, the conclusion that agriculture depends largely on a good working environment, in conditions of climate vulnerability and agriculture suffers. Applied research is needed in a range of areas both for the development of new practices and technologies for climate change mitigation and adaptation, and for improving existing practices and technologies in forest areas with a key role in mitigating the risks of agricultural production in their vicinity with conservation is vital.

References

Aldag Tecimen, H.B.: Land use effect on nitrogen and phosphorus fluxes into and from soil. Eurasian J. Forest Sci. **5**(1), 8–12 (2017)

Andrei, A.S., Robeson, M.S., Barics, A., et al.: Contrasting taxonomic stratification of microbial communities in two hypersaline meromictic:lakes. ISME J **9**, 2642–2656 (2015). https://doi.org/10.1038/ismej.2015.60

Aznar-Sánchez, J.A., Piquer-Rodríguez, M., Velasco-Muñoz, J.F., Manzano-Agugliaro, F.: Worldwide research trends on sustainable land use in agriculture. Land Use Policy **87**, 104069 (2019)

European Commission.: Forging a climate-resilient Europe—the new EU Strategy on Adaptation to Climate Change, Brussels: COM 82 (2021)

Huang, J., Chen, Y., Pan, J., Liu, W., Yang, G., Xiao, X., Zhou, L.: Carbon footprint of different agricultural systems in China estimated by different evaluation metrics. J. Clean. Prod. **225**, 939–948 (2019)

OECD.: Managing Risk in Agriculture, Policy Assessment and Design, s.l.: OECD Publishing (2011). https://read.oecd-ilibrary.org/agriculture-and-food/managing-risk-in-agriculture_9789264116146-en#page2

Popescu, L., et al.: Considerations regarding improving the ecological and environmental performance of agriculture. In: ICBE2020, Proceedings of the International Conference on Business Excellence, vol. 14, Issue 1. Business Revolution in the Digital Era (2020). https://doi.org/10.2478/picbe-2020-0066. Accessed 27 Jul 2020

Ramírez, P.B., Calderón, F.J., Fonte, S.J., Bonilla, C.A.: Environmental controls and long-term changes on carbon stocks under agricultural lands. Soil Tillage Res. **186**, 310–321 (2019)

Ruttan, V.W.: Sustainable agriculture and the environment: Perspectives on growth and constraints. CRC Press (2019)

Sauerbeck, D.R.: CO 2-emissions from agriculture: sources and mitigation potentials. Water Air Soil Pollut. **70**(1–4), 381–388 (1993)

Singh, N., Joshi, E., Sasode, D.S., Chouhan, N.: Conservation Agriculture. Biotica Res. Today **2**(5), 156–158 (2020)

Szymczyk, S.: Influence of the type of soil dewatering and land use on the dynamics of concentrations and volume of nitrogen discharged from agricultural areas. J. Elementol. **15**(1), 189–211 (2010)

Tilman, D., Fargione, J., Wolff, B., D'Antonio, C., Dobson, A., Howarth, R., Schindler, D., Schlesinger, W.H., et al.: Forecasting agriculturally driven global environmental change. Science **292**, 281–284 (2001)

Tuomisto et al.: The role of dung beetles in reducing greenhouse gas emissions from cattle farming Scientific reports, (2016) https://www.nature.com/articles/srep18140

Westhoek, H., Van Zeijts, H., Witmer, M., Van den Berg, M., Overmars, K., Van der Esch, S., Van der Bilt, W.: Greening the CAP. An analysis of the effects of the European Commission's proposals for the common agricultural policy, 2020 (2014)

Westhoek, H., et al.: Food choices, health and environment: effects of cutting Europe's meat and dairy intake. Global Env. Change **26**, 196–205 (2014). http://www.sciencedirect.com/science/article/pii/S095937801400338

COVID-19 Crisis Among Romanian Entrepreneurs—Challenges and Support Measures

Mihaela Cornelia Dan, Simona Irina Goia Agoston, and Shahrazad Hadad

Abstract There is consensus on the fact that COVID-19 crisis goes far beyond its medical roots and poses great challenges for entrepreneurs all over the world. Not only specific business models are provoked, but also the underlying principles of our economic and social system, based on human interaction and consumption as prerequisite for growth, prosperity and well-being. The pressure on the public system, national and international institutions and authorities, which were urged to come with rapid solutions and answers in order to bail-out and support the entrepreneurial ecosystem is enormous. The present paper aims at investigating the main challenges faced by Romanian entrepreneurs and the support schemes and public measures, as they were perceived by the interviewed persons, owners of small and medium-sized enterprises from Romania. After delivering a brief introduction of the specific features of the Romanian entrepreneurial environment and the impact of COVID-19 crisis on Romanian companies, the paper focuses on the policies and support measures offered by the state. Different categories of support schemes are depicted and critically analyzed within the broader European context, but also from the perspective of several Romanian entrepreneurs, target population of the aforementioned measures, participants in a qualitative study, based on in-depth interviews, conducted in the time framework July- December 2020.

Keywords Romanian entrepreneurs · COVID-19 crisis · Support measures · Mall and medium-sized enterprises · Challenges · In-depth interviews

M. C. Dan · S. I. G. Agoston · S. Hadad (✉)
The Bucharest University of Economic Studies, Piata Romana 6, 010374 Bucharest, Romania
e-mail: shahrazad.hadad@fabiz.ase.ro

M. C. Dan
e-mail: mihaela.dan@fabiz.ase.ro

S. I. G. Agoston
e-mail: simona.goia@ase.ro

M. Busu (ed.), *Digital Economy and New Value Creation*, Springer Proceedings in Business and Economics, https://doi.org/10.1007/978-3-031-07265-9_18

1 Introduction

At the beginning of the COVID-19 pandemic, the National Institute of Statistics (NIS 2020) of Romania conducted several ad hoc statistical surveys, Trends in the evolution of economic activity during March–April 2020, The impact of SARS-COV2 on the volume of exports and imports of goods in March and April 2020, to quantify the impact of the health crisis on economic activity and the volume of foreign trade, the results of which have been of real interest to policy makers, the media and users in general. The implications of the pandemic on economic activity and natural movement population are the basis of new statistical data requirements at national and European level on the impact of preventive measures on sectors of national economies. In the coming months, the National Institute of Statistics will make flash estimates to measure the impact of COVID-19 on the economy and society, complemented by a collection of innovative practices that European Union countries have implemented to meet key challenges, including the use of new data sources.

The most common measures taken by companies to reduce the negative effects of the pandemic on their business, at the aggregate level, are the partial reduction of activity, suspension of activity or sending into unemployment some of the staff or all employees. On the other hand, the least popular measures are recourse to insolvency proceedings or judicial reorganization proceedings, suspension of the payment of bank installments or dismissal of some or all employees.

According to a study conducted by HotNews (2020), on the background of the crisis generated by the COVID-19 pandemic, companies identify a worsening tendency of the main factors influencing their activity. They are of the opinion that the economic situation at national level, the situation of the sector to which they belong and the economic and financial situation of their own company have deteriorated compared to the previous year. However, the development prospects, the relationship with business partners (suppliers or customers) or the relationship with the financial institutions with which they work have remained largely unchanged.

According to NIS (2020), the problems that companies have felt are the most pressing in the last six months are production costs or labor costs, the availability of well-trained labor and lack of demand. Access to finance is not one of the pressing issues for companies, with 60% of them considering this factor to have little or no impact on them.

Moreover, in the context of the COVID-19 pandemic, almost a quarter of companies were significantly affected by the reduction in sales (a decrease of over 75%). From the point of view of measures taken by companies to reduce the negative effects of the pandemic on their business, the most common are partial reduction of activity, suspension of activity or sending to unemployment some of the staff or all employees.

The main sources of financing used by companies remained domestic, which became more important during this period than external resources, mainly due to the substantial advance recorded by the reinvestment of profits and the sale of assets. Bank loans remain a much less used option, and the number of companies opting for commercial loans and bank overdrafts continued to decrease during the analyzed period.

The payment discipline of business partners is a factor that is not assessed to be pressing for 49% of companies. About one in 5 companies consider that they are affected by the payment discipline of business partners. Corporations are more affected by this factor compared to small firms. However, lax payment discipline is a structural vulnerability of the Romanian economy, given the amounts of late payment to non-banking trading partners, which represents about 9% of gross domestic product, according to financial data of companies in 2018 (HotNews 2020).

2 Literature Review

As the COVID-19 crisis became a world-wide phenomenon, its associated economic impact has triggered chain reactions across multiple nations sending many countries into economic downturn, sanitary crises and social turmoil. The macroeconomic shocks translated into alterations of supply, demand, trade, investment, supply chain, financial stability, economic growth, exchange rates etc.

It is true that the world economy had already deteriorated widely over the past year, with GDP rising by only 2.9% in 2019, the lowest rate in 10 years, amid endless trade disputes and policy uncertainties, high rates of poverty and inequality, increasingly pressing climate risks, ubiquitous disparities and interactions (Băhnăreanu 2020).

In the first half of 2020 the anti-pandemic measures led to significant declines in almost all economic sectors, the collapse of stock indices, the collapse of the labor market, negative oil prices, disruption of world trade, etc. (McKibbin and Fernando 2021). It seems that this global health problem and the associated economic and social crisis will lead to the deterioration of most macroeconomic indicators in the coming years, which will call into question the activity and architecture of the economic and financial system and medium and long term development prospects (Padhan and Prabheesh 2021).

At global level, there have been conducted different reports investigating the effects of the corona virus outbreak over various economies. In a mild scenario, researchers show that GDP growth would take a hit, ranging from 3–6% depending on the country (Fernandes 2020), while for other countries GDP can fall more than 10%, and in some countries, more than 15%. The main service-oriented economies subject to negative effects (Baker et al. 2020), and have more jobs at risk, such examples could be represented by countries like Greece, Portugal, and Spain that are more reliant on tourism (more than 15% of GDP) will be more affected by this crisis. An interesting result suggests that on average, each additional month of crisis costs 2.5–3% of global GDP, therefore the need to limit the crisis.

The economic implications of this crisis which is deemed to be a Black Swan (Petro 2020) have been emphasized by massive lock-downs which interred with the free movement of people, capital and goods conducing to the temporary impairment of business (Barua 2020).

Aligning to most of the countries within the European Union and also globally, the Romanian Government has adopted various national measures in an attempt to alleviate the impact of the COVID-19 crisis over the economy. The main categories of actions relate to: Economic and budgetary-fiscal measures to support the business environment, Measures to support jobs in the economy and Social protection measures. Of course the measures also gained traction on the one hand, and were very severely criticized, on the other hand, as it will be discussed in the research results section where the mostly affected players will also turn out to be the most vocal. The following section delves into some of the most important measures within the specific categories and also indicate, where available, the legal documents that support their particular enactment and provisions (Romanian Government 2021).

2.1 Economic and Budgetary-Fiscal Measures to Support the Business Environment

- Implementation of a multi-annual program to support SMEs, IMM Invest, by guaranteeing loans and subsidizing interest for these loans (Governance Emergency Ordinance—GEO no. 29/2020 supplemented by GEO no. 42/2020, GEO 89/2020, GEO no. 143/2020 and the application norms adopted by GD (Government Decision) No. 282/2020, amended and supplemented by GD No. 326/2020 and GD No. 458/2020).
- Starting a program (about 8 billion lei) to support large companies and SMEs with a turnover of more than 20 million euros and which can benefit from state guarantees for investment loans and working capital up to 80%—program that will be run through Eximbank.
- SME Leasing guarantee scheme regarding the granting of state guarantee facilities for financial leasing type financing for the purchase of new or second-hand movable goods necessary for carrying out the activities of SMEs, as well as affiliated enterprises with an equal or greater number of 250 employees (GEO 118/2020).
- Approval of a commercial credit guarantee program supplier—SME Factor Program, to ensure the guarantee of payment facilities between merchants—guarantee offered by the state in proportion of 50%. The expected results are the avoidance of the risk of non-payment and the accumulation of arrears.
- Ensuring the necessary financing for contracting the projects submitted within the Regional Operational Program 2014-2020, Priority Axis 2—Improving the competitiveness of small and medium enterprises for financing applications whose total non-reimbursable value exceeds the value of the financial allocation.
- Extension of the payment term for taxes on buildings, on land, on means of transport, from March 31 to June 30, 2020, with the preservation of the bonus established by the Local Council (art. V of GEO 29/2020).

- Granting powers to local councils to reduce the building tax and exempt from paying the building tax, in the case of buildings used by taxpayers who have completely ceased their activity as a result of the establishment of the state of emergency, under certain conditions. (GEO no. 69/2020).
- Deferral of payment for utility and rent services in the case of small and medium-sized enterprises as well as national sports federations/sports clubs experiencing financial difficulties, with the fulfillment of specific conditions (art. X of GEO 29/2020 and GD no. 312/2020).
- Exemption of the holders of broadcasting licenses and licenses for the use of radio frequencies in digital terrestrial system (cf. Law 504/2002) from the payment of the spectrum use tariff for 2020.
- Suspension of seizures and forced executions for budget receivables, except for those established by court decisions in criminal matters (art. VII of GEO 29/2020, amended by GEO no. 99/2020).
- The granting of the fiscal facility consisting in the cancellation of some accessory obligations, related to the main budgetary obligations that were extinguished until 31.03.2020, for all categories of debtors (PF and PJ), with certain conditions. (GEO no. 69/2020 and the related norms from OMFP 2100/2020).
- Establishment of a specific method for taxpayers who apply the system of declaration and payment of annual income tax, with advance payments made quarterly (art. VIII of GEO 29/2020). Granting bonuses for companies that pay their taxes (art. 1 GEO 33/2020)
- Postponement of the deadline for submitting requests for restructuring of outstanding budget receivables on March 31, 2020 until December 15, 2020, with prior notification of the tax authorities until September 30, 2020 (art. I of GEO no. 90/2020).
- Extension of the deadline for submitting the declaration regarding the legal beneficiary until November 1. 2020, and during the state of emergency being suspended the submission of the declaration and the completion of the documents (art. XI of GEO 29/2020, amended by GEO no. 70/2020).
- Postponement of the deadline for submitting the annual financial statements for 2019 until July 31, 2020 (art. XIV of GEO 48/2020).
- Postponement of VAT payment for imports of medical equipment and medical supplies (including protective mask production machines) necessary to combat the SARS-CoV-2 epidemic for the period of emergency, plus another 30 days after its cessation (GEO no. 33/2020 completed of GEO No. 48/2020).
- Extension of the period of validity of licenses and authorizations for the operation of games of chance, which expire during the state of emergency, which may be exchanged within 90 days and suspension of payment obligations related to these authorizations, during the state of emergency (GEO no. 48/2020).
- Exemption from the payment of fees related to gambling operating authorizations provided in GEO no. 77/2009.
- Exemption from the payment of the specific tax for 2020, for a period of 90 days, by all taxpayers obliged to pay it, according to Law no. 170/2016.

- Reimbursement of VAT related to negative amount returns with the subsequent performance of the tax inspection—a measure applicable until October 25, 2020, provided certain conditions are met. (GEO no. 48/2020, amended by GEO no. 99/2020).
- Dissemination, for a fee, of public information campaigns on measures to prevent and limit the spread of SARS-CoV-2 and how to resume economic and social activities after the cessation of the state of emergency and alert (GEO no. 63/2020, supplemented by GEO No. 86/2020).
- Supporting, through a state aid scheme with an estimated total budget of 291 million euros, large energy consumers in 15 branches of industry, Romanian producers who have directly and indirectly employed over 100,000 employees, where there is a very high risk relocation due to the transfer of the cost of greenhouse gas emissions to the price of electricity (GEO 85/2020).
- Measures to support the cultural events industry—for festivals/events from March 8 to September 30. 2020, through which vouchers can be issued for customers who have purchased tickets for these canceled events (GEO no. 70/2020).
- Extension until June 15 of the term in which the Ministry of Economy (MEEMA) can issue, upon request, Emergency Situation Certificates (GEO no. 70/2020).
- Granting facilities for the payment of rent for the period related to the state of emergency (Law no. 62/2020).
- Regulation of a special way of extinguishing the fiscal and budgetary obligations administered by institutions of the central public administration, of the economic operators that have capacities for the production of electricity or electric and thermal energy in cogeneration connected to the National Electricity System.

2.2 *Measures to Support Jobs in the Economy*

- Payment of technical unemployment for employees of economic operators (including SMEs) whose individual employment contract was temporarily suspended, at the initiative of the employer, during the state of emergency and as a result of the effects of the coronavirus epidemic (GEO no. 30/2020, completed by GEO No. 32/2020).
- Establishment of an indemnity for other professional categories that do not have the quality of employers and interrupt the activity totally or partially, including lawyers affected by the coronavirus crisis according to the established criteria (GEO no. 30/2020, completed by GEO no. 32/2020 and GEO no. 53/2020).
- Regulating Romania's preparation for participation in the temporary Support to mitigate Unemployment Risks in an Emergency (SURE) Program, intended for Member States to finance support measures for companies and employees affected by the coronavirus crisis, by amending the legal framework on public debt, to allow Romania to participate as a Member State in guarantee mechanisms and European Union funding (GEO no. 77/2020).

- Providing active support measures for employees and employers, in order to support the resumption of economic activity (GEO no. 92/2020).

2.3 *Social Protection Measures*

- Regulation of the payment of social assistance benefits during the establishment of the state of emergency (GEO no. 30/2020 completed by GEO no. 32/2020).
- The further granting of the insertion incentive for a period of 3 months for the parents who returned to work faster from the parental leave are in the situation of termination or suspension of the employment contract due to the epidemic (GEO no. 30/2020 completed of GEO No. 32/2020).
- Granting the monthly allowance for the beneficiaries of the leave for the care of the child with disabilities up to the age of 7 during the whole period of the state of emergency and in the situation when the child has reached the age of 7 during this period.
- Maintaining the salary rights held in the month prior to the decree of the state of emergency for the staff employed in social services such as residential care and assistance centers for the elderly, residential centers for children/adults, with and without disabilities, for which the preventive isolation measure is established 14 days at work/specially arranged areas, followed cyclically by home isolation (GEO no. 59/2020).
- Suspension/non-initiation of foreclosure proceedings for all monthly fees paid through pension funds (budgetary, fiscal, financial and commercial claims that apply to these entitlements), for the period of state of emergency to which are added another 60 days (GEO No. 32/2020).
- Non-taxation of benefits in kind granted to employees who hold positions considered essential for carrying out the activity and are in preventive isolation at work or in special areas to which persons from outside do not have access. The granting period is established by the employer, not exceeding the period of establishment of the state of emergency (GEO no. 48/2020).
- Establishing a favorable fiscal regime, during the state of emergency, for the amounts representing incentives/bonuses granted by the employer from the salary fund to the individuals who realize incomes from salaries, these being subject only to the income tax (GEO no. 69/2020).
- Extension of the deadline for submitting the Single Declaration on income tax and social contributions due by individuals, as well as the payment deadline (June 30, 2020), as well as the granting of bonuses (10%) in case of payment of income tax and social contributions mandatory for 2019, as well as the submission of the Single Declaration by electronic means until June 30, 2020 (GEO no. 69/2020).
- Granting some days off to parents for the supervision of children (GEO no. 30/2020 completed by GEO no. 32/2020, GEO no. 41/2020, GEO 147/2020, GEO 182/2020 and GEO 198/2020)

3 Methodology of the Empirical Research

In order to assess the challenges and support coming from authorities, as perceived by respondents, in the time period July-December 2020 have been conducted in-depth, semi structured interviews with twelve Romanian entrepreneurs, owners of small and medium-sized enterprises from different fields of activity. The in-depth interview is a one to one interview, which is conducted by the researcher with each participant in the study, separately. Usually the in-depth interview might last between 30 and 90 minutes and in most of the cases the discussion is based on an interview guide which contains several questions, from which the conversation might divert also to other aspects that might be considered relevant for the topic researched by the interviewer. It is also very important that the focus lies always on the interviewee and the interviewer tries to minimize his own talking and provides so much time as possible to the respondent, in our case the business owner, to express his/her opinions about the various business topics related to the impact of COVID-19 crisis on their enterprises (Schiffman and Wisenblit 2015; Greener 2008). In the present study, despite the pandemic situation, the interviews were conducted face to face, as the non-verbal communication was considered also important. The average duration of an interview was around 80 minutes and the trained interviewer has provided the interviewee with the interview guide before starting the interview, so that both of them could follow the questions. The researcher who conducted the interview started with a short introduction about the scope of the research, the utilization of obtained data as well as reinforcing the expected duration of the interview.

The interviews have not been recorded because this approach was considered intimidating for the interviewees and the aim of the research was to obtain sincere, honest answers which could reveal insights of the business and also less formal business approaches of the COVID-19 situation. As recommended, the interviews were transcribed in the next period after the interview, in order to avoid that important and relevant information, which might not necessarily directly refer to the questions provided in the guide and rather to the personal impressions of the interviewer, gets lost and some contextual notes shade over time. Data obtained was analyzed using text mining and content analysis, based mainly on the experience of the research team.

According to Boyce and Neale (2006) in-depth interviews are recommended when the researcher seeks for detailed information about a person's thoughts, ideas, behaviors, actions and want to explore new issues in detail. This is exactly the situation in our research: COVID-19 crisis is a completely new situation considering the current paradigm of doing business, based on consumption and social, human interaction. Besides the pragmatic impact of the pandemic, we should not underestimate the emotional one, which highly impacts business behavior and decisions. Emotions are difficult, if not impossible, to capture through other data collection methods, such as surveys. During interviews, people feel more comfortable and are more willing to provide details about their personal and professional situation, about how the current crisis impacted them, about the challenges, threats and opportunities during these times and also some insights about the support received from authorities. Therefore, considering the aforementioned aspects we considered that we could obtain a more complete picture of the business situation of the small and middle-sized Romanian enterprises by conducting qualitative research.

The interview guide contains both open-ended and closed-ended questions and it is divided into five main sections: in the first two parts demographic questions and questions about the company are addressed. Next section is dedicated to evaluating the impact of COVID-19 crisis on the company's activity. An important series of questions refers to the challenges faced by small and medium-sized companies from Romania during COVID-19 pandemic. The last section refers to the support measures provided by Romanian authorities and their effectiveness for companies, as well as some general predictions related to the evolution of the economic crisis. The present paper focuses on the last part of the questionnaire, support measures offered by the state. Next, demographic data about respondents and data about their companies is presented, followed by the analysis of the questions related to the aid received from the authorities, as perceived by the interviewed entrepreneurs.

4 Results and Discussions

In the following table it is presented the demographic profile of the respondents. As it results from the table, we have an equal distribution of genders, half of the respondents being females and half males. 75% of them are between 31 and 50 years old. With respect to the educational level, 75% of respondents have an academic degree, in the field of activity or not. Eleven out of the twelve respondents are either married or in a stable relationship, and more than a half have at least one minor dependent child in the household (Table 1).

With respect to the profile of the companies, we can state that they are active in various fields of activity, all of them being impacted by the COVID-19 crisis to a certain extent (see Fig. 1): HoReCa (one third), Constructions/Real estate, Beauty, Photography, Event management, Sports, Medical services. As it results from Table 2, over 50% of the companies have a history of at least 6 years when the COVID-19 pandemic emerged, while most of them reported to have between 1–5 employees. With respect to the turnover, half of the companies have an annual turnover (2019) between 10.001 and 50.000 Euro, with just one company below the value of 10.001 Euro and the rest above 50.001 Euro.

In terms of the impact of COVID-19 pandemic on the company's activity as perceived by respondents, almost 85% of them considered that starting with 16th of March 2020 (when the state of emergency was declared on the territory of Romania) the activity of their businesses has been impacted to a large and very large extent. Just 2 respondents, representing a company from the HoReCa industry (production and selling of ice-cream) and a company from the construction sector stated that they have being impacted to a small and medium extent. The explanations behind these two answers refer to capturing actually some business opportunities: the ice-cream

Table 1 Profile of respondents

Gender		Age	
Male	50%	21–30 years	16.70%
Female	50%	31–40 years	50%
		41–50 years	25%
		51–60 years	8.30%
Education		Civil status	
Highschool degree	25%	Married	58.40%
Bachelor's degree	33.40%	Divorced	8.30%
Master's degree	25%	In a stable relationship	33.30%
PhD	8.30%		
Postgraduate degree	8.30%		
Minor children			
None	41.70%		
One child	25%		
Two children	25%		
Three children	8.30%		

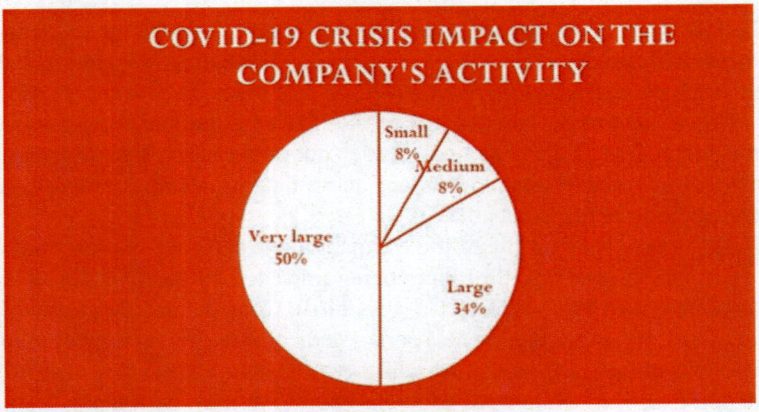

Fig. 1 COVID-19 impact on the company

manufacturer stated that they focused on deliveries and during crises people tend to indulge themselves with smaller things, such as a good ice-cream, and not with expensive ones, as they tend to be more cautious with their budget. The company from the construction sector captured the business opportunity of repairing old things rather than building new ones (in the case of private customers), as people tend also to be more careful with their expenses and focus on conserving existing buildings.

Table 2 Profile of respondents

Age of the company		Number of employees	
0–2 years	16.70%	1–5 employees	66.70%
3–5 years	25%	6–10 employees	8.30%
6–10 years	41.60%	11–20 employees	25%
Over 11 years	16.70%		
Turnover (Euro)			
0–10.000	8.30%		
10.001–50.000	50%		
50.001–100.000	8.30%		
100.001–200.000	16.70%		
200.001–500.000	16.70%		

The next section of the paper focuses on presenting the measures and support offered by Romanian authorities, as perceived by the interviewed entrepreneurs. In accordance with the next figure, which outlines the perception of the respondents regarding the appropriateness of the measures provided by the Romanian state, respectively the supporting measures offered by authorities they benefited from, it can be stated that the answers are balanced: half of the respondents consider that the measures taken by Romanian authorities in order to support the business environment during COVID-19 state of emergency/ alert are inappropriate and totally inappropriate, while the other half perceived them as neither inappropriate nor appropriate, and appropriate (8%). Therefore, the perception of the entrepreneurs is not negative, but neither positive, rather neutral. Concerning the support coming from the Romanian state through the adopted measures, more than 80% of respondents stated that they received no support or reduced support from authorities, only 16% declaring that they received moderated aid (Fig. 2).

Next there are provided details about the support measures implemented by the Romanian state, which respondents benefited from. According to the present

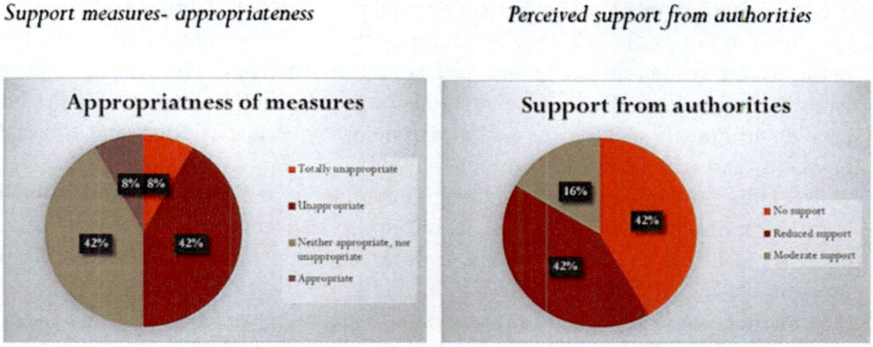

Fig. 2 Measures versus support from Romanian authorities

research the most popular support measure is represented by technical unemployment paid by the state (50% of respondents benefited from technical unemployment for their employees), while over 16% of respondents benefited from postponement of payments (taxes, social contribution) and 16% from tax reductions when paying before a due date. Only a reduced number of the participating entrepreneurs within the present research mentioned to have benefited from postponement of credits' payment and accelerated VAT refund. Most of the entrepreneurs interviewed explicitly complained about the conditions of accessing the Support Fund IMM Invest. They claimed to have unsuccessfully tried to access this Support Fund due to the very restrictive financing conditions and overwhelming bureaucracy while filling in the application file, contrary to the easy access and financing opportunities as presented and advertised in the general media. Therefore, one of the main remarks refers to the creation of a real, functional and accessible support fund- IMM Invest. Other remarks refer to the availability of non-refundable grants offered on time, not just postponement of different types of payments- an action which does not offer real help/ release of financial burden for companies facing financial difficulties. Another remark refers to the extension of technical unemployment for all categories of personnel under contract (not just own employees), as many companies work with various types of personnel (such as beauty salons, fitness centers etc.). Representatives of the HoReCa industry claim that restrictions have been extended over too long periods of time, making therefore the recovery of the business very difficult- softer economic limitations could have reached the target, if corroborated with logical, consistent and coherent medical measures. One of the respondents confessed to have hesitated to apply for support measures, being afraid that controls coming from authorities will intensify, if accessing the funding. Another questioned entrepreneur claimed that the support measures should be functional and of real help for the business environment, just as presented in media, and entrepreneurs should not have to face so many obstacles while trying to access them. For instance, besides the general remark referring to the functionality and accessibility of the IMM Invest Fund, the partial reimbursement of VAT (instead of full reimbursement) was also mentioned. The general conclusion refers to the fact that the business environment would have needed a more prompt response from authorities, with coherent measures, easy to access and of real help for the various industries. The general opinion reveals that the support schemes were presented in the media in a very good light, which does not necessarily reflects the reality faced by the representatives of business environment, while trying to benefit from them. However, there were stressed some very useful measures for the business environment, such as the technical unemployment, postponement of certain payments and accelerated VAT refund.

Regarding the information channels, as depicted in the following figure, the TV and the Online press rank first according to the answers received. On the second place we can find the Social Media, followed by Friends, acquaintances. Written press and official communication of public authorities are the least used methods of information, while Radio and Professional associations and networks are used by approximately half of the respondents (Fig. 3).

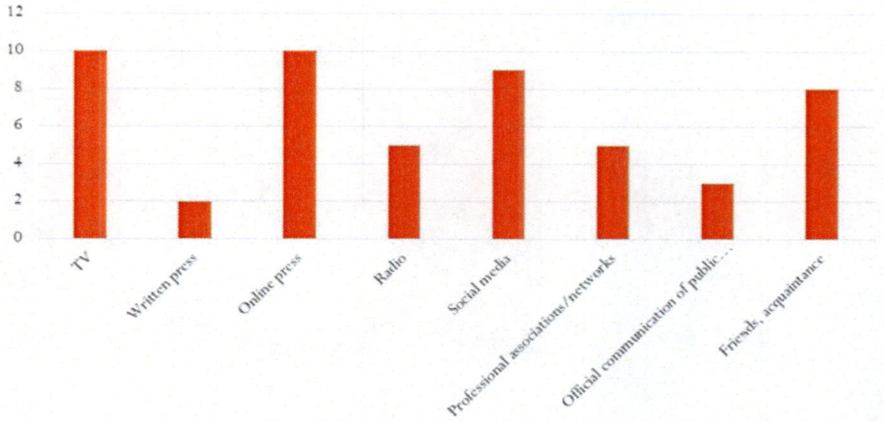

Fig. 3 Channels of information

An important aspect refers to the effects of the crisis: all of the respondents with only one exception consider that the effects of the present crisis generated by COVID-19 will be severe (66,7%) and very severe (25%), lasting over 2 years starting from 2020 (50%). One quarter cannot estimate the duration of the crisis, while one quarter are more optimistic, estimating a shorter crisis, which will end by the end of 2021. When asked to compare the effects of the economic crisis in Romania with other European countries, 75% of the respondents consider that in Romania the consequences will be more severe (due to the bad public management and administration of investment, lack of coherence of public measures), while 25% consider them similar to other European states. However, despite the crisis, 50% of respondents plan to extend their business or take advantage of new opportunities in related fields of activity. It is worth mentioning that it can be observed a positive correlation between the level of education and the identification of new business opportunities: the higher the level of education, the more opportunities have been identified. Similarly, it can be observed a connection between the capacity and desire to capture certain business opportunities during these challenging times and an academic background in the field of Business Administration (Fig. 4).

5 Conclusion

The COVID-19 crisis is a world-wide phenomenon and its associated economic impact has triggered chain reactions across multiple nations sending many countries into economic downturn, sanitary crises and social turmoil. The macroeconomic shocks translated into alterations of supply, demand, trade, investment, supply chain, financial stability, economic growth, exchange rates etc.

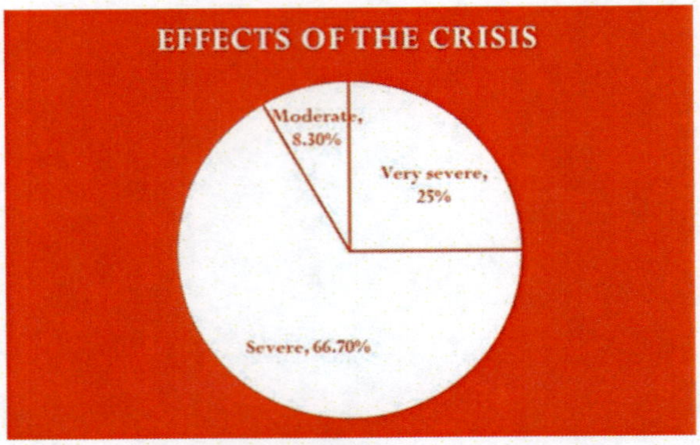

Fig. 4 Predictions related to the effects of the crisis

The main challenges faced by companies have been transformed, to certain extent, into measures meant to alleviate the damage caused by the COVID-19 crisis. In order to achieve an optimum balance between the necessities of the business environment and the measures the government has to take there should take place common meetings for designing the proper framework with a particular focus on the businesses. Moreover, flexibility and promptitude are of the essence for preventing further negative consequences of unaddressed challenges by creating corresponding measures.

As rendered by our study, specific actions need to be taken with the purpose of either improving public management or the perception of companies with respect to public management and efforts are invited to facilitate coherence of public measures (they should be properly aligned across all institutions and stakeholders).

Another major conclusion we can draw is that since TV and Online press rank first according to the answers received, policy makers and supervisory bodies should pay close attention to content of the materials distributed within the framework of these channels for ensuring that the population/businesses get properly informed.

The general conclusion refers to the fact that the business environment would have needed a more prompt response from authorities, with coherent measures, easy to access and of real help for the various industries. The general opinion within our research reveals that the support schemes were positively altered in the media, which turned out to be havoc when the representatives of business environment tried to access them. Still, some particularly useful measures for the business environment were identified: technical unemployment, postponement of certain payments and accelerated VAT refund.

Disclaimer

The present paper does not represent the position of the Romanian Government on the subject.

References

Băhnăreanu, C.: Impactul economic al pandemiei de COVID-19 la începutul anului 2020. Impact Strategic **2**(75), 87–98 (2020)

Baker, S.R., Bloom, N., Davis, S.J., Kost, K., Sammon, M., Viratyosin, T.: The unprecedented stock market reaction to COVID-19. Rev. Asset Pricing Stud. **10**(4), 742–758 (2020)

Barua, S.: Understanding coronanomics: the economic implications of the coronavirus (COVID-19) pandemic (2020)

Boyce, C., Neale, P.: Conducting in-depth interviews: a guide for designing and conducting in-depth interviews for evaluation input. Pathfinder Int. Tool Series Monit. Eval. **2** (2006)

Fernandes, N.: Economic effects of coronavirus outbreak (COVID-19) on the world economy. Available at SSRN 3557504 (2020)

Greener, S.: Business Research Methods, Dr. Sue Greener and Ventus Publishing ApS (2008)

HotNews.: Main challenges faced by companies. (2020) https://economie.hotnews.ro/stiri-finante_banci-24255508-masuri-luat-firmele-pentru-proteja-efectele-pandemiei-cele-mai-presante-probleme-ale-companiilor.htm. Accessed 21 Nov 2021

McKibbin, W., Fernando, R.: The global macroeconomic impacts of COVID-19: Seven scenarios. Asian Econ. Papers **20**(2), 1–30 (2021)

NIS COVID-19-Economic impact.: https://insse.ro/cms/ro/covid-19-impactul-economic (2021). Accessed 21 Mar 2021

Padhan, R., Prabheesh, K.P.: The economics of COVID-19 pandemic: a survey. Econ. Anal. Policy **70**, 220–237 (2021)

Petro, G.: The coronavirus tsunami: what's to Come for U.S. retail. forbes (2021). https://www.forbes.com/sites/gregpetro/2020/03/20/the-coronavirus-tsunami-whats-to-come-for-us-retail/#128e4da6609d. Accessed 21 Mar 2021

Romanian Government Official Website.: Official documents attesting the measures adopted by Romania against the COVID-19 Pandemic (2021). https://gov.ro/ro/masuri. Accessed 21 Nov 2021

Schiffman, L., Wisenblit, J.: Consumer Behavior, 11th edition, Global Edition, Pearson Education, England (2015)

Entrepreneurship and Short-Term Labor Market Sustainable Functioning—The Case of Romania

Speranţa Pîrciog, Adriana Grigorescu, Cristina Lincaru, and Cătalin-Corneliu Ghinăraru

Abstract If contractual work (waged) is traditional employment, "the new work" is represented by entrepreneurship and self-employment and to a certain extent by various forms of "subsistence economy". We develop an instrument to understand how new work shapes the short-term labor market's sustainable functioning. The dynamic interaction between labor market macro-aggregates related to employment status is treated in a systemic approach. We analyze the Granger Causality (GC) relations between contractual/dependent work, non-contractual/independent work and the work's social protection. A total of 14 VAR models have been analyzed following the Toda and Yamamoto (J. Econ. 66: 225–250, 1995) procedure as detailed by Giles (Testing for granger causality. Econometrics beat: dave giles, 2011), using 2 datasets (1999Q1-2018Q1 and 2008Q1-2018Q1). The main conclusion is that only a sustainable entrepreneurship mechanism could make the Romanian labor market suitable for the actual challenges. A powerful call for a new type of entrepreneurial behavior becomes evident in the case of Romania.

Keywords Sustainable entrepreneurship · Waged employment · Self-employment · Unemployment · Innovation

1 Introduction

Changing types of employment relations spurred by shifting demographics, a more dynamic labour force in a global scale, the rise of individual choice, the technological revolution and "client sophistication" are the root causes of an increasing pressure felt by all stakeholders in the "world of work" (World Economic Forum 2018). Bergmann in its book (2019) and Interview held by Ramsburg (Ramsburg 1999) pointed out "the 'Job System' for organizing work has only existed for around 200 years—since the

S. Pîrciog · C. Lincaru · C.-C. Ghinăraru
Institute for Labor and Social Protection, National Scientific Research, Bucharest, Romania

A. Grigorescu (✉)
National University of Political Studies and Public Administration, Bucharest, Romania
e-mail: adriana.grigorescu@snspa.ro

M. Busu (ed.), *Digital Economy and New Value Creation*, Springer Proceedings in Business and Economics, https://doi.org/10.1007/978-3-031-07265-9_19

industrial revolution". The Job System is now dying while the next system, labeled conveniently, "New Work" has to be created. Production in the New Economy driven by intelligent technologies is ubiquitous and hence accessible to each person."New work" is intrinsically entrepreneurial and this particular feature will only become evident as time passes.

Work is becoming increasingly "non-traditional", generating thus a demand for specific "policy innovations" capable both of sustaining higher assure living standards as well as providing a number of social benefits to which the workforce has grown accustomed. If we are therefore to consider contractual work (waged employment) as "traditional employment", then "the new work" will be represented by entrepreneurship, self-employment, and, in certain peculiar circumstances, by what we will call subsistence economy (the national accounts sector of "household production for its own final consumption").

Labour market functioning revolves, in our opinion, around three main pillars: (1) self-employment and entrepreneurship, (2) contractual employment (waged employment) and (3) unemployment (the system of unemployment protection). The dynamic interaction of these pillars characterizes the sustainability of the labor market assuring the quality of life for each person. A sustainable labour market means a higher rate of job creation, coming from more successful entrepreneurship. This should be coupled with fast enough labour force allocation and reallocation at an "optimal" security level (the "flexicurity" country-specific "model"). The present paper aims at analysing the interactions between the above components as triggering factors for public policy initiatives. Following an "evolutionary approach", the job creation and destruction process may have been seen as having a "contractual dimension" as per below:

(a) *Self-employment may evolve into self-employment with employees thus generating contractual employment*, or conversely generating (registered) unemployment with respect to social protection and safety nets, if *employment is deemed as not sustainable*;
(b) *Contractual employment* may evolve into *self-employment* as a spill-over effect, or conversely may also lead to unemployment;
(c) *Unemployment* may evolve into "contractual employment", or into *self-employment/independent* employment.

The labor market sustainable functioning "assessment instrument", described by the present paper, uses the Granger Causality (GC) relationship between contractual, non-contractual work and the social protection system (as expressed by the variable: rate of unemployment/unemployment rate) in Romania between 1999–2018. The analysis looks at 14 VAR models following the methodological approaches pioneered by Toda and Yamamoto (1995) and Giles (2011). Due to the large number of GC we decided to analyse only one region for 20 years and a replication of the study could be done for any other region.

2 Literature Review

Currently, the research field on the relationship between self-employment/entrepreneurship and unemployment is well developed, modeled by the following theories: GC1—the "refugee" or "unemployment push" theory and GC2—"the Schumpeter" or "entrepreneurial" theory.

GC1 states that a higher level of unemployment will have a positive lagged effect on start-up activities in a given spatial area. Ghavidel et al. (2011) explained that "unemployment reduces the amount of entrepreneurial activity", especially when at very high levels." This particular statement is in line with the work performed by Audretsch and Fritsch (1994, 1995, 2001).

However, the "refugee effect" is found to be reversed. This is what we call the "demand-driven" leverage: A higher level of unemployment will have a negative lagged effect on start-up activities in a given spatial area as a mere result of depressed demand. GC2 by contrast states that a higher level of start-up activity will have a negative lagged effect on unemployment in a given spatial area (Cole 2018).

Literature showed a validation of both, GC1 and GC2. Cole (2018) found a "circular relationship" between unemployment (unemployment rate) and new firm start-ups, and Audretsch et al. (2001) found a relationship between unemployment and the number of business owners. Audretsch et al. (2005) and Thurik et al. (2008) both studies confirm distinct relationships for 23 OECD countries for the period 1974–2002, where the "entrepreneurial" effects are considerably stronger than "refugee" effects. Faria et al. (2008) concluded that "the Granger-causality tests show there is bidirectional causality between unemployment variation and entrepreneurship for countries that have the most flexible labor markets, such as Ireland, Germany, the United States and Australia. Unemployment variation causes business creation in Italy and Japan, and for France, the causation runs from entrepreneurship to unemployment."

At the same time as GC1 is rejected, unemployment does often lead to entrepreneurship (Lasch et al. 2007) for the case of France, with "agglomeration economies" displaying no significant effect. Halicioglu and Yolac (2015) establishes that the results for H1 are ambiguous as increased unemployment may lead to increased self-employment activities (results reveal the existence of a long-run relationship for Belgium, Canada, Sweden and the UK). The increase in unemployment rates may however work so as to decrease human capital endowment and entrepreneurial talent thus causing a rise in unemployment rates (observed empirically in Luxembourg, Greece and Portugal). However, these tendencies materialize only in the short term (Garba et al. 2013), which confirms that in the case of Nigeria with unemployment positively influencing entrepreneurship.

Also, as GC2 is rejected, a one-way validation (Bokhari et al. 2012) points to the fact that the impact of entrepreneurial activity on the unemployment rate is significantly negative in the case of 31 selected countries during the period 2008–2010, as supported also by data from Global Entrepreneurship Monitor (GEM). Dilanchiev (2014) confirms this relationship for Georgia in 2013 and Verheul et al. (2006) did the same for Spain for the period 1972–2004. Spain has become convergent with the OECD model only recently and following labor market reforms with the effect of increasing the quality of business ownership. Baptista and Thurik (2007) using COMPENDIA data 2000.1 (1974–1998) and COMPENDIA Data 2002.1 (1972–2002), noted that Portugal is an outlier when compared to the OECD average because of the "*subsistence entrepreneurship*", similar to other Southern European countries such as Italy and Greece. For seven developing and 23 OECD countries, during the period 1995–2007, the Schumpeter effect is established definitively, but the refugee effect is not clear, suggesting that the causal effect of the unemployment rate on entrepreneurship is first positive and then negative (Ghavidel et al. 2011).

Opolskie Voivodship (PL), during the Q1 2007 to Q2 2013 period, showed that neither GC1 nor GC2 are validated (Szewczyk et al. 2013). "Countries with a high rate of entrepreneurial activity have a low rate of unemployment (developed countries). While on the other hand countries with a high rate of unemployment have a low rate of entrepreneurial activities (under-developed or developing countries). Therefore, the interaction between entrepreneurship and unemployment is essentially determined by the country's position in the labor market" (Audretsch et al. 2005).

The Granger causality model was applied to Romania mainly to study the relationship between education and economic growth (Dănăcică 2011), the employment rate of older workers and the shadow economy, inflation and unemployment (Dilanchiev 2014), and GDP and entrepreneurial activity (Simionescu 2015; Şipoş-Gug and Bădulescu 2013; Şipoş-Gug et al. 2015). Another causal link evidenced by literature is the particular relationship between entrepreneurship, economic growth and employment as studied for the case of Taiwan (RoC–Republic of China) (Chen 2013). The Granger causality test indicates that there is a two-way relationship between the employment rate and entrepreneurship. Entrepreneurship is one of the mechanisms facilitating the spillover of knowledge, alongside research and development, and human capital, promoting thus economic growth alongside job creation ("job-rich growth") (Acs et al. 2012). Between the employee and the company the relationship could be influenced by the motivation (Uka and Prendi 2021) or the satisfaction or happiness (Eckhaus 2021).

3 Systemic Approach of Waged Employment, Self-Employment and Unemployment

The main theoretical argumentation of the paper is that of providing a valuable input for policy makers in support of "job-rich growth". The dynamic perspective of Waged Employment, Self-Employment, and Unemployment (Constant and Zimmermann 2014) from the labor market transitions in relation to the business cycle. Bennett and Rablen (2015) looked at the self-employment, wage employment, and informality in a developing economy concluding on "the importance of underlying macroeconomic conditions in determining the effects of education and training on transitions of individuals from one labor market status to another". Poschke (2010) considers some dimension of FLEXICURITY such as high hiring costs or low matching efficiency (simultaneous with high job destruction rate) as being indicative of the presence of high friction on the labor market" causing not only high unemployment rates, but also high rates of self-employment in poor countries". We should also consider the Atkinson model about flexibility Atkinson (1984) that considers the pressure of change from job creation, technical/production, institutional and supply factors.

Building upon the previous GC1 and GC2 theories as well as upon other research as previously referenced, our approach for the current paper considers the three main pillars as mentioned before: contractual employment, self-employment and unemployment rate. Two-way relations and the dynamic between groups were under analysis. The research hypothesis starts with H1 and H2 used in the theories GC1 and GC2 adding to these H3, H4, H5, and H6 as per below.

H1—Social Protection does not cause Non-Contractual Employment.

H2—Non Contractual Employment does not cause Social Protection.

H3—Non Contractual Employment does not cause Contractual Employment.

H4—Contractual Employment does not cause Non-Contractual Employment;

H5—Social Protection does not cause Contractual Employment.

H6—Contractual Employment does not cause Contractual Employment.

The systemic approach can be easily understood by looking at Fig. 1.

The hypothesis reflects the dynamics (in our case the causality relationship GC) between labor markets' considered statuses, for the period 1999Q1–2018Q1 for all the variables, with exception of self-employment without employees, not in agriculture (NACE, Rev 2), available since 2008Q1, and excluding the case of subsistence entrepreneurship:

Contractual employment involves WR—employee's rate (share of waged employees).

Non-Contractual Employment[1] represented by:

(a) SE1R—self-employment with waged employees rate;
(b) SE0R—self-employment without waged employees rate, marked by "*";
(c) SE0NAR—self-employment without employees, not in agriculture (NACE, Rev 2), marked by "**";
(d) CFR contributing family workers rate marked by "**".

Social protection status is represented by the unemployment rate. In Fig. 2, we present the short-term labor market sustainable functioning assessment instrument as detailed by the explored relationship. We use supplementary notations "#" for the second time period 2008Q1–2018Q1.

These relationships reflect the intersection of various policies (Pinelli 2015) growth and job creation and unemployment reduction (Bokhari et al. 2012), entrepreneurship and sustainable entrepreneurship; labor market policies (LMP);

[1] We do not consider the case of cooperative members while, since 2011, RO-national statistics office (The National Institute for Statistics of Romania/NIS-RO) no longer reports the number for the category"Members of an agricultural holding or of a cooperative".

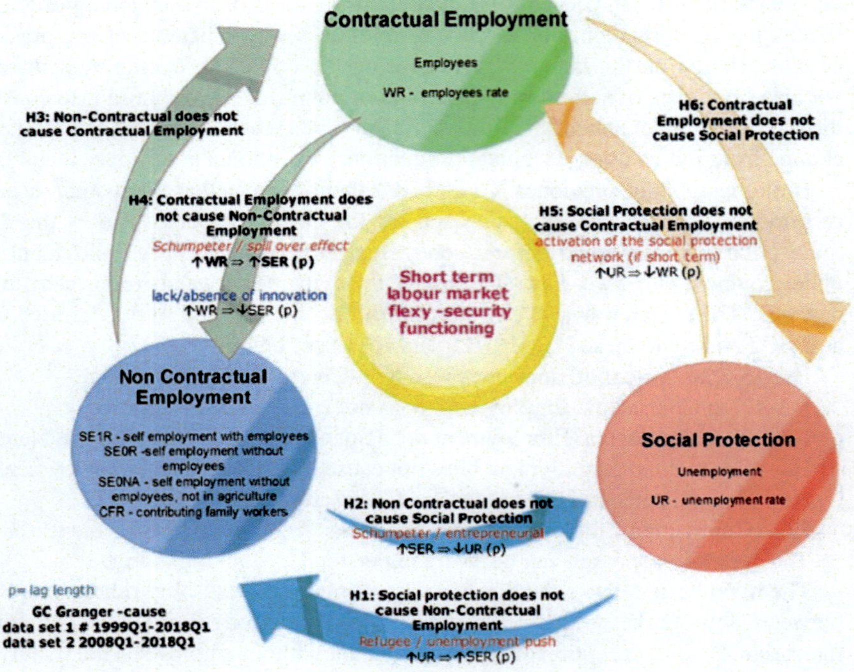

Fig. 1 The short-term labor market sustainable functioning assessment instrument

measures such as direct job creation and start-up incentives; regional development; economic performance, and enhancing productivity and competition, etc., in opposition to the undeclared work (Xavier and Badea 2018). Non-contractual employment may refer to entrepreneurship (Babaginda and Semasinghe 2013), new firm startups; self-employed persons with or without employees; small business ownership; innovators (Bokhari et al. 2012), and social enterprise as the precursor to entrepreneurial non-profit organizations (Borzaga et al. 2010). The literature considers non-contractual employment and unemployment as being "considerably ambiguous" with respect to their causal relationship (Audretsch et al. 2001).

Entrepreneurship in Romania during the period 2008–2014 presents some worrying features regarding its sustainability restrictions Paradoxically, it first increases, but its hope of survival decreases after 2 years, with business activity decreasing to a historical minimum. Following the "great recession" of 2008–10, entrepreneurship has apparently failed to generate employment growth. Based on business demography characteristics in Romania, as described by short-term indicators, we apply the model of Toda-Yamamoto.

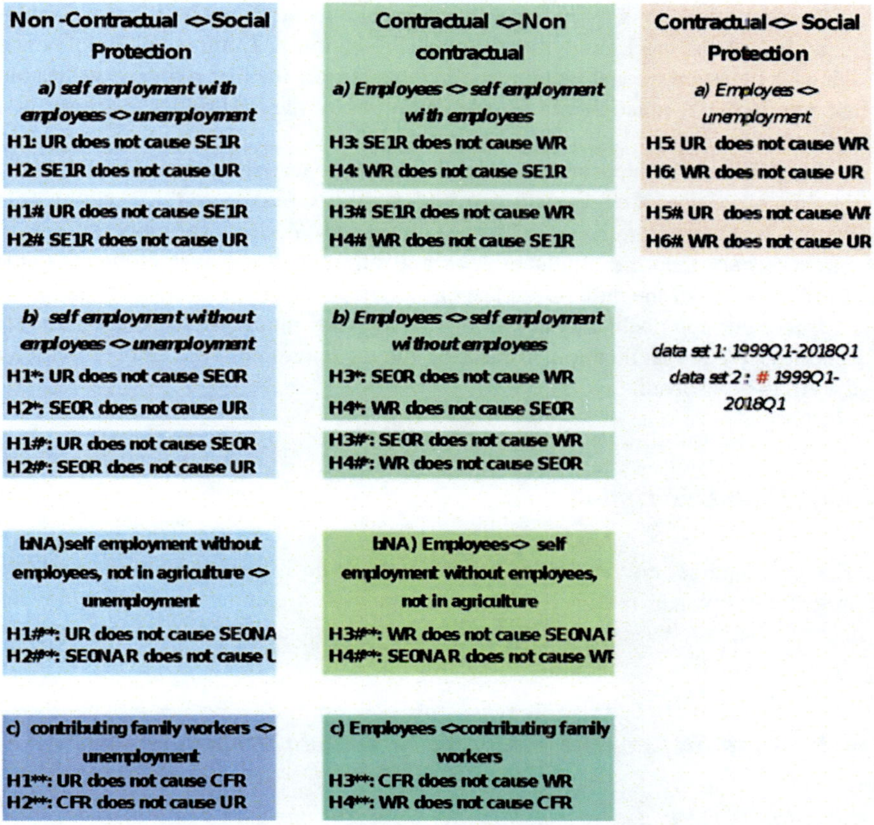

Fig. 2 The short-term labor market sustainable functioning assessment instrument detailed for their allocated hypothesis (<> relationship)

4 Methods and Research Data

Following Toda and Yamamoto (1995), Giles (2011) and Bennett and Rablen (2015) the GC model, we applied the following VAR model equations:

$$Y_t = a_0 + a_1 Y_t - 1 + \ldots + a_p Y_t - p + b_1 X_t - 1 + \ldots + b_p X_t - p + u_t \quad (1.)$$

$$X_t = c_0 + c_1 X_t - 1 + \ldots + c_p X_t - p + d_1 Y_t - 1 + \ldots + d_p Y_t - p + v_t \quad (2.)$$

Then, testing H0: b1 = b2 = ... = bp = 0, against HA: 'Not H0', is a test that X does not Granger-cause Y.

Similarly, testing H0: d1 = d2 = ...= dp = 0, against HA: 'Not H0', is a test that Y does not Granger-cause X. p lag numbers

Social Protection & Non-Contractual Employment; Non Contractual Employment & Contractual Employment; Social Protection & Contractual Employment. For each equation we test that the coefficients of only the first *p lagged* values of X are zero in the Y equation and reverse. Rejection of the H0 implies the presence of Granger causality.

The Giles (2011) approach assures the causality testing be done properly, when the time series used are non-stationary and possibly co-integrated (Lütkepohl 2005). We use Wald test from Eviews 7 to test GC in the VAR in level model. If data are non-stationary then the "statistics does not follow its usual asymptotic chi-square distribution under the null"(Giles 2011).

Under our approach, the endogeneity of the VAR model is controlled in the step "checking VAR specification" increasing the lag length until the autocorrelation of the errors is resolved.

4.1 Data Description

The first data set covers 1999Q1–2018Q1 and the following working variables: waged employment rate; self-employment rate with employees; self-employment rate without employees; contributing family workers, and unemployment rate.

4.2 Main Steps of the Toda and Yamamoto Procedure

(a) *Check the time-series stationarity for each time series and their respective orders of integration.*

If the average and dispersion of a series are variable in time, then the series is non-stationary (Jula 2011).

The augmented Dickey-Fuller (ADF) unit root tests on WR, SE1R, and SE0R indicate that all these variables are non-stationary at the 1%, 5% and 10% levels for all three exogenous models. They are: *model 1* with Constant, *model* 2 with Constant and Linear trend, and *model 3* when none have the maximum order of integration, m = 1. Only the unemployment rate is stationary at the second difference level and therefore has m = 2. The stationarity in the ADF test follows the H0, and the series has the unit root (nonstationary). Regarding Schwartz test, we consider the variable to be stationary if in at least one of the three models, ADF for H0 is rejected (Jula 2011).

Consider the maximum order of integration for the group of the time-series as "m". If there are two time-series and one is found to be I(1), and the other is I(2) then, "m" = 2. If one is I(0) and the other is I(1) then, "m" = 1, etc. (Giles 2011).

(b) *Unrestricted VAR Equations Setting.*

We set up a 5*2-equation (unrestricted VAR type, co-integrated or without any long-run association) VAR model at various levels of the data, including an intercept in each equation. We assume that the variables are of short-time association only.

(c) *Optimal Lag Selection.*

We determine the appropriate maximum *lag length* (*p*) for the variables in the VAR using the usual information criteria methods, such as AIC, SIC (Giles 2011). For each pair of series, we apply the lag length criteria for the VAR build for levels with exogenous variables, and the statistics are determined to be at the 5% level.

(d) *Checking the VAR Specification.*

We check that there is no serial correlation in the residuals: "If the majority of probability values are larger than 5%, then there is no serial correlation at lag order. If more than half of p values are >0.05, then no autocorrelation is observed" (Toda and Yamamoto 1995). The errors are independent (Giles 2011). We apply the LM test/EVIEWS 7 = VAR residual serial correlation LM tests under the "Null Hypothesis: no serial correlation at lag order h."

(e) *Checking the Dynamic Stability of the Models.*

Checking the dynamic stability of the models presents the inverse roots of the characteristic AR polynomial (Lütkepohl 2005). The estimated VAR is stable (stationary) if all roots have modules less than one and lie inside the "unit" circle. If the VAR is not stable, certain results (such as impulse response standard errors) are not valid. There will be roots at the number of endogenous variables with the largest lag. If we estimated a VEC with co-integrating relations, the roots should be equal to the "unit" (Lütkepohl 2005).

We confirmed the lag 5 for the VAR estimated models, WR SE0R, SER1 UR, and SE0R UR, with lag 7 for WR UR, and raised the lag from 3 to 7 for WR SER1. Equations are dynamically stable based on the fact that all roots have a modulus less than one and lie inside the unit circle.

(f) *Checking the Cointegration of the Time-Series Using Johansen's Methodology.*

In literature is considered that "if there is a linear stationary combination between random non-stationary variables, then the combined variables are co-integrated" (there is a common non-stationary dynamic) (Jula 2011). Both tests of unrestricted co-integration rank, the Johansen's Trace Test and Max. Eigenvalue, indicate the presence of 1 co-integrating equations at the 0.05 level for each model (WR SE0R, WR CFR, WR UR, SE0R UR and CFR UR, with the exception of WR SE1R and SE1R UR models.

(g) *Finalize VAR models models.*

We finalize the VAR models by adding in "m" additional lags of each of the variables into each of the equations (Giles 2011). As m = 1, we now re-estimate the levels of VAR with one extra lag of each variable in each equation. To "trick" EViews into doing what we want when we test for causality, rather than declaring the lag interval for the 2 endogenous variables to be from 1–7 (the latter being p + m), we leave the interval at 1–6, and declare the extra (7th.) lag of each variable to be an "exogenous" variable. The coefficients of these extra lags will then not be included when the subsequent Wald tests are conducted. If we just specified the lag interval to be from 1–7, then the coefficients of all seven lags would be incorrectly included in the Wald tests. In this case, the Wald test statistic would not have its usual asymptotic chi-square null distribution (Giles 2011).

In our VAR models we finalize the equations and add in m additional lags as exogenous variables. According to Giles (Giles 2011) "it's essential that you don't include the coefficients for the 'extra' m lags when you perform the Wald tests. They are there just to fix up the asymptotic".

(h) *Applying the Granger Non-Causality Test.*

We test for Granger non-causality by testing the hypothesis that the coefficients of (only) the first p lagged values of the unemployment rate are zero in equations, using a standard Wald test for the first data set (Fig. 3) and for the second data set (Fig. 4).

Relationship **Non -Contractual <> Social Protection**	**Direction (based on Impulse)**
• ***Refugee /unemployment push***	
H1#** reject: UR does cause SE0NAR, after 2 lags 2*Q=2*3 months=6 months | *2008Q1-2018Q1*	↗UR⇨↗SE0NAR
• ***Schumpeter / entrepreneurial effect***	
H2** reject: CFR does cause UR, after 5 lags 5*Q=5*3 months= 15 months | 1999Q1-2018Q1	↗CFR⇨↗UR (5Q to 8Q), then ↗CFR⇨↘UF
H2# reject: SE1R does cause UR, after 3 lags 3*Q=3*3 months= 9 months | *2008Q1-2018Q1*	**↗SE1R⇨↘UR**
H2# reject: SE0NAR does cause UR, after 2 lags 2*Q=2*3 months=6 months | *2008Q1-2018Q1***	↗ SE0NAR⇨↗UR
Contractual <> Non contractual	
• **Dependent employment creation in emerging & expansion sectors / job creation policy**	
H3 reject: SE1R does cause WR, after 7 lags 7*Q1=7*3 months= 21 months | 1999Q1-2018Q1	↗SE1R⇨↗WR
Contractual<> Social Protection	
• ***ALMP case / searching and matching on labor market***	
H5 reject: UR does cause WR, after 7 lags 7*Q1=7*3 months= 21 months | 1999Q1-2018Q1	↗UR⇨↗WR
• *Gift exchange* (Akerlof 1984) *& involuntary unemployment* (Pettinger, 2017)	
H6 reject: WR does cause UR, after 7 lags 7*Q1=7*3 months= 21 months | 1999Q1-2018Q1	↗WR⇨↗UR

Then we apply the standard Wald test for the coefficients of the lagged values of the same equations. We conduct an evaluation on the effects induced by various impacts on system variables (the impulse response function). Each variable affected by its own innovations, as well as by the innovations of the other variables (Jula 2011) the response to Cholesky One S.D. Innovations (Fig. 5).

Non-Contractual - Social Protection

a) Non-Contractual: self employment with employees

Included observations 70

H1			
Dependent variable: SE1R			
Excluded	Chi-sq	df	Prob.
UR	4.6941	5	0.4543
All	4.6941	5	0.4543
H2			
Dependent variable: UR			
Excluded	Chi-sq	df	Prob.
SE1R	4.4391	5	0.4881
All	4.4391	5	0.4881

b) Non-Contractual: self employment without employees

Included observations 70

H1*			
Dependent variable: SE0R			
Excluded	Chi-sq	df	Prob.
UR	7.9116	5	0.1612
All	7.9116	5	0.1612
H2*			
Dependent variable: UR			
Excluded	Chi-sq	df	Prob.
SE0R	6.4183	5	0.2676
All	6.4183	5	0.2676

c) contractual - contributing family workers

Included observations 70

H1**			
Dependent variable: CFR			
Excluded	Chi-sq	df	Prob.
UR	5.9248	5	0.3136
All	5.9248	5	0.3136
H2**			
Dependent variable: UR			
Excluded	Chi-sq	df	Prob.
CFR	16.262	5	0.0061
All	16.262	5	0.0061

Contractual - Non contractual

a) Employees- self employment with employees

Included observations 69

H3			
Dependent variable: WR			
Excluded	Chi-sq	df	Prob.
SE1R	19.82402	7	0.006
All	19.82402	7	0.006
H4			
Dependent variable: SE1R			
Excluded	Chi-sq	df	Prob.
WR	5.112598	7	0.6462
All	5.112598	7	0.6462

b) Employees - self employment without employees

Included observations 71

H3*			
Dependent variable: WR			
Excluded	Chi-sq	df	Prob.
SE0R	7.266605	5	0.2016
All	7.266605	5	0.2016
H4*			
Dependent variable: SE0R			
Excluded	Chi-sq	df	Prob.
WR	8.572562	5	0.1274
All	8.572562	5	0.1274

c) contractual - contributing family workers

Included observations 71

H3**			
Dependent variable: WR			
Excluded	Chi-sq	df	Prob.
CFR	6.9465	5	0.2246
All	6.9465	5	0.2246
H4**			
Dependent variable: CFR			
Excluded	Chi-sq	df	Prob.
WR	6.399927	5	0.2692
All	6.399927	5	0.2692

Contractual - Social Protection

a) Employees - unemployment

Included observations 68

H5			
Dependent variable: WR			
Excluded	Chi-sq	df	Prob.
UR	27.53901	7	0.0003
All	27.53901	7	0.0003
H6			
Dependent variable: UR			
Excluded	Chi-sq	df	Prob.
WR	25.0565	7	0.0007
All	25.0565	7	0.0007

Fig. 3 VAR Granger Causality/Block exogeneity Wald Tests. Sample: 1999Q1 2018Q1

Non -Contractual <> Social Protection

a) Non -Contractual: self-employment with employees <> unemployment

Included observations 37

H1#: UR does not cause SE1R

Dependent variable: SER1

Excluded	Chi-sq	df	Prob.
UR	4.455	3	0.2163
All	4.455	3	0.2163

H2#: SE1R does not cause UR

Dependent variable: UR

Excluded	Chi-sq	df	Prob.
SER1	21.565	3	0.0001
All	21.565	3	0.0001

b) Non -Contractual: self employment without employees <> unemployment

Included observations

H1#*: UR does not cause SE0R

VAR does not satisfy the stability condition.

H2#*: SE0R does not cause UR

VAR does not satisfy the stability condition.

bNA) Non -Contractual: self-employment without employees, not in agriculture <> unemployment

Included observations

H1#: UR does not cause SE0NAR**

Dependent variable: SE0NAR

Excluded	Chi-sq	df	Prob.
UR	15.259	2	0.0005
All	15.259	2	0.0005

H2#: SE0NAR does not cause UR**

Dependent variable: UR

Excluded	Chi-sq	df	Prob.

Contractual <> Non contractual

a) Employees <> self-employment with employees

Included observations

H3#: SE1R does not cause WR

VAR does not satisfy the stability condition.

H4#: WR does not cause SE1R

VAR does not satisfy the stability condition.

b) Employees <> self employment without employees

Included observations

H3#*: SE0R does not cause WR

VAR does not satisfy the stability condition.

H4#*: WR does not cause SE0R

VAR does not satisfy the stability condition.

bNA) Employees<> Non -Contractual: self-employment without employees, not in agriculture

Included observations

H3#: WR does not cause SE0NAR**

VAR does not satisfy the stability condition.

H4#: SE0NAR does not cause WR**

VAR does not satisfy the stability condition.

Contractual<> Social Protection

a) Employees <> unemployment

Included observations

H5#: UR does not cause WR

VAR does not satisfy the stability condition.

H6#: WR does not cause UR

VAR does not satisfy the stability condition.

Fig. 4 VAR Granger Causality/Block Exogeneity Wald Tests. Sample: 2008Q1–2018Q1

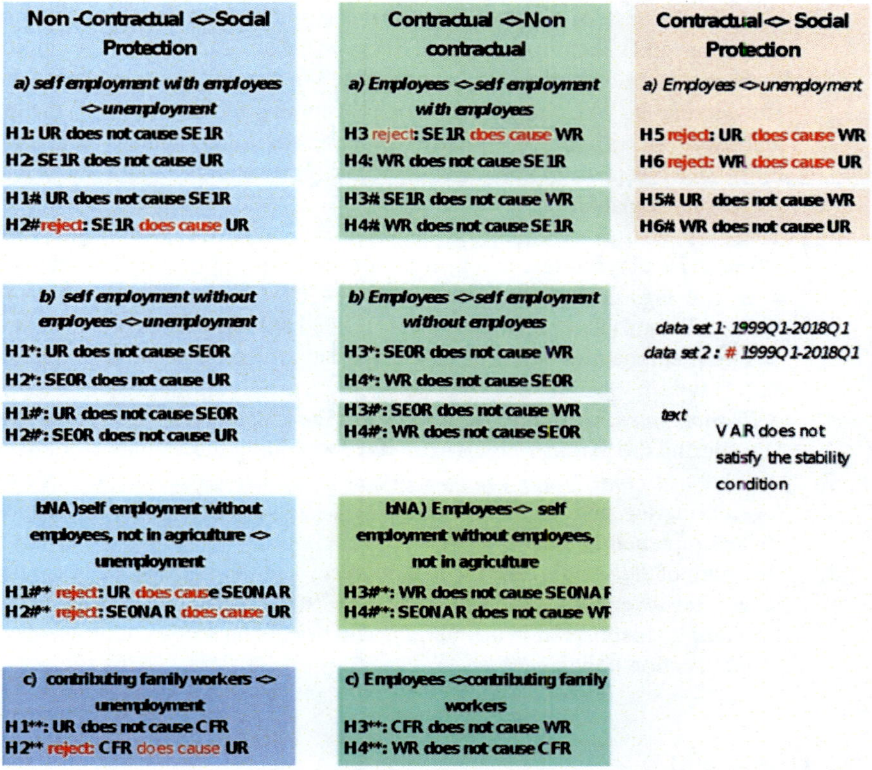

Fig. 5 Waged employment, self-employment, and unemployment relationship in Romania: A flexy-security profile in the short run (H with no reject means H accepted)

5 Results

Based on the fact that we apply a Toda-Yamamoto procedure detailed by Giles (Giles 2011) adding but not testing the extra "m" lags and we use only the Wald test "our causality test results will be meaningless, even asymptotically". The VAR equation coefficients are not significantly caused by the multiple methodological breaks in series. We run the model multiple times for the trimmed variable series (to avoid the broken series), but the results are not statistically robust. The GC is robust, but the equations are not appropriate for forecasting.

The Granger causality test indicates that there is robust evidence for.

(a) A two-way relationship between the share of wage employment and *unemployment rate* after a lag of 7 quarters (reject H5 and H6) for the period between 1999Q1–2018Q1, confirming the presence of social protections for contractual employment, respectively wage employment. Unemployment growth causes wage employment growth after making due consideration for the effectiveness of the ALMPs, especially the information, consultation, searching, and matching services of SPOs on the labor market. Wage employment increases determine ("Granger sense") a rise of unemployment when considering the "Akerlof gift exchange" according to Akerlof 1984) theoretical situation, with its consequence of the presence of involuntary unemployment. This is especially notable with the background of the labor force deficits present in Romania.

(b) There is also a relationship between the share of **self-employment without employees** (excluding **agriculture**) and the **unemployment rate** after two lags during the period 2008Q1–2018Q1. Additionally, both the "refugee" and "unemployment-push" (H1 and H2 rejected) and the "Schumpeter/entrepreneurial" hypotheses are validated. There is a one-way relationship between:

(c1) *Self-employment* with *employees share* and the *unemployment rate*, after three lags, during the period 1999Q1–2018Q1, presenting the Schumpeter effect;

(c2) the share of *contributing family workers* and *unemployment rate* after five lags during the period 1999Q1–2018Q1, as a "subcase" of the Schumpeterian/entrepreneurial effect; and

(d) the share of *self-employment with employees* and the share of *wage employment* after seven lags during the period 1999Q1–2018Q1, indicating dependent employment creation in emerging and expansion sectors or the existence of job creation policies.

6 Discussion

The discussion of the results is focused on their relevance to the characterization of the unemployment, waged employment, self-employment and unemployment relationships in Romania. Performance assessment is considered in the short-run with a focus on entrepreneurial sustainability as the engine of job creation coupled with social policies innovation. Based on the demography of enterprise and entrepreneurship indicators in Romania for the period 2008–2014, we have identified several features leading to a preliminary conclusion of a rather poor sustainability of entrepreneurship and entrepreneurial job creation processes. A purely quantitative increase of our chosen sustainability metric matched by a decrease in the enterprise survival expectancy (two years following enterprise start-up); business activity decreases to its historical minimum in 2014. In a context marked by a severe recession during 2009–2011, entrepreneurship failed to play its expected role as job generator and thus made little contribution to unemployment rate reduction. Our result, for a developed country, is the reverse where "both immigrants and Germans use self-employment for transition between different employment status" (Constant and Zimmermann 2014).

Pineiro-Chousa and López-Cabarcos (Akerlof 1984) endorses reputational management as a driver of entrepreneurial sustainability resulting from an original multidisciplinary analysis based on the "game theory of organizational realities".

Agriculture in Romania's entrepreneurship profile is still important and therefore covers a specific profile of self-employment supported by the fragmented properties of the small producers. The high level of self-employment in agriculture (70% of total) indicates the presence of a *subsistence-type entrepreneurship* (Baptista and Thurik 2007), akin to the ones identified in the economies of Portugal, Italy and Greece. The quality of Romanian entrepreneurship and reputational management falls broadly in line with the one found in Portugal. Its large share in the overall employment is poorly supported by a corresponding quality, especially with respect to innovation. (it is to be noted that Romanian enterprises rank lowest in the EU-28 on the innovation metrics). This may also find an explanation in the low levels of SME survival rates, which prevent these entities from engaging into innovative production processes.

Romanian unemployment metrics in the meantime, are to a certain extent deficient. This deficiency basically translates into an incomplete, partial statistical capture of the phenomenon. Adding the information provided by the underemployment EUROSTAT indicators ([lfsi_sup_a]) is important to illustrate that in 2015 there were 625 thousand underemployed persons in Romania—11 thousand more than in 2014. In reference to their behavior on the labour market, in 2015 there were 357 thousand underemployed persons willing to work but who were not looking for jobs (decreasing by 19 thousand persons compared with 2014) and 268 thousand "part-time worker underemployed persons" (increasing by 30 thousand persons compared with 2014) (Pineiro-Chousa et al. 2016).

The model uses the ILO (harmonized) definition of the unemployment rate. This is not entirely relevant as when looking to social protection of the unemployed and the coverage into the corresponding system then, one should take rather the administratively calculated unemployment rate.[2] Romania, as a former communist country, has a low risk acceptance level, reflected by low levels of independent employment search. In 2015, from 624 thousand unemployed persons, only 82.5 thousand were independently searching for or finding employment [lfsa_ugates]. In relative terms, the share of unemployed persons expressing readiness for independent employment increased from 7% in 2014 to 13.3% in 2015. In the context of relatively high rates of youth unemployment in Romania (three times higher than the national average), the youth unemployed persons aged 15–24 years displayed a positive tendency towards entrepreneurship acceptance, as a form of independent employment. However, this remains less intense when compared with the "adult" unemployed persons (25–55 years of age). The share of young (aged 15–24 years) unemployed persons expressing a readiness for self-employment in the total number of unemployed persons was 5.9% in 2014; this share doubled in 2015, reaching 9.4% (Pineiro-Chousa et al. 2016).

Contractual employment is in a two-way relationship with unemployment (indicative of its strong relation with contractual/wage employment), indicating the intra and inter sector allocation of the labor force, simultaneously with **gift exchange**

[2] The unemployment rate as calculated by the Romanian PES- the national agency for employment (ANOFM, see for more at: www.anofm.ro).

(Jula 2011) and the presence of "involuntary unemployment". *Non-contractual work* generates *contractual work* only in one way for self-employment with employees after seven lags, indicating the absence of spill over from dependent employment. This relationship confirms the **gift exchange** choice even if productivity is not increasing at the same rate as wages; there is low innovation presence, and a low rate of economic sector diversification.

A *Non-contractual employment type* increase causes an *unemployment increase* thus being indicative of the presence of *social protecti*on arrangements for the self-employed (agriculture excluded) (for the period 2008Q1–2018Q1). It is remarkable that the effect of the Schumpeter model is present after 2008Q1 in the case of increasing self-employment with employees causing a decrease in unemployment. The absence of a relationship between self-employment without employees (including the self-employed in agriculture) and unemployment indicates the vulnerability and low level of access to social protections for this category of workers. Between contributing family workers and contractual employment there is no relationship; the only relationship with this category is established via the "social protection system". The functioning of the Romanian labor market suffers from multiple imbalances, confirming the dual labor market paradigm illustrated by"the high instability of jobs in non-contractual compared to contractual wage employment, regardless of the type and period, lower by more than 2 times and even 3 times as the lag—interpreted as the speed of exiting from employment". The jobs are segmented (Jula 2011) between a "primary sector" (i.e.: characterized by job stability, low exit rates, good working conditions, regular promotion and career perspectives, acquisition of skills and reasonably high salaries) and a "secondary sector" (high exit rates, low or no chance of promotion or acquisition of skills, and poor pay)".

We analyse the Non-contractual employment from the perspective of evolutionary entrepreneurship by quantity and quality in a limited spectrum (self-employed with employees, without employees, without employees not in agriculture, and contributing family workers), ignoring social enterprises, cooperatives, etc. as they are practically absent from official statistics. While "institutional and organizational variety increases characterizing advanced economic systems" (Borzaga et al. 2010), in Romania this tendency is inverted. The systemic model proposed by the authors and the results of the analysis for 1999–2018 data presented in Fig. 6.

The analysis is subject to the following limitation:

*given by the concept of variables:

(1) The VAR equation coefficients are not significantly caused by the multiple methodological breaks in series. The CG is robust, but the equations are not appropriate for predictions.

(2) *Entrepreneurship definition*: Self-employment is not comparable with the OECD methodology. "Limits of comparability self-employment are defined in COMPENDIA **as the number of non-agricultural self-employed** (including both unincorporated and incorporated firms) as a share of the labor force" (Verheul et al. 2006). Our entrepreneurship definition is exhaustive, based on the EU-LFS Methodology, comparable at the EU28 level. This model needs to

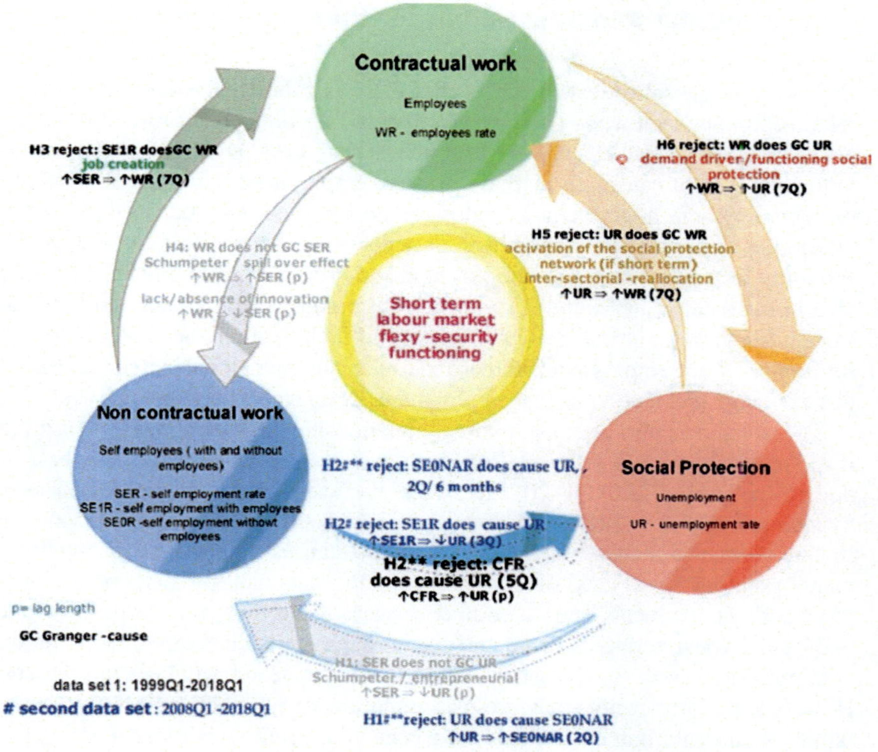

Fig. 6 Short-term labor market flexy security monitoring tool—2018 results overview

be applied on EU28 countries in order to validate the model better according to international literature, especially when applied to developed countries (OCED).

(3) The unemployment rate is the one calculated according to the ILO methodology, and the variable with administrative unemployment should also be checked.

*Given by the model: Wage employment rates, the entrepreneurship rates/contributing family workers, and the unemployment rate relationship are only a short run causality as the result of the application of the Toda–Yamamoto procedure. "If the Granger causality analysis requires that the lagged coefficients of the independent variable in each equation are jointly statistically significant, using Wald statistics, Toda-Yamamoto has the advantage of no pre-testing the integration order of variables and also of the cointegration rank in the VAR system" (Pîrciog and Lincaru 2017).

7 Conclusions and Suggestions for Future Research

The Granger causality test indicates that there is robust evidence for a two-way relationship between *wage employment rate* and *unemployment rate* after a lag of seven quarters (reject H5 and H6) for the period 1999Q1–2018Q1. An increase in unemployment causes an increase in wage employment, a comparable situation to unemployment acting as a stimulus for entrepreneurship, but in this case, the "unemployment stimulus" applied to contractual work. Various forms of "subsistence economy" categories (e.g., self-employed in agriculture, contributing family workers also in subsistence agriculture) for which sustainable employment and especially sustainable entrepreneurship are at the most difficult to attain, become thus eligible for support. Wage employment increase causes an unemployment increase, indicating gift exchange and involuntary unemployment presence.

The test shows also a robust, two-way relationship between the *self-employment without employees not in agriculture rat*e and *unemployment rate*, after two lags during the period 2008Q1–2018Q1, with "refugee" and "unemployment-push" hypotheses and the "Schumpeter/entrepreneurial" hypothesis partially validated. In this case, the increase in entrepreneurship causes an increase in unemployment, pointing to a lack of entrepreneurship sustainability determined by its "founder's endowments and its competitiveness", and thus, there will be winners and losers. These results indicate that even though there are entrepreneurial efforts, entrepreneurial success is not easily reached. Future research will be in this direction. Howard and Martinez note that "from an evolutionary approach, process and context (strategy and environment) interact in a recursive continuous process, driving the fate of entrepreneurial efforts" (Davidescu 2015). This model's results are partially concordant with the results of Şipoş-Gug and Bădulescu (2015), for the level of unemployment and entrepreneurship. They point out that there exists a two-way relationship between the variables analyzed: "Entrepreneurship is a cause in the Granger sense for unemployment level with a 3 quarters lag, while unemployment is a Granger cause for entrepreneurship with a 13 quarters lag".

The following one-way relationships have were confirmed by the Ganger test:

(c1) *Self-employment with employee rate* and *unemployment rate* after three lags, confirming the Schumpeter effect during the period 2008Q1–2018Q1. The increasing entrepreneurship rate (self-employment with employees) causes a decrease in unemployment.

(c2) *Contributing family's workers rate* and *unemployment rate*, after five lags, during the period 1999Q1–2018Q1, as a subcase of the Schumpeterian/entrepreneurial effect but with a reverse effect; the CFRs rate increase causes a decrease in unemployment in the short run.

(c3) *Self-employment with employees rate* causally affects *wage employment rate* after seven lags during the period 1999Q1–2018Q1, indicating the dependent (wage) employment role in job creation in emerging and expanding sectors/job creation policies as a case of inter-sectorial reallocation of the labor force.

These results, especially with regard to the general mechanism of employment generation by the self-employed category, are in line with (Verheul et al. 2006), self-employment becoming an opportunity and not only an employment of last resort (Grebel et al. 2001). Another explanation for this atypical result could be found in the conclusion of Peric and Vitezic (2016) for the Croatia case that "during recession large and medium-sized firms seem to possess higher growth potentials than do small ones, which is inconsistent with previous research in this field". Faria et al. (2008) pointed that between unemployment and entrepreneurship is a cyclical relation, a fact that explains the variability of results in time.

The absence of a relationship between self-employment without employees (inclusive of the self-employed in agriculture) with contractual and social protections reflects the presence of segmentation on the Romanian labor market, thus preserving the dual labor market paradigm. Segments of Self-employed without employees as well as contributing family workers are still vulnerable workers, trapped in the "in-work poverty cycle".

The low level of entrepreneurial success (in a global economy) reflects the low competitive performance of Romanian labor market. This may prove indicative of the low entry labor costs manifested by the firm's low rate of adoption of new and advanced technologies (Poschke 2010). On the other side, the significant presence of contributing family workers in the rural area, confirm the model of rural–urban migration model driven by the low agricultural earnings, labeling the persistent rural urban divide with poverty spots (Faria et al. 2009). Based on Günther and Launov (2012) findings, the Romanian informal sector too, presents a heterogeneous structure, which could be the result of "competitive market forces or labor market segmentation". The lack of "flexisecurity type" arrangements expressed by low transition levels from entrepreneurship to waged employment on Romanian labor market reflects the important presence of informality, specific to developing countries. This conclusion is in line with Bennett and Rablen (2015) whereby for a governmental strategy aiming at the reduction of informality, reducing the "costs" of informality makes for a prerequisite..

The main conclusion is that the "current performance and (functioning model" of the Romanian labor market is not sustainable while the entrepreneurship efforts are still not successful. The dual labor market paradigm preserved according to Salem and Bensidoun (2012).

The jobs in **non-contractual work** are two and even three times more unstable compared to contractual work, regardless of their type and period, and are also less covered by social protection (for self-employed without employees, especially those in agriculture where any type of social protection, with the voluntary type aside, is entirely absent).

The Granger causality relation identification between its macro-aggregates referencing *status in employment* (contractual work, non-contractual work and social protection) provides for a rapid labor market performance assessment instrument with an emphasis on **"entrepreneurial sustainability" as an alternative engine for job creation**. (Fig. 6). There is a certain evident need to enhance the role of novel opportunities for entrepreneurship such as social enterprises, cooperatives etc., rooted in innovation including the high-end, high tech innovation processes (Borzaga et al. 2010; Salem and Bensidoun 2012).

The results indicate that Romania is advancing towards a "developed country" status but suffers from a lack of "entrepreneurship sustainability". The absence of innovation with respect to the activities undertaken by entrepreneurs is an important cause of a de-balanced "flexysecurity type arrangements" on the labor market, coupled with the lack of social and environmental orientation of entrepreneurs. Kuckertz and Wagner (2010) pointed out that the new economic success implies a new entrepreneurial behavior explained by the positive relationship between an individual's "sustainability orientation" and entrepreneurial intentions.

The issue of successful and sustainable entrepreneurship in Romania opens further inroads into this research area, with a particular regard to:

- The high relevance of this topic for ALMPs and other policies stressing the ALMP's impact on entrepreneurship as well as other public policies designated to support entrepreneurship, e.g., the "Start-up Nation" National Programme launched in June 2017 in Romania. We suppose that the target groups of these policies are more system-following than system-building entrepreneurs (?) (Kuckertz and Wagner 2010);
- The need to introduce quality criteria (and not only quantitative criteria) in entrepreneurship policies regarding job creation and independent employment and the creation of a procedure for accelerating innovation adoption, including eco-innovation, considering eco-innovation an instrument to ensure sustainable competitive advantages (Klein Woolthuis 2010);
- The need to further explore the relationship between entrepreneurship and unemployment with a distinct glance towards immigration inflows: "The economic variables examined here have an enormous explanatory capacity for migration flow (GDP per capita, GDP growth, and unemployment rate), fundamentally in the periods of economic crisis and recovery. According to the regression coefficients, the variables related to GDP attract immigration to a high degree; however, the rate of unemployment is a decreasing factor for immigration, but has a much smaller impact." (Albort-Morant et al. 2017);
- The need to further explore entrepreneurship as "an important tool to enhance knowledge diffusion and promote economic growth" (Grau and Ramírez López 2017). The authors applied the Granger test for OECD countries which "confirms that causality goes in the direction from entrepreneurs to growth". Şipoş-Gug and Bădulescu (2015) finds **that entrepreneurship is not a Granger cause for GDP growth in** Romania. Also, there is no evidence regarding entrepreneurship and knowledge diffusion in the Romania analysis. Entrepreneurship represented by startups as a mechanism for knowledge spillover that produces economic growth (Acs et al. 2012);
- The need to explore the profile of sustainable entrepreneurs (Kuckertz and Wagner 2010), entrepreneurs seem to have to be rather "networkers" than inventors. Their products might be less central to their entrepreneurial actions; it is the embedding of the products in new structures and the reinvention of the "entrepreneurial game"

that seems to be key in these entrepreneurial processes. A "game with new players" and new relationships amongst them, and, obvious enough, maybe new rules.

A flexible, secure and well-functioning labor market is possible only through its integration with innovation processes. However, "sustainable innovation" is not possible without a robust and equally sustainable social protection system.

Acknowledgements This work was supported by a grant from the Romanian Ministry of Research and Innovation, Programme NUCLEU, 2019–2022, project no: Structural changes in the labor market and innovative change response policies, cod PN 19_13_01_02.

References

Acs, Z.J., Audretsch, D.B., Braunerhjelm, P., Carlsson, B.: Growth and entrepreneurship. Small Bus. Econ. **39**(2), 289–300 (2012). https://doi.org/10.1007/s11187-010-9307-2

Akerlof, G.: Gift exchange and efficiency-wage theory: four views. Am. Econ. Rev. **74**(2), 79–83 (1984). https://econpapers.repec.org/article/aeaaecrev/v_3a74_3ay_3a1984_3ai_3a2_3ap_3a79-83.htm

Albort-Morant, G., Henseler, J., Leal-Millán, A., Cepeda-Carrión, G.: Mapping the field: a bibliometric analysis of green innovation. Sustainability **9**(6), 1011 (2017). https://doi.org/10.3390/su9061011

Aldrich, H.E., Martinez, M.A.: Many are called, but few are chosen: an evolutionary perspective for the study of entrepreneurship. In: Cuervo, Á., Ribeiro, D., Roig, S. (eds.), Entrepreneurship (pp. 293–311). Springer Berlin Heidelberg (2007). https://doi.org/10.1007/978-3-540-48543-8_14

Atkinson, J.: Flexibility, uncertainty and manpower management (Report 89). Brighton: Institute of Manpower Studies (1984)

Audretsch, D.: Innovation, growth and survival. Int. J. Ind. Organ. **13**(4), 441–457 (1995). https://econpapers.repec.org/article/eeeindorg/v_3a13_3ay_3a1995_3ai_3a4_3ap_3a441-457.htm

Audretsch, D.B., Carree, M.A., Thurik, A.R.: Does Entrepreneurship reduce Unemployment? Tinbergen Institute, Erasmus University Rotterdam, TI 2001–074/3. https://pdfs.semanticscholar.org/7b33/3bfff107ca29fcff6bfbb491ab698324d735.pdf (2001)

Audretsch, D.B., Carree, M.A., Van Stel, A.J., Thurik, A.R.: Does Self-Employment Reduce Unemployment? Papers on Entrepreneurship, Growth and Public Policy. https://www.econstor.eu/obitstream/10419/19998/1/2005-07.pdf (2005)

Audretsch, D.B., Fritsch, M.: The geography of firm births in Germany. Reg. Stud. **28**(4), 359–365 (1994). https://doi.org/10.1080/00343409412331348326

Babaginda, M., Semasinghe, D.M.: Entrepreneurship and unemployment: a litrature review. Int. Conf. Bus. Inf. (2013). https://www.academia.edu/5322204/Entrepreneurship_and_Unemployment_A_Litrature_Review

Baptista, R., Thurik, A.R.: The relationship between entrepreneurship and unemployment: Is Portugal an outlier? Technol. Forecast. Soc. Chang. **74**(1), 75–89 (2007). https://doi.org/10.1016/j.techfore.2006.04.003

Bennett, J., Rablen, M.D.: Self-employment, wage employment, and informality in a developing economy. Oxf. Econ. Pap. **67**(2), 227–244 (2015). https://doi.org/10.1093/oep/gpu047

Bergmann, F.: New work new culture: work we want and a culture that strengthens Us (Kindle Edition). https://www.goodreads.com/book/show/41396901-new-work-new-culture (2019)

Bokhari, A.A.H., Alothmany, N.S., Magbool, S.S.: Entrepreneurship and Unemployment in The Kingdom of Saudi Arabia. 24. https://www.researchgate.net/profile/Nazeeh_Alothmany/publication/235751134_Entrepreneurship_and_Unemploymentin_The_Kingdom_of_Saudi_Arabia/links/0912f51319eda567ef000000/Entrepreneurship-and-Unemployment-in-The-Kingdom-of-Saudi-Arabia.pdf (2012)

Borzaga, C., Depedri, S., Tortia, E.C.: The growth of organizational variety in market economies: the case of social enterprises. SSRN Electron. J. (2010). https://doi.org/10.2139/ssrn.1622155

Braunerhjelm, P., Acs, Z.J., Audretsch, D.B., Carlsson, B.: The missing link: Knowledge diffusion and entrepreneurship in endogenous growth. Small Bus. Econ. **34**(2), 105–125 (2010). https://doi.org/10.1007/s11187-009-9235-1

Chen, C.C.: Entrepreneurship, economic growth, and employment: a case study of Taiwan. Hitotsubashi J. Econ. **55**(19), 71–88 (2013). https://doi.org/10.15057/26817

Cole, I.M.: Unemployment and entrepreneurship in the mid-Atlantic region of the United States: a spatial panel data analysis. Rev. Reg. Stud. **48**(3), 347–375 (2018). https://ideas.repec.org/a/rre/publsh/v48y2018i3p347-375.html

Constant, A.F., Zimmermann, K.F.: Self-employment against employment or unemployment: markov transitions across the business cycle. Eurasian Bus. Rev. **4**(1), 51–87 (2014). https://doi.org/10.1007/s40821-014-0005-x

Dănăcică, D.-E.: Causality between school education and economic growth in Romania. Argumenta Oeconomică **1**(26), 55–72 (2011). https://www.researchgate.net/publication/261991200_Causality_between_school_education_and_economic_growth_in_Romania

Davidescu, A.A.: The relationship between shadow economy and unemployment rate. a ardl causality analysis for the case of Romania. Romanian Stat. Rev. **63**(4), 46–62 (2015). https://ideas.repec.org/a/rsr/journl/v63y2015i4p46-62.html

Dilanchiev, A.: Relationship between entrepreneurship and unemployment: the case of georgia. J. Social Sci. **3**(2), 5 (2014). http://oaji.net/articles/2016/2903-1455537612.pdf

Eckhaus, E.: The fourth dimension of happiness and work satisfaction. Manag. Market. Chall. Knowl. Soc. **16**(2), 118–133 (2021).

Faria, J.R., Cuestas, J.C., Gil-Alana, L.A.: Unemployment and entrepreneurship: a cyclical relation? Econ. Lett. **105**(3), 318–320 (2009). https://doi.org/10.1016/j.econlet.2009.09.004

Faria, J.R., Cuestas, J.C., Mourelle, E.: Entrepreneurship and unemployment: A nonlinear bidirectional causality (2008/6; NBS Discussion Papers in Economics). Economics, Nottingham Business School, Nottingham Trent University. https://ideas.repec.org/p/nbs/wpaper/2008-6.html (2008)

Garba, A., Djafar, F., Mansor, S.A.: Evidence of opportunity and necessity driven entrepreneurship in Nigeria. J. Entrepr. Manag. Innov. (2013) https://doi.org/10.13140/2.1.3484.0963

Ghavidel, S., Farjadi, G., Mohammadpour, A.: The Relationship Between Entrepreneurship and Unemployment in Developed and Developing Countries. Int. J. Econ. Behav. IJEB **1**(1), 71–78 (2011). https://econpapers.repec.org/article/butijebfa/v_3a1_3ay_3a2011_3ai_3a1_3ap_3a71-78.htm

Giles, D.: Testing for granger causality. Econometrics beat: dave giles' blog*:* https://davegiles.blogspot.com/2011/04/testing-for-granger-causality.html (2011)

Grau Grau, A., Ramírez López, F.: Determinants of immigration in europe. the relevance of life expectancy and environmental sustainability. Sustainability **9**(7), 1093 (2017). https://doi.org/10.3390/su9071093

Grebel, T., Pyka, A., Hanusch, H.: An evolutionary approach to the theory of entrepreneurship. Volkswirtschaftliche Diskussionsreihe **206**, 22 (2001). https://www.econstor.eu/bitstream/10419/70016/1/335892833.pdf

Günther, I., Launov, A.: Informal employment in developing countries. J. Dev. Econ. **97**(1), 88–98 (2012). https://doi.org/10.1016/j.jdeveco.2011.01.001
Halicioglu, F., Yolac, S.: Testing the impact of unemployment on self-employment: evidence from OECD Countries. Procedia. Soc. Behav. Sci. **195**, 10–17 (2015). https://doi.org/10.1016/j.sbspro.2015.06.161
Harris, J.R., Todaro, M.P.: Migration, unemployment and development: a two-sector analysis. Am. Econ. Rev. **60**(1), 126–142 (1970). https://www.jstor.org/stable/1807860
Jula, D.: Econometrie [Course]. http://www.postdoc.acad.ro/data/files/ECONOMETRIE_-_prof.Jula.pdf (2011)
Klein Woolthuis, R.J.A.: Sustainable entrepreneurship in the Dutch construction industry. Sustainability **2**(2), 505–523 (2010). https://doi.org/10.3390/su2020505
Kuckertz, A., Wagner, M.: The influence of sustainability orientation on entrepreneurial intentions—Investigating the role of business experience. J. Bus. Ventur. **25**(5), 524–539 (2010). https://doi.org/10.1016/j.jbusvent.2009.09.001
Lasch, F., Gundolf, K., Kraus, S.: The Impact of Unemployment on Entrepreneurship: Empirical Evidence From France. ResearchGate. https://www.researchgate.net/publication/281904438_The_Impact_of_Unemployment_on_Entrepreneurship_Empirical_Evidence_From_France (2007)
Lütkepohl, H.: New Introduction to Multiple Time Series Analysis (Springer). Springer Berlin Heidelberg (2005). https://doi.org/10.1007/978-3-540-27752-1
Peric, M., Vitezic, V.: Impact of global economic crisis on firm growth. Small Bus. Econ. **46**(1), 1–12 (2016). https://doi.org/10.1007/s11187-015-9671-z
Pineiro-Chousa, J., Vizcaíno-González, M., López-Cabarcos, M.: Reputation, game theory and entrepreneurial sustainability. Sustainability **8**(11), 1196 (2016). https://doi.org/10.3390/su8111196
Pinelli, M.: Can entrepreneurship solve the youth unemployment crisis? | World Economic Forum. https://www.weforum.org/agenda/2015/09/can-entrepreneurship-solve-the-youth-unemployment-crisis/ (2015)
Pîrciog, S., Lincaru, C.: Evaluarea eficienței și impactului Strategiei naționale de ocupare a forței de muncă 2014–2020 la jumătatea perioadei de implementare. Study/Evaluation of the efficiency and impact of the National Employment Strategy 2014–2020 at the mid-term. Study (INCSMPS Studies) (2017)
Poschke, M.: The regulation of entry and aggregate productivity. Econ. J. **120**(549), 1175–1200 (2010). https://doi.org/10.1111/j.1468-0297.2010.02367.x
Poschke, M.: Wage employment, unemployment and self-employment across countries, F-32205-ETH-1. Int. Growth Centre **78** (2018). https://www.theigc.org/wp-content/uploads/2018/05/Poschke-2018-Working-Paper.pdf
Ramsburg, J.: An Interview with Frithjof Bergmann: Rethinking Work on a Global Scale. J. Int. Inst. **6**(2) (1999). http://hdl.handle.net/2027/spo.4750978.0006.203
Salem, M.B., Bensidoun, I.: The heterogeneity of informal employment and segmentation in the Turkish labour market. J. Asia Pacific Econ. **17**(4), 578–592 (2012). https://doi.org/10.1080/13547860.2012.724546
Shepherd, D.A., Patzelt, H.: The new field of sustainable entrepreneurship: studying entrepreneurial action linking "what is to be sustained" with "what is to be developed." Entrep. Theory Pract. **35**(1), 137–163 (2011). https://doi.org/10.1111/j.1540-6520.2010.00426.x
Simionescu, M.: Kalman Filter or VAR Models to Predict Unemployment Rate in Romania? Naše Gospodarstvo/our Econ. **61**(3), 3–21 (2015). https://doi.org/10.1515/ngoe-2015-0009
Şipoş-Gug, S., Bădulescu, A.: Granger causality analysis of the relationship between gross domestic product, unemployment and entrepreneurial activity in Romania. Emerging Markets Economics and Business. 4th Conference of Doctoral Students in Economic Sciences. http://steconomiceuoradea.ro/wp/wp-content/uploads/2011/09/Volumul-conferintei-doctoranzi-2013.pdf (2013)
Şipoş-Gug, S., coord. Bădulescu, A.: Antreprenoriatul tinerilor. Motivaţii, dimensiuni şi condiţionări / youth entrepreneurship. Motivations, dimesnions and conditionings. [Universitatea Din Oradea

Facultatea de Ştiinţe Economice Şcoala Doctorală de Ştiinţe Economice Domeniul Economie]. https://www.uoradea.ro/display8183 (2015)

Szewczyk, M., Widera, K., Parvi, R.: The relationship between unemployment and entrepreneurship: a case of opolskie voivodship. Res. Gate 156–159 (2013). https://www.researchgate.net/publication/281869880_M_SZEWCZYK_K_WIDERA_R_PARVI_The_Relationship_between_Unemployment_and_Entrepreneurship_A_Case_of_Opolskie_Voivodship_ARSA_2013_EDIS_-_Publishing_Institution_of_the_University_of_Zilina_2013_156-159_ht

Thurik, A.R., Carree, M.A., van Stel, A., Audretsch, D.B.: Does self-employment reduce unemployment? J. Bus. Ventur. **23**(6), 673–686 (2008). https://doi.org/10.1016/j.jbusvent.2008.01.007

Toda, H.Y., Yamamoto, T.: Statistical inference in vector autoregressions with possibly integrated processes. J. Econ. **66**(1–2), 225–250 (1995). https://doi.org/10.1016/0304-4076(94)01616-8

Uka, A., Prendi, A.: Motivation as an indicator of performance and productivity from the perspective of employees. Manag. Market. Chall. Knowl. Soc. **16**(3), 268–285 (2021).

Verheul, I., van Stel, A., Thurik, R., Urbano, D.: The relationship between business ownership and unemployment in Spain: a matter of quantity or quality?/La relación entre el autoempleo y el desempleo en España: Una cuestión de cantidad o de calidad? Estudios de Economia Aplicada **24**, 435–457 (2006). https://econpapers.repec.org/article/lrkeeaart/24_5f2_5f5.htm

World Economic Forum.: The Future of Jobs Report 2018. World Economic Forum. https://www.weforum.org/reports/the-future-of-jobs-report-2018/ (2018)

Xavier, A., Badea, P.: EU Labour Market Policies: How active are we and how do we respond to unemployment? European Commission, Employment, Social Affairs & Inclusion. https://ec.europa.eu/social/main.jsp?langId=en&catId=1196&newsId=2389&furtherNews=yes (2018)